The Chemistry of Excitation
at Interfaces

The Chemistry of Excitation at Interfaces

J. Kerry Thomas

ACS Monograph 181

American Chemical Society

Washington, D.C.   1984

Library of Congress Cataloging in Publication Data

Thomas, J. Kerry, 1934–
    The chemistry of excitation at interfaces.
    (ACS monograph, ISSN 0065-7719; 181)

    Bibliography: p.
    Includes index.

    1. Radiation chemistry. 2. Photochemistry.
3. Surface chemistry.

    I. Title. II. Series.

QD636.T46   1984   541.3'8   84-9365
ISBN 0-8412-0816-6

ACS Monographs

Marjorie C. Caserio, *Series Editor*

Advisory Board

FOREWORD

ACS MONOGRAPH SERIES was started by arrangement with the interallied Conference of Pure and Applied Chemistry, which met in London and Brussels in July 1919, when the American Chemical Society undertook the production and publication of Scientific and Technological Monographs on chemical subjects. At the same time it was agreed that the National Research Council, in cooperation with the American Chemical Society and the American Physical Society, should undertake the production and publication of Critical Tables of Chemical and Physical Constants. The American Chemical Society and the National Research Council mutually agreed to care for these two fields of chemical progress.

The Council of the American Chemical Society, acting through its Committee on National Policy, appointed editors and associates to select authors of competent authority in their respective fields and to consider critically the manuscripts submitted. Since 1944 the Scientific and Technologic Monographs have been combined in the Series. The first Monograph appeared in 1921.

These Monographs are intended to serve two principal purposes: first to make available to chemists a thorough treatment of a selected area in form usable by persons working in more or less unrelated fields to the end that they may correlate their own work with a larger area of physical science; secondly, to stimulate further research in the specific field treated. To implement this purpose the authors of Monographs give extended references to the literature.

ABOUT THE AUTHOR

J. KERRY THOMAS earned his doctorate in 1957 from the University of Manchester, England. He was awarded an Honorary Doctor of Science at Manchester in 1969, and the Research Award of the Radiation Research Society in 1974. He developed pulse radiolysis and laser-flask photolysis studies while at Harwell in England and Argonne National Laboratory in Illinois. He is presently the Father Julius A. Nieuwland, C.S.C., Professor of Science at the University of Notre Dame. His interests lie in radiation and photochemistry, in particular fast-pulsed methods, in the chemistry of surfaces, and in colloidal systems.

CONTENTS

PREFACE

ADVANCES IN BIOCHEMISTRY and biophysics have encouraged more chemists to direct their attention to these areas of research. Perhaps two areas of study stand out: the role played by organized assemblies such as membranes or enzymes in controlling material transport and catalysis selection reactions, and the role played by such assemblies in the interaction of light, as in vision or photosynthesis.

Direct chemical and physical studies usually are made in vivo or on extracted parts of the organism. However, it is also possible to start with simple chemical systems and by gradually increasing their complexity, arrive at model systems that provide information to the biologist. This text deals with the latter approach to colloids. Their structure and the role that they play in radiation-induced reactions are investigated.

I would like to thank T. Gautier, J. Inglish, and P. McCormack, for converting my hand-written text into something that could be read by others. I would also like to thank my research group T. Wheeler, J. Kuczynski, R. Stramel, P. Hite, D. Chu, S. Hashimoto, and T. Nakamurai for checking the manuscript. The inspiration for the text is due to those who provoked and guided me over the years, the late J. H. Baxendale, J. L. Magee, M. Burton, and E. J. Hart.

I would also like to thank various research agencies who have supported my work over the years: the U.S. Atomic Energy Commission, for studies in radiation chemistry; the National Science Foundation, for photochemistry and micellar studies; the Army Research Office, for studies in inorganic colloids; and the Petroleum Research Fund of the American Chemical Society, for helping me start many important projects.

The text is dedicated to that greatest of all organized assemblies, my family: Ronald, Rebecca, June, Delia, Roland, and Roger.

J. KERRY THOMAS
University of Notre Dame
Notre Dame, Indiana

January 1984

1

Introduction

MUCH OF THE CHEMISTRY TAUGHT TO STUDENTS, or even the scientific community at large, concerns itself with events in homogeneous media. However, chemical and physical processes at surfaces and interfaces are of fundamental importance in nature and in the technological society developed by humanity. Numerous examples of such events are available in medicine, biology, radiobiology, industry, etc., as well as in the new development of solar energy storage. A few choice examples, which follow, illustrate and stress the many important features of interfacial chemical events.

Biological Systems

A living cell is distinguished by possessing, among other things, a nucleus and a cell wall or membrane. The vital nature of the nucleus has been well impressed on the thinking of the community by many communications at all levels of sophistication. By contrast, the cell membrane or wall has received little notoriety and is often dismissed as a convenient "hold all" for the organism. Yet, through this structure the cell receives nutrients, eliminates waste products, and recognizes its environment. A failure in the cell membrane could be disastrous to the well-being of an organism. This lightly dismissed boundary contains the most sophisticated physicochemical apparatus that communicates information of vital importance to the cell. It may also participate in the decision-making faculty of the organism.

Medical science has realized the awesome consequences of "cell contact inhibition" (1). The outer membrane of a healthy cell contains molecules that recognize the situation where cells are in physical contact and cell multiplication has reached its limit. This information is interpreted by the cell as a signal to cease cell division. A damaged cell, one that is transformed virally or chemically, does not possess this critical faculty. Cell-to-cell contact is not recognized, and cell division proceeds ceaselessly leading to an altered life form or tumor. The damaged cell apparently has an alteration in the outer membrane that stops

0065-7719/84/0181-0001$06.00/1
© 1984 American Chemical Society

it from recognizing its surroundings accurately, and hence, leads to catastrophy.

The cell membrane also plays a crucial role in the treatment of cancer by high-energy radiation. In the early days of radiobiology oxic cells were found to be more sensitive to high-energy radiation than anoxic cells. Thus, increasing the oxygen content of tumor cells as much as possible is desirable so that lethal damage to the tumor is maximized. Oxygen is used up during the cell irradiation by chemical free radical species $R^{\cdot}$ and ions e,

$$R^{\cdot}, e + O_2 \rightarrow RO_2^{\cdot} \text{ and } O_2^{-}$$

The long-lived peroxy species $RO_2^{\cdot}$ and O_2^{-} strongly promote damage to the cell. In a series of experiments with pulsed accelerators, high radiation doses decreased the cell oxygen level and maximum cell killing only occurred if sufficient time elapsed between radiation doses so that oxygen could penetrate through the cell membrane and into the cell (2, 3). The site of radiation damage in cells is the nucleus; however, the rather innocuous role of oxygen transport across the cell membranes plays a major part in the final result.

The role of a barrier given to a membrane is realized immediately in photosynthesis, where the primary step, the adsorption of a light quantum by the chloroplast chromophore, leads to electron transfer:

$$\text{chromophore} \xrightarrow[\text{acceptor}]{\text{light}} (\text{chromophore})^+ + (\text{acceptor})^-$$

For this process to occur efficiently, the donor chromophore and acceptor must be arranged in close proximity so that efficient electron transfer occurs. A phospholipid membrane system is used to provide a matrix for the photosystems that are activated in the initial event. Even more important is the role played by the membrane in guaranteeing that back electron transfer, a process that would decrease the yield of the photoreaction, does not occur. The photoevents of vision are particularly complicated, because the information of the photons has to be communicated to the brain as electrical impulses passed along the nervous system. Numerous interfacial chemical processes take place along the nervous system, processes that are not well documented. However, the interfacial photon systems of the eye are still under vigorous investigation. The photosensory device of the eye is a rod or cone, which contains about 1000 closed bilayer membranes packed with rhodopsin molecules. Excitation of rhodopsin by absorption of light leads to isomerization of 11-*cis*-retinal to all-*trans*-retinal. This isomerization leads to a closing of the sodium ion channels in the plasma membrane, which, in turn, lead to signals that are passed by the nervous system to the brain. Nature often uses an interface to achieve efficiency in a desired process. Those uses given in this section are pertinent to later considerations in this text.

Industrial Effects of Surfaces

Most articles in everyday use are covered by a layer of organic material. This layer serves both to protect the object from corrosion by the environment, and to provide a vehicle for color and for pleasant handling of the object. For the most part, an organic-based paint is used, and it provides a thin membranelike layer between the object and its environment. The analogy between the paint and the membrane ends there because the paint layer has little of the ordered and sophisticated properties of a membrane. However, diffusion of materials, in particular oxygen, through the paint layer and photolytic damage of the paint by light are reminiscent of similar effects in membranes.

The joining of two surfaces by an adhesive is an important industrial technique. The surfaces of the objects to be bonded together have to readily receive the adhesive, which in most cases is a foreign material. The subsequent setting or polymerization of the adhesive has to proceed smoothly, in spite of the solid structures into which it has diffused. The chemistry and physics of such processes are complex and are modified by the narrow object–adhesive interface. The diametrically opposite effect of lubrication also operates at a thin interface of object and lubricant.

Catalysis

In the examples just discussed, the interface between two materials is important, mainly because of the specialized physical properties imparted to the system by the interface. Often, the chemistry at interfaces is altered or different from the norm, that is, from that experienced in homogeneous media. The unique properties of an interface are used in catalysis, and many outstanding solid-state catalysts that promote reactions of great value are available in industry. Catalysis is due to adsorption of the reactants on the catalyst surface. The nature of the adsorption may give rise to low levels of active species (4, 5). For example, H_2 probably is adsorbed on some surfaces as atoms:

$$H_2 + 2M \rightarrow 2HM$$

where M denotes an active site on the catalyst. Other molecules may be adsorbed at an active state without dissociation, for example, H_2S, or CO.

$$H_2S + M \rightarrow M \overset{\displaystyle HSH}{\diagup}$$

$$CO + 2M \rightarrow M—C \overset{\displaystyle M}{\underset{\displaystyle O}{\diagup \atop \diagdown}}$$

The adsorbed molecules are in a higher energy state than free molecules.

Reactions are catalyzed because the activation energy of the reaction is lowered. The reactants are also organized on the catalyst in a suitable form for reaction to occur, and are maintained in close proximity on the catalyst during the course of reaction. The rapid three-dimensional, rotational, and translational motion of free reactants is eliminated in favor of a more rigid situation where reactants are brought together slowly but for long periods of time. The relative importance of lowering the activation energy compared to optimum alignment of molecules is not known with industrial catalysts. This concept will be discussed further in the section on colloidal systems.

Photosystems

Photochemistry in heterogeneous systems is of particular importance in the photoimaging industry, where phototechniques are used in the production of literature and textiles and in the preparation of printed electronic circuits (5, 6). Most photographic imaging is carried out with silver-based systems. However, in the future efficient nonsilver systems would probably be based on heterogeneous emulsion systems. Many immediate difficulties are encountered in the development of such systems (7), but some colloidal surfactant systems show promise. Heterogeneous chemistry across two phases is an essential feature of all systems.

Solid-state solar devices, which convert solar energy to electrical energy, have become important because of the diminishing world stocks of oil. These systems consist of solid materials that eject electrons to a conductor when light impinges on them. The event is again photochemistry at an interface. Developments in this area have led to the design of low cost solar cells and the increase in their response to the solar spectrum (8, 9). Most of the systems considered to be possible avenues for solar energy research (10) involve the use of membranes or colloids, that is, chemistry at interfaces.

Radiation as a Catalyst

Chemistry initiated by the absorption of radiation may be looked upon as a catalyzed reaction. In thermal catalysis the reactants are strained by the catalyst and become energized for efficient rapid reactions. Absorption of radiation also leads to energized molecules and promotes reaction. However, the energies involved can be much larger than those involved in thermal catalysis; usually, the action of radiation is to produce chemistry that does not occur thermally. This method is at odds with thermal catalysis where chemical reaction does occur slowly in the absence of the catalyst. The convenience of a radiation-induced reaction cannot be overemphasized, however, as efficient and unusual reaction is promoted readily without introducing a foreign body into the reactants. Radiation-induced processes depend rather critically on the nature of the medium in which they are carried out and tend to be difficult to control or marshall into desired reaction pathways. This area is where the value of a directing system comes into play, that is, the introduction of a catalyst that promotes one of the radiation-induced reactions at the expense of the others. Perhaps the term catalyst is a little misleading because the chemical reactions already occur as a result of the action of radiation absorption; just guidance is required. Interfaces introduced into the chemical systems via colloidal particles provide starting routes for the desired effect of control of the radiation-induced reaction.

Photochemistry and radiation chemistry have been carried out in solution and, to benefit from the large body of work already available, catalysts that can be used in fluid media must be studied. Colloidal chemistry provides many such systems from organic micelles to colloidal metals. For the most part, these systems consist of small particles (radii of 20–2000 Å) suspended in liquids such as water or hydrocarbons. Other systems may utilize other liquids, but little is known about them. The interface or surface exists between the particle and the bulk fluid medium. The nature of the interface (i.e., surface charge and surface type) may be changed readily to promote optimum conditions for reaction. These systems have provided useful vehicles for the catalysis of thermal reactions (11, 12), and can promote radiation-induced reactions (12–15). In many instances, the knowledge gained from studies of the radiation-induced reactions can be used to comment on the structure of the colloidal particle. It is useful to consider the colloidal particle as an *organized assembly* that organizes the reactions taking place at the particle.

Reactions at Colloidal Surfaces

Figure 1 depicts four types of processes that are directed by or occur as a direct consequence of colloidal particles. The particles are represented as spherical objects with the reactants A, B, or C in different locations with respect to the particle. The nature of the site for A, B, or C will be discussed subsequently.

At least one of the reactants has to reside for a major portion of its time with the particle. In Case I (Figure 1) both reactants A and B reside primarily in the particle and not in the aqueous phase, due to their hydrophobic character. Excitation by light of one of the reactants leads to reaction when the excited species meets with the other reactant. This reaction is promoted more strongly than that in homogeneous

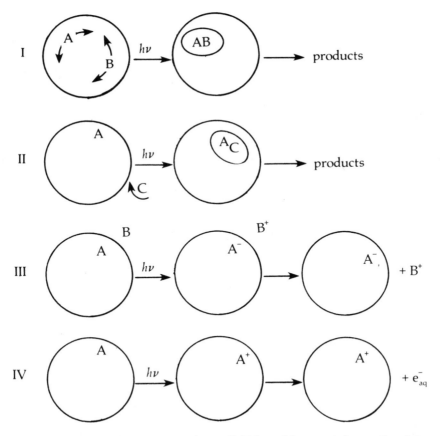

Figure 1. Four possible situations where colloidal particles may influence the photo-induced reaction of A, B, and C.

solution because of the close proximity of the reactants. Several examples of this proximity effect will be discussed later.

In Case II one of the reactants A, which is hydrophobic, is with the particle, while the other, which is hydrophilic, is in the water phase. Excitation of A leads to the excited state A* and, for reaction to occur, C must penetrate the particle to A*. Usually A* need not exit the particle to react with C. Unlike Case I no catalysis of reaction is observed, but frequently a retardation is experienced and the excited molecule actually is protected from damage by impurities such as C in the aqueous phase. This retardation is particularly true if C and the particle have like charge, when electrostatic repulsion keeps C away from the particle containing A*.

Case III systems are a derivation of Case I systems. Both A and B are with the particle with the stipulation that one reactant, say B, is only weakly solubilized by the particle. Excitation by light may lead to electron transfer from A to B, forming A^- and B^+, or B to A depending on the system. This electron transfer is the proximity effect of Case I. However, if the surface charge of the particle is positive or the same as that of B^+ then repulsion of B^+ from the surface occurs. Repulsion leads to efficient separation of A^- and B^+ (or A^+ and B^- if the particle is anionic) and prevents the ion neutralization $A^- + B^+ \rightarrow A + B$. Thus, light energy is stored efficiently as chemical energy of the ions.

Case IV is a variation of Case III where the light-induced electron transfer process involves transfer of e^- to the aqueous phase to give the hydrated electron e_{aq}^-. Again an anionic surface prevents the back neutralization of A^+ and e_{aq}^-. The energy required is reduced in micellar colloidal systems compared to the gas phase. Such effects make Cases III and IV of value in solar-energy storage.

Methods for photochemical reactions in aqueous micellar systems are quite useful in the search for new energy sources (10, 16). Aqueous, micellar systems aid in the production of ions in photochemical reactions, and ionic chemistry is a suitable starting point for subsequent production of electrical power, or of hydrogen gas via further breakdown of the aqueous medium. These types of processes mimic the photosynthetic system where light is absorbed in chlorophyll pigment and leads to charge separation, which is promoted by the membrane system of the chloroplast. The induced charged separation by absorption of light and the subsequent separation of ions can all be modeled in simple micellar systems.

Literature Cited

1. Pollack, R. E.; Burger, M. M. *Proc. Nat. Acad. Sci.*, **1969**, *62*, 1074.
2. Adams, G. E.; Michael, B. D.; Asguith, J. C.; Shenoy, M. A.; Watts, M. E.; Whillans, D. W. In "Radiation Research"; Nygaard, O. F., Adler, H. I., Sinclair, W. K. Eds.; Acad. Press: New York, 1975; p. 478.

3. Epp, E. R.; Weiss, H.; Ling, C. C.; Djordjevic, B.; Kessaris, N. D. In "Fast Processes in Radiation Chemistry and Biology"; Adam, G. E.; Fielden, E. M.; Michael, B. D., Eds.; J. Wiley & Sons: London, 1973; p. 341.
4. Bond, G. C. "Heterogeneous Catalysis"; Clarendon Press: Oxford, 1974.
5. Thomas, J. M.; Thomas, W. J. "Introduction to the Principles of Heterogeneous Catalysis"; Acad. Press: London, 1967.
6. Jacobson, K. I.; Jacobson, R. E. "Imaging. Systems."; Halsted Press: New York, 1976.
7. *Chemistry in Britain*, 16 (1980).
8. Wrighton, M. S. *Chem. Eng. News* **1979**, *57*, 29.
9. "Interfacial Photoprocesses: Energy Conversion and Synthesis"; Ed. Wrighton, M. S. ADVANCES IN CHEMISTRY SERIES No. 184 (1980) American Chemical Society: Washington, D.C.
10. Almgren, M. p. 1 in Vol II of Solar Energy Project Results. Eds. S. Claesson and L. Engstrom.
11. Cordes, E. H. "Reaction Kinetics in Micelles"; Plenum Press: New York, 1973.
12. Fendler, J. H.; Fendler, E. J. "Catalysis in Micellar and Macromolecular Systems"; Acad. Press: New York, 1975.
13. Thomas, J. K. *Acc. Chem. Res.* **1977**, *10*, 133.
14. Turro, N. J.; Grätzel, M.; Braun, A. M. *Angew. Chem.* **1980**, *19*, 675.
15. Thomas, J. K. *Chem. Rev.* **1980**, *80*, 283.
16. Proceeding of the Third International Conference on Photochemical Conversion and Storage of Solar Energy, 1980.

Basic Principles of Excitation Processes

THE BASIC PHYSICAL PROCESS in both photochemistry and radiation chemistry is the transfer of electromagnetic energy to the electrons of the system. In photochemistry this event occurs at relatively low energies (*see* box). Mercury lamps are usually employed for steady-state irradiation and lasers for pulsed irradiation. Almost any wavelength of the UV–visible spectrum can be produced by the use of dye lasers. However, the output of these lasers is low, particularly after frequency doubling to achieve UV light. This condition limits the use of these lasers in pulsed experiments where short-lived intermediates are observed by absorption measurements. In photochemistry, low-energy quanta of light are absorbed in a specific resonance process that nearly always appertains to the solute molecules. This situation is in marked contrast to radiation chemistry where the high energies of the radiation, often 10^6 eV/quantum or greater, are well in excess of the bond energies of molecules. Therefore, the component of the system in excess, that is, the solvent, is excited.

Table I shows several common types of high-energy radiation and their salient properties. With each type of radiation, energy is transferred to the electrons of the medium in a nonresonant fashion. A component of the system will receive a fraction of the total energy lost to the system that is proportional to its electron fraction in that system. Thus, the energy is almost entirely deposited in the solvent if the system is a dilute solution. By contrast, in photochemistry, a solute at low concentration, that is, small electron fraction, may still absorb 100% of the total light absorbed by the system. With high-energy radiation the initial chemistry is that of the solvent and subsequently it may involve a dissolved solute. This difference between photochemistry and radiation chemistry is the key in understanding the chemistry resulting from the two different modes of excitation.

0065-7719/84/0181-0009$07.25/1
© 1984 American Chemical Society

Scan of Wavelengths Employed in Photochemistry: Quanta Energies, and Some Bond Energies

Energy/Quantum (eV)	1.26	3.20			4.20		6.30
Energy/Quantum (kcals/mole)	29	74	Hg	Hg	97	Hg	145
Wavelength, λ	10,000 Å	4000 Å	3650	3130	3000 Å	2537	2000 Å

Lasers:　Ruby　　　　　6943 and 3471 Å
　　　　　Neodymium　　10,600, 5300, and 2650 Å
　　　　　N_2　　　　　　3371 Å
　　　　　Excimer　　　　3080, 248, 193, and 157

$$\lambda \text{ (Å)}$$

Selected Reactions:
$$H_2O \rightarrow H + OH \qquad < 2000$$
$$H_2S \rightarrow H + SH \qquad < 2500$$
$$I^- \rightarrow I + e_{aq}^- \qquad < 2300$$
$$Fe^{2+} \rightarrow Fe^{3+} + e_{aq}^- \qquad 537$$

$$CH_3 - C \overset{O}{\underset{\underset{CH_3}{|}}{\big\Vert}} \rightarrow CH_3^{\cdot} + \underset{\underset{CH_3}{|}}{C} = O \; < 3200$$

biacetyl $\rightarrow 2(CH_3CO)$　　　sunlight (< 5000 Å)
iodobenzene $\rightarrow I^{\cdot}$ + radical　　　(2800 Å)
$HCOOH \rightarrow H_2 + CO + H_2O +$　　　(2537 Å)
CO_2

Table I.
Types of High-Energy Radiation

Radiation	LET	Energies	Chemical Effects
Fast electrons or β-rays	Low	>0.5 MeV	Low yields of molecular products, e.g., H_2 and H_2O_2 from water; large yields of free radical and ionic fragments
Slow electrons	High	<1000 eV	Large yields of molecular products
α-Particles	Very high	Up to 100 MeV	
Deuterons	Very high	Up to 100 MeV	
Neutrons	Very high	Up to 100 MeV	
γ-Rays, X-rays	Low	>1000 eV	Low yields of molecular products

Photochemistry

Photochemistry requires that the system must absorb light so that chemical change can take place (Draper's law). This property is ascertained readily by taking the absorption spectrum of the molecule in the environment of interest. The next step requires the selection of a suitable light source to illuminate the molecule in its absorption band. The absorption of light arises directly from the interaction of electrons of the molecule with the oscillating electric field of the light. The molecule gains energy from the light precisely as one quantum of light of frequency v is absorbed by the molecule. The energy of the excited state formed is given by $E = hv$ where E is the energy involved and h is Planck's constant (1, 2). Suitable rearrangement of this equation produces the form

$$E = \frac{12.6 \times 10^3}{\lambda}$$

where E is in electronvolts, and the wavelength λ is in angstroms. The number of molecules, n, decomposed by m quanta of light is used to define the quantum yield ϕ of a photochemical process by $\phi = n/m$. The resulting chemistry is derived from reaction of the excited state of the molecule A^*, which is formed from the ground state A, by absorption of radiation:

$$A \xrightarrow{hv} A^* \longrightarrow \text{chemical reactions}$$

This reaction is the basic premise of Draper's law, a concept that is not difficult to accept.

Various types of excited states can be formed. In simple organic molecules such as benzene and acetone, the interaction of the light's electric field with electrons of the molecules produces a transitory dipole into the molecule, and a node or nodes are introduced into the electronic wavefunction. Energy is abstracted from the electromagnetic radiation by the electronic system of the molecule. These events are visualized readily in Figure 1 by considering the interaction of light with the 1s orbital of a hydrogen atom. The ground state H atom is represented in the 1s state with a positive nucleus that is surrounded by a spherical cloud of negative charge produced by the electron. The vertical electric field of the light tends to force the electronic charge into two concentrated regions. The positive nucleus is at the node between them and thus produces a 2p excited state in the H atom.

Table II lists some of the possible excited states formed in photochemistry, together with examples of actual systems, the photochemical

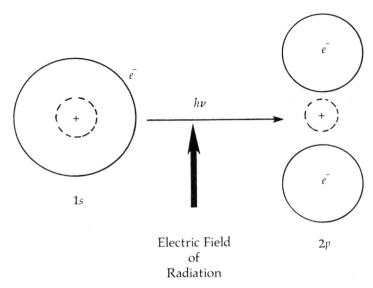

Electric Field
of
Radiation

$2p$

Figure 1. Diagrammatical representation of the interaction of radiation with a 1s electron to produce an excited state with a 2p electron.

Table II.
Examples of Excited States in Photochemistry

Chromophore	λ (Å)	E (eV)	System
$\eta-\pi^*$	<3200	>4.0	acetone $\rightarrow$ excited singlet $\rightarrow$ triplet $\rightarrow$ chemistry (e.g., $CH_3^{\cdot}$ + CO)
	<4000	>3.2	benzophenone $\rightarrow$ singlet $\rightarrow$ triplet $\rightarrow$ chemistry (BQ^T + RH $\rightarrow$ $BQH^{\cdot}$ + $R^{\cdot}$)
$\pi-\pi^*$	2700	4.7	benzene $\rightarrow$ singlet $\rightarrow$ triplet singlet $\rightarrow$ excimers + exiplexes
$\sigma-\sigma^*$	2000	6.3	$CH_3Cl \rightarrow CH_3^{\cdot}$ + Cl
Charge transfer	2400	5.3	$I_{aq}^- \rightarrow I + e_{aq}^-$ $I \rightleftharpoons I_2^-$ (similar reactions for CN_5, Cl^-, and Br^-)
	2600	4.9	$Fe_{aq}^{2+} \rightarrow Fe_{aq}^{3+} + e_{aq}^-$
	5000	2.5	tris(bipyridyl)ruthenium $\rightarrow$ excited triplet $\swarrow$ acceptor Ru(III) + another acceptor other organometallic compounds, e.g., tetraphenylporphinezinc
	5000	2.5	CdS $\rightarrow$ hole-pair $\rightarrow$ emission $\downarrow$ e^- donation to acceptor (also other semiconductors, e.g., TiO_2)

wavelengths used, and energies of these states. In nonconjugated molecules, absorption of light results in the promotion of a single electron from a σ, π, or η orbital of the ground state to an antibonding π^* or σ^* orbital in the excited state. One of the most common molecular transitions is an electronic transition from a bonding π orbital to an antibonding π^* orbital ($\pi \rightarrow \pi^*$ transition). The well-known transitions of benzene fall into this category.

If the molecule contains a heteroatom such as oxygen or nitrogen, then the highest filled orbitals are nonbonding or η orbitals; the lowest unfilled orbitals are π^* in character. Thus, absorption of light, which promotes electronic transitions of lowest energy in ketones, aldehydes, nitrogen heterocycles, quinones, etc., is due to an $\eta \rightarrow \pi$ transition. Other transitions such as $\pi \rightarrow \sigma^*$ or $\sigma \rightarrow \sigma^*$ occur in alkanes and alkyl halides.

An interesting electronic transition occurs when an electron is transferred from a molecule or ion to another molecule, or to the surrounding solvent. These charge-transfer transitions often lead to the formation of ions. The complex between N,N-dimethylaniline and chloranil exhibits an absorption band in the visible region due to an electron transfer from the amine to chloranil:

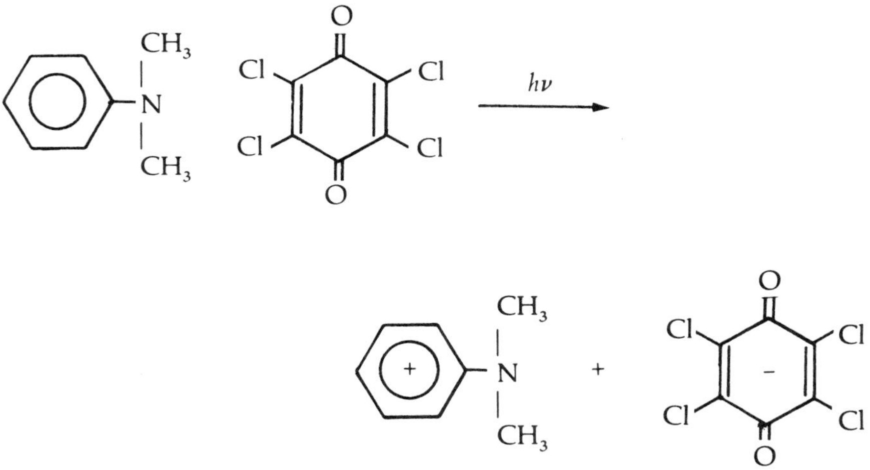

Many examples are available (1, 3, 4). Iodide ion in polar solvents has a strong absorption in the UV region around 2300 Å. Excitation of iodide ion in water leads to the formation of hydrated electrons e_{aq}^- and I atoms:

$$I_{aq}^- \xrightarrow[\lambda, \sim 2300 \text{ Å}]{h\nu} I + e_{aq}^-$$

This process is analogous to gas-phase photoionization, but occurs at much reduced energies in solution compared to the gas phase.

The absorption bands due to the electronic transitions just described show marked solvent effects (see Table II) and may be used to identify the nature of the transition. For example, the $\eta \rightarrow \pi^*$ transition of acetone shows a blue shift with increasing solvent polarity because polar solvents are more strongly hydrogen bonded to the ground state than to the excited state. Increasing solvent polarity continually decreases the energy of the ground state and, hence, increases the difference in energy between this state and the excited state.

A red shift is observed with increasing polarity for a $\pi \rightarrow \pi^*$ transition. Therefore, the dipole moment of the excited state is stronger than that of the ground state, and solvent interaction lowers the energy of the excited state more than that of the ground state. Increasing polarity then leads to a decrease in the energy difference between ground and excited states.

The environment of the molecule in an organized assembly can be understood by knowing how a solvent effects the spectroscopic properties of a molecule. Such probing techniques are very useful in micellar chemistry, and they will be discussed in detail later.

A common belief in spectroscopy and photochemistry is that excitation of a molecule into an excited state higher than the first excited state leads rapidly ($\sim 10^{-12}$ s) to the first excited state via the internal conversion of the states. This concept may also apply to radiation chemistry where the high-energy radiation suggests that high-energy loss events should be predominant on excitation with ionizing radiation.

Energy Loss with High-Energy Radiation

The concept of linear energy transfer (LET) is used widely in radiation chemistry (5–8). This term means that the energy is lost along the track of the radiation. Fast electrons and β-, γ-, and X-rays are radiation of low LET and transfer small amounts of energy at intervals along their respective radiation tracks. This process spaces out the chemical events produced by the radiation. Radiation of high LET (e.g., α-particles and slow electrons) deposits large amounts of energy over short distances of traverse, thus crowding together the radiation-produced chemical events.

For the most part, radiation chemistry is carried out with radiation of low LET. Fast electrons from accelerators are the most useful source of radiation for pulse radiolysis experiments. For completeness, the mode of energy loss by several types of radiation is important to consider after stating the common units of measurement of radiation work.

High-Energy Units. High-energy radiation is not measured in quanta as in photochemistry, but in rads or electronvolts. The rad finds use mostly in radiobiology, and the electronvolt finds use in radiation chemistry. A *rad* is the loss of 100 ergs of energy to 1 g of material. The radiation chemical yield is defined by the symbol G, which is the number of molecules changed per 100 eV of radiation absorbed.

Energy Loss by Electrons. Electrons or β-rays lose energy to matter by two processes, via Bremsstrahlung, and via elastic and inelastic collisions. An electron is decelerated on passing close to the nucleus of an atom, a process that gives rise to emission of radiation, usually X-rays or Bremsstrahlung. The subsequent energy loss by the X-ray to the medium gives rise to chemical reactions. Bremsstrahlung is the major mode of energy loss for electrons in the energy range from 10 to 100 MeV. Another energy loss by radiation emission is Cerenkov radiation.

Electrons of lower energy (<10 MeV), which are used primarily in chemical studies, also lose energy by elastic and inelastic collisions with electrons of the medium. This energy loss also gives rise to chemical reactions. Bethe produced an equation to describe this process and it emphasizes important differences between radiation chemistry and photochemistry. The energy loss E (ergs cm^{-1}) along the length of the track of the radiation x is given by:

$$\frac{dE}{dx} = \frac{2\pi N e^4 Z}{M_o V^2} \left[\ln \frac{M_o V^2 E}{2I^2(1 - \beta^2)} - \ln 2 \{2\sqrt{(1 - \beta)^2} - 1 + \beta^2\} + 1 - \beta^2 + \frac{1}{8} \{1 - \sqrt{(1 - \beta^2)}\}^2 \right]$$

where N is the number of atoms per milliliter, e is the electronic charge, Z is the atomic number of the medium, M_o is the rest mass of the electron, V is the velocity of the electron, I is the mean excitation potential of the materials, and $\beta = V/c$. The equation illustrates that the LET increases as the energy decreases, that is, as the radiation traverses the material. The LET is directly proportional to the electron density of the medium; therefore, the resonance effects, as in photochemistry, are not important. The energy lost via elastic collision is large and gives rise to ionization of the medium.

Energy Loss by Heavy Particles. Fast helium ions or α-particles lose energy to the electrons of the medium through which they pass by inelastic collision. Thus, the radiation events are distributed along the track of the particle, and only a small fraction of the α-particle energy is lost per collision. Deuterons—energetic deuterium ions—and fast protons lose energy similarly to α-particles. These radiations tend to produce tracks densely covered by free radicals and ions because the LET is high.

Neutrons. Neutrons normally react with the nucleii of the medium they pass through to produce energetic fragments, such as protons of other heavy positive ions. These particles then traverse the medium and produce chemistry characteristic of their LET. Thus, the radiation chemistry of neutrons is that of the energetic particles produced by the interaction of the neutron with the nucleii of the medium.

Electromagnetic Radiation. High-energy X-rays or γ-rays, which have very low LETs, are convenient sources of radiation because they irradiate large objects quite uniformly; consequently, a fairly uniform dose of energy is given to a sample. Two main forms of energy loss occur, via the photoelectric effect and the Compton process.

PHOTOELECTRIC EFFECT. In the photoelectric effect the entire energy of the photon is transferred to an electron of the medium, which is then ejected from the atom at high energy. The precise energy E is given by the difference in the energy of the photon E_p and the binding energy of the electron in the atom or molecule E_s by $E = E_p - E_s$. The ionized electron often has sufficient energy to cause further energy loss, and, thus, behaves as a low-energy β-ray. The photoelectric effect is most prominent in mixtures of high atomic number, and is important at low photon energies (i.e., below 60–100 keV).

COMPTON PROCESS. For most steady-state radiation chemical work, ^{60}Co γ-rays are used, where the photon energy is in excess of 1 MeV. With this type of radiation, the Compton process is the important mechanism of energy loss. The photon is considered to exhibit the properties of a particle, and only a portion of the γ-ray photon energy is imparted to an electron of the medium; the scattered photon retains the remaining energy. This process is most important at energies above 100 keV or so, and is very effective at these energies in material of low mass.

Cerenkov Radiation. Incident β-rays and ejected electrons can possess velocities in excess of the velocity of light in the medium. If this happens, intense radiation, called Cerenkov radiation, can be emitted by the electron. The deep blue color observed in reactor assemblies where uranium rods are immersed in a water matrix and in liquid samples irradiated by ^{60}Co γ-rays is a result of this radiation. The Cerenkov radiation extends over the whole optical spectrum, from red to UV, and ends at the cutoff absorption of the solvent. The Cerenkov radiation in water can extend down to vacuum UV below 2000 Å. The amount of photon energy expended as Cerenkov radiation is small (below 1%), but it can have useful and significant effects. Because Cerenkov radiation behaves as a source of white light from the IR to UV regions, it can be used as a monitoring light source for observation of short-lived species in radiolysis (9). Cerenkov radiation can also be used to excite molecules photochemically. Pulses of Cerenkov radiation, produced via pulses of fast electrons from accelerators, are sometimes more convenient to use than light pulses produced by conventional photochemical methods. Even small energy losses, such as those produced by Cerenkov radiation, may have some significance in radiation biology, where the biodamage, although small, seriously may affect a selected function of a complicated organism such as the cell nucleus.

Optical Approximation

The nature of the energy loss in radiolysis necessitates that one or all of the various possible modes of excitation of the solvent are activated; yet another popular name for the forementioned radiation is ionizing radiation. This name suggests that the chemistry of high-energy radiation is the chemistry of ions. Experimental data, particularly in the gas phase, show that this statement is largely true. However, extensive yields of excited states are observed in the radiolysis of several liquids, particularly arenes such as benzene and toluene. These excited states also may arise from very rapid ion neutralization following the initial formation of solvent ions (10–13).

Ions apparently are formed in large yields during the radiolysis of many liquids. However, evidence does not indicate that excited states are formed directly. Direct excitation may lead to excited states of a molecule by high-energy radiation. A simple exposé of this approach may be approached as follows (14, 15). The primary yield of a product G_1, expressed as molecules changed per 100 eV absorbed, is given by the proportionality $G_1 \propto f_1/E_1$, where f_1 is the oscillator strength of a transition leading to excitation, and E_1 is the energy of that transition. The yield G_T for all excitation produced, including ionization, is given by a corresponding expression. An immediate problem of this approach is the summation of the oscillator strengths and energies E for each transition of a molecule. The Thomas–Reiche–Kuhn rule gives an approximation that states that the total oscillator strength f_T is equal to the actual number of electrons in the molecule; the average energy E_T of all transitions also can be approximated. Thus the fraction F_1 of the energy lost in a particular transition is given by:

$$F_1 = \frac{f_1}{f_T} \cdot \frac{E_T}{E_1}$$

For example, in benzene the following values may be used: $f_T = 42$ and $E_T \sim 20$ eV; and the strong $E_{1u} \leftarrow A_{1g}$ transition to the third excited state is given by $f \approx 1$, and $E_1 = 6.5$ eV.

Thus, F_1 is $1/42 \times 20/6.5 \approx 7\%$. Therefore, the yield G_3 of the E_{1u} state of benzene in molecules per 100 eV of energy is unity. No evidence shows that the E_{1u} state is formed directly in the radiolysis of liquid benzene or toluene, although the system has been interpreted in this way (16).

The treatment just described is quite rough and extremely approximate for condensed systems. However, in the gas phase such processes do occur with electrons of energies ~ 100 eV (17).

Spatial Location of Events

For each type of radiation (α, β, γ, neutrons, etc.), the chemical events produced by passage of the radiation through the medium are strung out along the radiation track. Figure 2 illustrates this effect for radiation of low and high LET. The subsequent chemistry depends critically on the LET of the radiation. For example, radiolysis with radiation of low LET, such as fast electrons and γ-rays, leads to isolated chemical events along the radiation track in the medium. Radiolysis at high LET pushes the events together on a very short track and leads to chemical events clustered closely together in short cylinders. Under the conditions of high LET, the radiation produces chemical species at high concentrations, and intratrack reaction is greater than track–solute interactions. With radiation of low LET, track–solute interactions are more important. This picture manifests itself quite strangely in the observed chemistry; for example, radiolysis of water with α particles gives rise to H_2, O_2, and H_2O_2, due to intratrack interaction of the radiation produced species (e_{aq}^-, H, and OH):

$$H_2O \rightarrow e_{aq}^- + H^+ + OH$$
$$OH + OH \rightarrow H_2O_2$$
$$e_{aq}^- + H^+ \rightarrow H$$
$$H + H \rightarrow H_2$$
$$H + OH \rightarrow H_2O$$

Low LET (β Rays)

Short Tracks

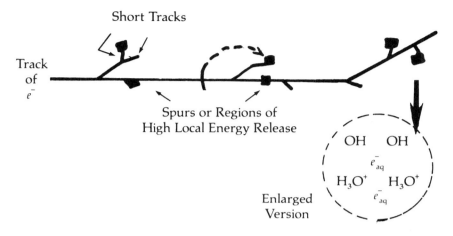

Track of e^-

Spurs or Regions of High Local Energy Release

Enlarged Version

High LET (α Particles)

Short Track

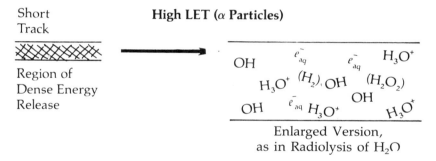

Region of Dense Energy Release

Enlarged Version, as in Radiolysis of H_2O

Figure 2. Illustration of radiation events with radiation of low and high LET.

Radiolysis of water with low intensities of low LET radiation gives rise to vanishingly small yields of products, such as H_2 and H_2O_2, because the products, although at low concentration, act as solutes and pick up the radicals formed (5–8).

$$H_2O_2 + OH \rightarrow H_2O + HO_2$$
$$H + H_2O_2 \rightarrow H_2O + OH$$
$$H_2 + OH \rightarrow H_2O + H$$

Radiolysis of water with very high intensities of low LET radiation, (e.g., pulsed electron beams), tends to simulate the effects of α-particles (i.e., high LET). The low LET tracks overlap and the free radical concentration is increased; these events lead to radical–radical reactions similar to those in single tracks of high LET radiation (*18, 19*).

Energy loss along the tracks causes multiple excitation or ionization events, and as much as 100 eV is lost in one location of the irradiated medium. Strings of blobs or spurs of reactive species are produced along the radiation track (*20*). The free radicals and ions in these blobs interact to form products, and they diffuse away to react with solutes in the bulk of the medium.

Summary of Radiolysis

Radiation of low LET is used almost exclusively to irradiate samples of interest. High intensities of fast electrons are used to carry out pulse radiolysis experiments; any undesirable effects of high radical or ion concentrations are avoided by suitable adjustment of the system. Reactions in the blob areas of the track, or spur events, are short-lived and are not important in a consideration of the chemical events of the system studied.

Formation of Excited States. In photochemistry the absorption of a light quantum by molecule leads directly to the formation of an excited state, which is usually a singlet excited state. A small probability exists for the direct formation of an excited triplet state. Excited singlet and triplet states are formed with relatively the same efficiency in the radiolysis of condensed materials; the mode of their formation is still a matter for debate.

PHOTOCHEMICAL EXCITATION. The ground states of most molecules are singlet in character; that is, the spins of the electrons in the molecule are all paired. In a triplet state two electrons of the molecule are not paired but do have parallel spins. Optical selection rules require that excitation of a singlet ground state of a molecule should lead to a singlet excited state. A very small probability exists that a singlet ground state may be excited directly to a triplet excited state. The yield of such a process is low, and chemistry resulting from such transitions is not observed readily. This situation is understandable from the concept of the oscillator strength of a particle transition that the molecule undergoes.

Oscillator Strength. The oscillator strength of a transition f, is defined by $f = 4.3 \times 10^{-9} \int \varepsilon d\bar{\nu}$ where ε is the extinction coefficient, and $\bar{\nu}$ is the energy in wave numbers of the absorption giving rise to the transition. The quantity $\int \varepsilon \, d\bar{\nu}$ is the area under the absorption curve of the molecule; ε is the ordinate and $\bar{\nu}$ as the abscissa. For a fully allowed transition, f approaches unity (Kuhn–Thomas rule); f may be as low as 10^{-10} for nonallowed transitions. The oscillator strength varies as the extinction coefficient of the absorption spectrum and is significant for singlet–singlet transitions. However for singlet–triplet and triplet–singlet transitions the oscillator strength is very low.

The singlet–triplet or triplet–singlet transitions may be enhanced in the presence of perturbing influences such as heavy atoms (I^-, Xe, and Tl^+) and magnetic molecules (O_2 or free radicals). These effects will be described later.

Triplet Excited States. Triplet excited states T_1 are conveniently formed by intersystem crossing from a singlet excited state $S_1{}^*$:

$$S_0 \xrightarrow{\;h\nu\;} S_1{}^* \xrightarrow{\;k_1\;} T_1$$

Reaction 1

This process occurs rapidly in most molecules, $k_1 \sim 10^8\ s^{-1}$, and also with high efficiency. Intersystem crossing competes with deactivation of the $S_1{}^*$ state by fluorescence emission: $S_1{}^* \rightarrow S_0$ + fluorescence, and also with other nonradiative processes. The number of molecules formed per quantum absorbed (quantum yield ϕ) for formation of T_1 by Reaction 1 can be close to unity for such molecules as benzophenone, and less than 0.1 for molecules such as *p*-terphenyl.

Higher Excited States. Excitation of a molecule into excited states beyond that of the first excited state usually leads to rapid internal conversion to the first excited state. Azulene is an exception to this rule, however, and the predominant fluorescence of this molecule arises from the second excited state; little fluorescence is observed from the first excited state. The transition from the azulene ground state S_0 to the first excited state $S_1{}^*$ exhibits an absorption maximum at 5850 Å. The coincidence of the zero–zero level bands of the $S_1{}^*$ fluorescence and S_0 absorption indicate strong Frank–Condon overlaps of the two states. This overlap leads to rapid relaxation of $S_1{}^*$ to S_0 and is in competition with the much slower fluorescence emission from $S_1{}^*$, which has a small quantum yield ($\phi < 10^{-4}$). The relaxation from the second excited state $S_2{}^*$ to $S_1{}^*$ is slower than normal due to the larger energy difference between the states. Thus, a reasonable opportunity is afforded the $S_2{}^* \rightarrow S_0$ transition with fluorescence from the $S_2{}^*$ state, $\phi(S_2{}^* \rightarrow S_0) \sim 0.02$.

In some cases chemical products are observed from higher excited states. A classic example is benzene, where excitation into the first excited state, B_{2u} ($\lambda = 2537$ Å), gives benzvalene. Excitation into the third excited state, E_{1u} ($\lambda = 1839$ Å), gives dewar benzene and benzvalene as products (21). However, internal conversion from $E_{1u} \rightarrow B_{2u}$ still takes place with a lower yield, which can be dependent on solvent (22).

The rapid rate of internal conversion previously was taken as an indication that fluorescence could not be observed from these states because the quantum yield would be too low. However, both steady-state (23, 24) and pulsed methods (25, 26) have produced accurate fluorescence spectra of higher excited states. Figure 3 summarizes the various processes discussed. Excitation of most molecules in this discussion into their higher excited states leads rapidly (10^{-12} s) to the first excited state S_1^*, and subsequent, less rapid ($10^{-8}-10^{-7}$ s) proc-

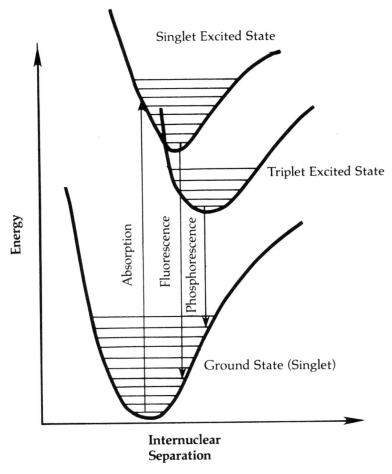

Figure 3. Potential energy diagram of ground state, singlet, and triplet excited states.

esses such as fluorescence and intersystem crossing lead to triplet states and luminescence.

HIGH-ENERGY RADIATION. The exact mechanism of formation of excited states in radiolysis is still a matter of concern. The occurrence of high-energy events in irradiated systems suggests that ionization is all important and that subsequent ion neutralization leads to excited states. The optical approximation suggests that direct excitation may be possible; however, the low oscillator strengths of the first excited states of most molecules of interest (e.g., benzene, toluene, acetone and alkanes) indicate that the only direct excitation of any significance must be to higher excited states (e.g., the third excited state E_{1u} in benzene). Radiolysis of aromatic liquids gives rise to large yields of excited singlet and triplet states that are produced very rapidly following excitation (<10–11 s) (27, 28). This effect is a direct contrast to photochemistry where only excited singlet states are formed directly. The production of both excited singlet and triplet states supports the evidence for an ionic origin for the excited states (29).

In alkanes the situation is similar; however, solute ions recombine to form excited solute states (30). This recombination is similar to other systems where slower ion neutralization in low temperature glasses (31) or steady-state mixing experiments (32) show that excited states arise from ion recombination. Excited triplet states have been observed in acetone (33), and singlet excited states in dioxane (34). The mechanisms for excited-state formation in these liquids are not known.

Table III summarizes the observed yields of excited states on radiolysis of various liquids of interest.

TWO-PHOTON EXCITATION. At high laser intensities it is possible to produce excited states by two-photon absorption (35) and by biphotonic processes (26, 36). In two-photon absorption a molecule normally transparent to the radiation can absorb, at high laser power, two photons simultaneously. Good examples are benzene (37) and toluene (38), both of which are transparent to light at the doubled frequency of a ruby laser, $\lambda = 3471$ Å. However, irradiation of these liquids at high laser intensity with light of $\lambda = 3471$ Å gives rise to fluorescence characteristic of excited monomer and excimer states:

$$\text{benzene} \xrightarrow{2h\nu} (\text{benzene})^* \text{ or } B_1^* \rightarrow (B)_2^*$$

The excimer states were also identified by absorption spectroscopy. The yield of excited states varies with the square of the laser intensity, a feature that indicates a two-photon process. A simple relationship that describes the number of two-photon excitations (N) per unit volume is

Table III.
Yield of Excited States upon Radiolysis of Solvents

Solvent	G (Triplet)	G (Singlet)
o-Xylene[a]	1.7	2.5
m-Xylene[a]	1.8	2.7
p-Xylene[a]	2.4	2.0
Pseudocumene[a]	1.8	1.6
Mesitylene[a]	1.8	1.6
Toluene[a]	2.4	2.1
Benzene[a]	3.8	1.6
Benzyl alcohol[b]	1.1	0.7
Benzonitrile[c]	1.4	1.2
Dimethylaniline[d]	3.1	0.9
Phenol[e]	0	0
Cyclohexane[f]	0.7	0.7
Tetrahydrofuran[g]	0.1	0.04
Dioxane[h]	—	1.03
Methanol[g]	0	0
Water[g]	0	0

NOTE: G = molecules/100 eV.
[a] From Ref. 59.
[b] From Ref. 7.
[c] From Ref. 8.
[d] From Ref. 60.
[e] From Ref. 61.
[f] From Ref. 9.
[g] From Ref. 62.
[h] From Ref. 63.

given by (39) $N = (e^2/mc^2) (\lambda^2/b^2 v)N_0 F^2$ where $e^2/mc^2 = 2.8 \times 10^{-13}$ cm, λ is the wavelength of excitation, b is the refractive index of the medium, v is the width of the higher state absorption band, N_0 is the number of molecules/unit volume, and F is the photon flux per unit area. This relationship indicates that the process is most efficient at high concentration, that is, in the pure state. Even at conventional laser intensities (200 MW), only about 10% of the laser power is absorbed in a 1-cm pathlength of pure liquid. Experimentally, this situation is similar to radiolysis, where the solvent absorbs energy that subsequently is passed on to a solute at low concentration.

BIPHOTONIC PROCESSES. The term biphotonic process describes the consecutive absorption of two photons of energy. This event can occur at high laser intensities and also at low light intensities if the intermediate species formed by the absorption of the initial photon is long-lived (26, 36). An example in micellar chemistry is the two-photon photoionization of pyrene (P) that occurs via absorption of a second photon by the excited

singlet state (P*), which is formed by absorption of the first photon (40, 41).

$$P_1 \xrightarrow[\text{3471 Å}]{h\nu} P_1{}^* \xrightarrow{h\nu} P_1{}^{**} \longrightarrow P^+ + e^-$$

The excited $P_1{}^*$ has a sufficiently long lifetime ($\tau > 300$ ns) so that it
can absorb a second photon to produce a super excited state for ionization.

If the intermediate formed by the first photon is long-lived, for
example, a triplet state, then it may subsequently be photolyzed. A
good example is N, N, N', N'-tetramethyl-p-phenylenediamine (TMPD),
which on photolysis in low temperature glasses forms triplet TMPD
that absorbs a second photon to give photoionization (36):

$$\text{TMPD} \xrightarrow{h\nu} \text{TMPD}_1{}^* \longrightarrow \text{TMPD}_3{}^* \xrightarrow{h\nu} (\text{TMPD})^+ + e^-$$

This process is quite prevalent in micellar systems where intermediates
such as triplet states are stabilized and have long lifetimes.

Formation of Ions. PHOTOCHEMICAL. Quanta with energies in excess of
8.0 eV ($\lambda < 2000$ Å) are sufficiently energetic to photoionize many
aromatic molecules; the ejected electron and parent cation are quite
reactive in subsequent reactions. A typical point of reference in discussing photoionization is to consider either the effect of phase on the
energy required to eject an electron from a molecule P or the photoionization potential or threshold. (The symbol I_g is used to denote the
ionization energy in the gas phase and I_s for the ionization energy in
solution.)

$$P \xrightarrow{h\nu} P^+ + e^- \ (I_g \text{ or } I_s)$$

The ions produced in photoionization are observed readily by sensitive
conduction methods that monitor both the electron e^- and the parent
cation P^+. For most molecules light quanta in the vacuum UV, that is,
$I_g > 7\text{--}8$ eV, have to be used to produce photoionization. A few molecules have lower I_g, for example, TMPD, where $I_g = 6.5$ eV.

The energy required for the photoionization process is lowered
markedly in the solution phase, because of interaction of the medium
with the cation and the electron. The ionization potential in the condensed phase, I_s, is related to that in the gas phase by the relationship (42):

$$I_s = I_g + P_+ + V_o$$

where P_+ is the polarization energy of the cation and V_o is the energy of e^- in the medium. For large aromatic molecules the polarization energies are typically -1.5 eV (43, 44); V_o may be as large as -2.0 eV for water (45). Thus, the net lowering of the ionization potential often is greater than -3.50 eV. Only 3.0 eV is required to eject an electron from TMPD in aqueous solution (46, 47).

The polarization energy P_+ may be expressed via the Born equation:

$$P_+ = \frac{-Z^2e^2}{2R}\left(1 - \frac{1}{\varepsilon}\right)$$

where ε is the fast dielectric constant of the medium and is taken to be the square of the refractive index, Ze is the electronic charge of the cation, and R is the radius of the cation in question. Table IV lists I_s, I_g, P_+, V_o, and R for several molecules of interest in micellar chemistry.

Both theory and data indicate a substantial decrease in the energy required to photoionize a molecule in solution when compared to the gas phase. The biphotonic excitation process can be used to generate sufficient energy for the e^- ejection process, either via high powered

Table IV.
Parameters Connected with Photoionization of Selected Molecules

Molecule	I_g (eV)	I_s (TMS) (10) (eV)	P_+ (eV)	P_+ (avg) (eV)	r (Å)	R (Å)
TMPD	6.25	4.27	-1.37	-1.36	2.51	3.73
		4.72	-1.36			
TMB	6.40	4.52	-1.27	-1.32	2.59	4.11
		4.87	-1.36			
Phenothiazine	6.96	4.80	-1.55	-1.57	2.18	3.87
		5.20	-1.59			
1-Aminopyrene	7.02	4.77	-1.64	-1.67	2.05	4.08
		5.15	-1.70			
3-Aminoperylene	6.94	4.75	-1.58	-1.62	2.11	4.32
		5.12	-1.65			
Pyrene	7.55	5.46	-1.91	-1.91	1.66	3.98
Perylene	6.92	4.82	-1.49	-1.52	2.25	4.23
		5.20	-1.55			
Anthracene	7.42	5.60	-1.21	-1.30	2.63	3.84
		5.87	-1.38			
Tetracene	6.88	4.92	-1.33	-1.43	2.39	4.18
		5.19	-1.52			

lasers or via steady-state irradiation in a system where the intermediate state is protected.

CHARGE-TRANSFER ABSORPTION. Many ions show charge transfer to solvent spectra. The halide ions are well-known examples, particularly iodide ion, which exhibits a strong UV spectral absorption at $\lambda < 2500$ Å. This absorption is attributed to an e^- charge transfer to solvent spectrum (48):

$$I^- \xrightarrow{h\nu} I + e_s^-$$

The solvated electron (e_s^-) is one product of this system.

Flash photolysis studies show that solvated electrons are formed in the photolysis of anions (X^-) in many polar solvents, and that an intermediate of the type X_2^- (i.e., I_2^- in the case of I^-) is also formed via the following reaction sequence (49):

$$X^- \xrightarrow{h\nu} X + e_s^-$$
$$X + X^- \rightleftharpoons X_2^-$$

The quantum yield for the e_s^- is low ($\phi \sim 0.2$) and is attributed to a back reaction of e_s^- and I in the primary solvent cage, as follows: $I + e_s^- \rightarrow I^-$. Similar data are reported for several ions, for example, CNS^-, $Fe(CN)_6^{4-}$, Fe^{2+}, SO_4^{2-}, and OH. All of these ions produce e_{aq}^- on photolysis in water with UV light (50).

HIGH-ENERGY RADIATION. High-energy radiation readily ionizes the medium through which it passes, a fact supported by conduction studies, and cloud- and bubble-chamber phenomena. The yield of ionic species [G (ions)] increases with solvent polarity and is very large in water (> 5) [G (ions) is the yield of ions produced per 100 eV of energy absorbed]. In benzene, G (ions) < 0.1. Table V lists the ion yields in several solvents. In nonpolar liquids such as cyclohexane, the so-called free-ion yield is quoted. For cyclohexane this yield is about 0.1. However, the initial ion yield at short times following the excitation process is much larger; G (ions) ~ 1.6 at 10^{-9} s (12). The free-ion yield is a measure of the ions that escape geminate ion recombination. In practice, free ions are still detected at 1 ms following a short radiation pulse.

The cation of the system is derived from the solvent and, in some cases, may retain its basic integrity. However, in many solvents, in particular proteated ones, ion–molecule reactions rapidly degrade the solvent cation to a proton and a free radical:

$$ROH \rightarrow ROH^+ \xrightarrow{ROH} RO^{\cdot} + ROH_2^+$$

Table V.
Yield of Ions on Radiolysis of Solvents

Solvent	G (Ions)	Solvent	G (Ions)
o-Xylene	~0.1	Benzonitrile	1.4
m-Xylene	~0.1	Dimethylaniline	<0.2
p-Xylene	~0.1	Phenol	3.0
Pseudocumene	~0.1	Cyclohexane	1.6
Mesitylene	~0.1	Tetrahydrofuran	0.66
Toluene	~0.1	Dioxane	0.12
Benzene	~0.1	Methanol	2.0
Benzyl alcohol	2.1	Water	3.5

RANGES OF ELECTRONS. The yield of ions that escape paired recombination depends entirely on the mobility of the ions in the solvent and on the initial separation of ions. The initial ionization events eject the electron away from the cation with excess energy. The energy is rapidly degraded to the solvent molecules which leads, within 10^{-12} s, to a thermal electron (51). Because most of the excess electron energy is lost during the very early stages of the electron motion, the separation of the cation and the thermal electron may be approximated by the range of low energies of the electrons in the solvent. The ranges of slow, low-energy electrons in many fluids have been measured and are given in Table VI.

Formation of Free Radicals. PHOTOPRODUCTION. Of the many examples of photolytic production of free radicals, a few are pertinent to the present discussion.

Carbon–Halide Bond Scission. Many organic halides give rise to radicals upon photolysis; particularly, aryl iodides readily decompose to aryl radicals and iodide atoms (52):

$$ArI \xrightarrow{h\nu} (ArI)^* \rightarrow Ar^{\cdot} + I$$

The dissociation process takes place through the triplet state of the halide that is formed rapidly via intersystem crossing from the excited singlet state. Sophisticated beam work at low pressures (53), shows that the rates of dissociation of several aryl and alkyl halides are very rapid (10^{-12}–10^{-11} s). The state involved is believed to be a σT_1^* triplet state. Thus, excitation is into the π–π^* excited state followed by intersystem crossing to the π–π^* and α–π^* triplet states. Pulse radiolysis studies (54) have shown that the rate of decomposition of an excited triplet aryl halide is temperature dependent and decreases upon lowering the temperature of the solution. This observation agrees with previous

Table VI.
Low-Energy Electron Ranges in Liquids

Liquid	Electron Energy (eV)	Range (Å)
n-Pentane	0.15	36
	0.23	36
Cyclopentane	0.52	71
	0.47	84
Neopentane	0.24	135
	0.23	132
	0.34	178
n-Hexane	0.43	36
	0.36	49
2-Methylpentane	0.30	68
	0.13	74
	0.22	47
3-Methylpentane	0.38	53
	0.33	48
	0.17	43
	0.20	47
2,2,4-Trimethylpentane	0.20	100
	0.24	103
Tetramethylsilane	0.25	186
	0.25	142
	0.12	160
	0.27	184
	0.20	198
	0.14	190
	0.08	176

(Reproduced from Ref. 64. Copyright 1972, American Chemical Society.)

photochemical work and also with the observed aryl halide phosphorescence in rigid glasses at 77 K. Decomposition of 1-iodonaphthalene at room temperature occurs with an activation energy of 5.9 kcal/mol, and decomposition of 2-iodonaphthalene is 4.8 kcal/mol. Therefore, a lifetime of 10^{-11} s is predicted for the triplet state at room temperature, a value that agrees with previous beam studies (53). Alkyl and aryl radicals may be prepared conveniently by photolysis of suitable halide; however, even more convenient methods are described later.

Decomposition of Peroxides. Many peroxides decompose on photolysis to yield oxy radicals. The simplest example is H_2O_2, which decomposes upon photolysis to give hydroxyl radicals (55–57).

$$H_2O_2 \xrightarrow{h\nu} OH + OH$$

The net decomposition of H_2O_2 upon photolysis has a quantum yield of about 1.0. However, this number is misleading because about 50% of the OH radicals recombine to form H_2O_2 in the solvent cage. A series of free radical reactions can lead to further H_2O_2 decomposition:

$$OH + H_2O_2 \rightarrow H_2O + HO_2$$
$$HO_2 + HO_2 \rightarrow H_2O_2$$

Further complications at low light levels are due to chain reactions of HO_2 and H_2O_2. The HO_2 radical may exist in two forms:

$$HO_2 \leftrightharpoons O_2^- + H^+$$

The rate of dimerization of O_2^- is very slow compared to that of HO_2. The OH radical is very reactive and can be used to generate further radicals.

Photolysis of Carbonyl Compounds. Photolysis of carbonyls of the form $R(C=O)R_1$ (e.g., acetone) leads to the formation of radicals. The initial excitation produces the $n\pi^*$ excited singlet state which rapidly converts to an $n\pi^*$ triplet state:

$$CH_3-\overset{\overset{\displaystyle O}{\|}}{C}-CH_3 \rightarrow [\, (CH_3-\overset{\overset{\displaystyle O}{\|}}{C}-CH_3)^*]^{\text{singlet}} \rightarrow [\, (CH_3-\overset{\overset{\displaystyle O}{\|}}{C}-CH_3)^*_3]^{\text{triplet}}$$

The triplet subsequently breaks down into two radicals:

$$[\, (CH_3-\overset{\overset{\displaystyle O}{\|}}{C}-CH_3)^*]^{\text{triplet}} \rightarrow CH_3-C=O + CH_3^{\cdot}$$
$$CH_3-C=O \longrightarrow CH_3^{\cdot} + CO$$

The triplet state is reactive in many media and gives rise to other radicals.

$$\overset{\displaystyle R_1}{\underset{\displaystyle R3}{\diagdown \diagup}} C=O^* + RH \rightarrow \overset{\displaystyle R_1}{\underset{\displaystyle R_3}{\diagdown \diagup}} C^{\cdot}-OH + R^{\cdot}$$

In some instances the activated C=O group can undergo hydrogen abstraction with the attendent alkyl groups to produce a short-lived biradical:

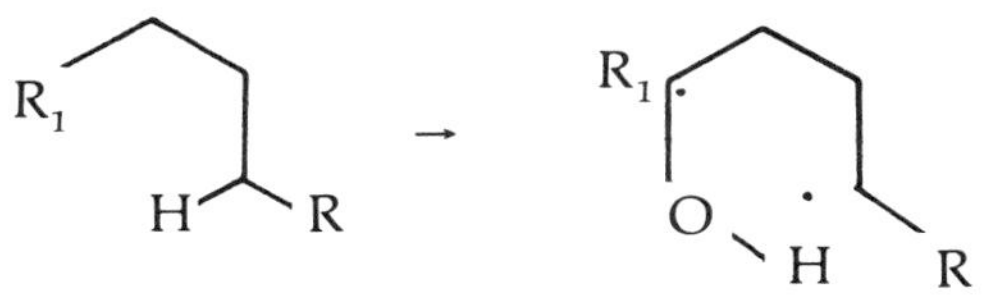

HIGH-ENERGY RADIATION. High-energy radiation may produce free radicals directly. However, a discussion of the radiolysis of a condensed system usually is confined to secondary production of free radicals. A typical example is the radiolysis of water:

$$H_2O \rightarrow H_2O^+ + e^-$$

The initial radiolytic event usually leads to the formation of the ions H_2O^+ and e^-; products such as OH radicals and solvated electrons are formed rapidly ($<10^{-11}$ s). The following series of steps may explain the observed chemistry:

$$e^- + H_2O \rightarrow e_{aq}^-$$
$$H_2O^+ + H_2O \rightarrow H_3O^+ + OH$$

The latter reaction has not been observed in liquid water but is projected from mass spectrophotometric data of water vapor (58). Hydrogen atoms and hydrogen peroxide are also observed in low yield and are explained by reactions of the primary species, that is, e_{aq}^-, H_3O^+, and OH, in regions of high-energy loss or spurs,

$$e_{aq}^- + H_3O^+ \rightarrow H + H_2O$$
$$OH + OH \rightarrow H_2O_2$$

Free radicals of various kinds have been reported in organic liquids (5–7). The exact origin of these species is uncertain, although plausible reaction schemes may be written to explain them.

It is customary to "tailor make" a particular free radical species by appropriate design of the chemical system. Precise details of these systems will be given later.

Literature Cited

1. Turro, N. "Modern Molecular Photochemistry"; Benjamin; New York, 1978.
2. Jaffe, H. H.; Orchin, M. "Theory and Application of Ultra-violet Spectroscopy"; J. Wiley: New York, 1962.
3. Birks, J. B. "Photophysics of Aromatic Molecules"; Wiley: London, 1970.
4. Mataga, N. In "Molecular Association"; Foster, R. Ed.; Acad. Press: London, 1979; Vol. 2, p. 2.
5. Spinks, J. W.; Woods, R. J. "An Introduction to Radiation Chemistry"; Wiley: New York, 1976.
6. Hughes, G. "Radiation Chemistry"; Clarendon Press; Oxford, 1973.
7. Swallow, A. J. "Radiation Chemistry"; Halstead Press: London, 1973.
8. Allen, A. O., "The Radiation Chemistry of Water and Aqueous Solutions"; Van Nostrand: New York, 1961.
9. Hunt, J. W.; Bronskill, M. J.; Wolff, R. K. *Adv. Radiat. Res. Phys. Chem.* **1973**, *1*, 271.
10. Thomas, J. K. *Int. J. Radiat. Phys. Chem.* **1976**, *8*, 1.
11. Salmon, G. A. *Int. J. Radiat. Phys. Chem.*, **1976**, *8*, 13.
12. Beck, G.; Thomas, J. K. *J. Phys. Chem.* **1972**, *76*, 3856.
13. Thomas, J. K. *Annu. Rev. Phys. Chem.* **1970**, *21*, 17.
14. Fano, U. *Phys. Rev.* **1960**, *118*, 451.
15. Platzman, R. L. In "Radiation Research"; G. Silini, Ed.; North Holland: Amsterdam, 1967; p. 20.
16. Gangwer, T. E.; Thomas, J. K. *Int. J. Radiat. Chem. Phys.* **1975**, *7*, 305.
17. Lassettre, E. N.; Skerbele, A.; Dillon, M. A.; Ross, K. J. *J. Chem. Phys.* **1968**, *48*, 5066.
18. Thomas, J. K; Hart, E. J. *Radiat. Res.* **1962**, *17*, 408.
19. Fricke, H.; Thomas, J. K. *Radiat. Res. Suppl.* **1964**, *4*, 35.
20. Mozumder, A.; Magee, J. L. *J. Chem. Phys.* **1966**, *45*, 3332; Mozumder, A. In "Advances in Radiation Chemistry"; Barton & Magee, Eds.; Wiley: New York, 1969; Vol. 1, p. 1.
21. Smith, D:; Gilbert, A.; Robinson, D. A. *Angew. Chem.* **1971**, *10*, 745; Kaplan, L.; Ransch, D. J.; Wilzbach, K. E. *J. Am. Chem. Soc.* **1972**, *94*, 8638; Ibid **1966**, *88*, 2881; **1966**, *89*, 1030, 1031.
22. Lawson, C. W.; Hiryama, F.; Lipsky, S. *J. Chem. Phys.* **1969**, *51*, 1590.
23. Gregory, T. A.; Hirayama, F.; Lipsky, S. *J. Chem. Phys.* **1973**, *58*, 4697.
24. Siomos, K.; Konronkis, G.; Christophorou, L. G. *Chem. Phys. Lett.* **1981**, *80*, 504.
25. Lin, H.-B.; Topp M. R., *Chem. Phys Lett.* **1977**, *48*, 251; Lin, H.-B.; Topp, M. R. *Chem. Phys. Lett* **1979**, *36*, 365.
26. Nickel, B. *Chem. Phys Lett.* **1974**, *27*, 84.
27. Beck, G.; Ding, A.; Thomas, J. K. *J. Chem. Phys.* **1979**, *71*, 2611.
28. Beck, G.; Thomas, J. K. In "Liquid Scintillation Counting"; Peng, Alpen, and Horrachs, Eds.; Acad. Press: New York, 1980; Vol. 5, p. 17.
29. Magee, J. L.; Huang, J. T. J. *J. Phys. Chem.* **1972**, *76*, 3801.
30. Thomas, J. K.; Johnson, K.; Klippert, T.; Lowers, R. *J. Chem. Phys.* **1968**, *48*, 1608.
31. Brocklehurst, B.; Russell, R. D. *Trans. Faraday Soc.* **1969**, *65*, 2159.
32. Weller, A.; Zachariasse, K. *J. Chem. Phys.* **1967**, *46*, 4984.
33. Arai, S.; Dorfman, L. M. *J. Phys. Chem.* **1965**, *69*, 2239; Rodgers, M. A. J. *Trans. Faraday Soc.* **1971**, *67*, 1029.
34. Baxendale, J. H.; Rodgers, M. A. J. *J. Phys. Chem.* **1968**, *72*, 3849.
35. Kaufman, K. J.; Rentzepis, P. M. *Acc. Chem. Res.* **1975**, *8*, 407.
36. Cadogan, K. D.; Albrecht, A. C. *J. Chem. Phys.* **1969**, *51*, 2710.
37. Richards, J. T.; Thomas, J. K. *Chem. Phys. Lett.* **1970**, *5*, 527.
38. Beck, G.; Thomas, J. K. *J. Chem. Soc. Faraday Trans.* **1976**, T, *72*, 2610.
39. Kaiser, W.; Garett, C. E. B. *Phys. Rev. Lett.* **1961**, *7*, 229.
40. Richards, J. T.; West, G.; Thomas, J. K. *J. Phys. Chem.* **1970**, *74*, 4137.

41. Thomas, J. K.; Piciulo, P. p. 97 in Advances in Chem. Series No. 184, M. Wrighton Ed. 1980.
42. Ray, B.; Jortner, J. *Chem. Phys. Lett.* **1969**, *4*, 155.
43. Holroyd, R. A.; Russell, R. L. *J. Phys. Chem.* **1974**, *78*, 2128.
44. Thomas, J. K.; Picuilo, P. *J. Am. Chem. Soc.* **1978**, *100*, 3239.
45. Barker, G. C.; Bottura, G.; Cloke, G.; Gardner, A. W.; Williams, M. J. *J. Electroanal. Chem. Interfacial. Electrochem.* **1974**, *50*, 323.
46. Richards, J. T.; Thomas, J. K. *Trans. Faraday Soc.* **1970**, *66*, 621.
47. Wu, K. C.; Lipsky, S. *J. Chem. Phys.* **1977**, *66*, 5614.
48. Platzman, R. L.; Franck, J. *Z. Phys.* **1954**, *138*, 411.
49. Dorfman, L. M.; Matheson, M. S. *J. Chem. Phys.* **1960**, *32*, 1870.
50. Calvert, J. G.; Pitts, J. N. "Photochemistry"; J. Wiley: New York, 1966; p. 271.
51. Holroyd, R. A.; Dietrich, B. K.; Schwarz, H. A. *J. Phys. Chem.* **1972**, *76*, 3794.
52. Levy, A.; Meyerstein, D.; Ottolenghi, M. *J. Phys. Chem.* **1973**, *77*, 3044.
53. Dzvonik, M.; Yang, S.; Bersohn, R. *J. Chem. Phys.* **1974**, *61*, 4408.
54. Grieser, F.; Thomas, J. K. *J. Chem. Phys.* **1980**, *73*, 2115.
55. Baxendale, J. H.; Wilson, R. *Trans. Faraday Soc.* **1957**, *53*, 344.
56. Baxendale, J. H.; Thomas, J. K. *Trans. Faraday Soc.* **1958**, *54*, 1515.
57. Hunt, J; Taube, H. *J. Am. Chem. Soc.* **1952**, *74*, 5999.
58. Tal'Rose, V. L.; Frankevich, E. L. *Zh. Fiz. Khim.* **1960**, *34*, 2709; **1959**, 33, 1093.
59. Gangwer, T.; Thomas, J. K. *Rad Res* **1973**, *54*, 192.
60. Land, E. J.; Richards, J. T.; Thomas, J. K. *J. Phys. Chem* **1972**, *76*, 3805.
61. Platzner, T.; Thomas, J. K. published work.
62. Allen, A. O. NSRDS-NBS 57, NBS Reports 1976.
63. Baxendale, J. H.; Rodgers, M. A. J. *J. Phys. Chem.* **1968**, *72*, 3849.
64. Holroyd, R. A.; Dietrich, B. K.; Schwarz, H. A. *J. Phys. Chem.* **1972**, *76*, 3794.

3

Chemistry of Intermediates

Both in photochemistry and radiation chemistry the initial act of energy loss gives rise instantaneously to excited states, ions, and/or free radicals. Rarely is this "trinity" of products balanced; one type of species tends to predominate. The species formed initially are not the final, stable result of the energy loss process unless the initial active species are stabilized in some way, for example, in low temperature rigid glasses. The final stable products of the radiation-induced processes result from the degradation of the excited states, ions, free radicals, or reactions of these species with other components of the system. The many and varied reaction sequences of the initially produced species are of vital concern to the present discourse.

Excited States

The various kinetic processes of excited states may be divided conveniently into two categories: energy-transfer processes whereby the energy of the excited state is transferred to another molecule (these processes are physical in nature); and genuine chemical reactions such as H atom abstraction, bond rupture, and electron transfer.

Energy Transfer Processes. Both singlet and triplet excited states undergo static and dynamic energy transfer. In static energy transfer, the energy of the donor excited state is transferred over a distance to the acceptor without actual physical contact between the two species. In dynamic energy transfer, both the acceptor (A) and the donor excited state (D) have to diffuse to within a short distance of each other, usually 3–5 Å, so that the reactants are in physical contact.

Dynamic Energy Transfer. The transfer of energy from D to A is usually efficient if the energy of the excited state of D is larger than that of A; therefore, the process is essentially exothermic. If the energies of the excited states of A and D, which are denoted as A* and D*, are comparable, then energy transfer becomes inefficient, and it is vanishingly small if the energy of D* is smaller than A*. Energy transfer is rapid provided the free energy change for the reaction is less than -40 kJ/mol (1). Consider the situation where the energy of D* is larger than A*, so that energy transfer is efficient. In this case the rate of energy transfer in solution is controlled by the rate of mutual diffusion of D* and A to each other, and is described by the Smoluckowski equation (2):

$$k_\gamma = 4\pi\gamma D \left(1 + \sqrt{\gamma/\pi D\tau}\right) \qquad (1)$$

where k_γ is the reaction rate constant, D is the sum of the diffusion constants of D* and A, γ is the interaction distance, and τ is the time after initiation of the reaction. Equation 1 may be used in various forms to present a convenient fit of data and experiment (3). However, for most purposes if τ is sufficiently large, that is, $\tau > 10^{-9}$ s for simple liquids such as water and hexane, then the second term in parentheses becomes small compared to unity and the equation reduces to $k = 4\pi\gamma D$. Using a fair degree of approximation, we can write the diffusion constants of D* and A in the form:

$$D_{D^*} = \frac{kT}{6\pi\gamma_{D^*}\eta} \quad \text{and} \quad D_A = \frac{kT}{6\pi\gamma_A\eta}$$

where k is the Boltzmann constant, T is the medium temperature, η is the viscosity of the medium, γ_{D^*} is the radius of D*, and γ_A is the radius of A.

For small reactants in liquids of low viscosity and at room temperature, k_γ is calculated as 5×10^9 to 5×10^{10} M^{-1}s^{-1}. If the reactants are ions then an additional correction is placed in the equation for k, for the electrostatic effects experienced by the ionic reactants (4):

$$k_{ions} = k\left(\frac{z_1 z_2 e^2}{\gamma\mu kT}\right)\left[\exp\left(\frac{z_1 z_2 e^2}{\gamma\mu kT}\right) - 1\right]^{-1}$$

where $z_1 e$ and $z_2 e$ are the charges on reactants 1 and 2, respectively, and μ is the dielectric constant of the medium. Sample calculations for the reaction of hydrated electrons with various solutes and comparison to experimental data are given here (5):

Reaction	γ (Å)	k (calculated) (LM^{-1}s^{-1})	k (observed) (LM^{-1}s^{-1})
$e_{aq}^- + O_2$	1.6	2.31×10^{10}	2.0×10^{10}
$e_{aq}^- + e_{aq}^-$	2.7	0.94	0.45
$e_{aq}^- + Cu^{2+}$	0.7	5.4	3.3

Table I contains a list of excited singlet and triplet reactions in several solvents. The simple theory just given describes the observed data quite well for many reactions. Other writings on this general subject also cover details of exciton energy transfer in solid systems (6–9).

OTHER CONSIDERATIONS IN DIFFUSION PROCESSES. The simple formulations for diffusion-controlled reactions indicate that the rate of reaction and rate constant vary inversely with the viscosity of the medium. Even if the reaction is truly diffusion controlled and no barrier exists to prevent reaction (i.e., the activation energy of the reaction is zero in the gas phase), the activation energy is still finite in solution. To a first approximation, as k varies as $1/\eta$, the activation energy of a diffusion-controlled reaction is simply the activation energy of the viscosity of the medium. Several reported reactions of the hydrated electron where $E_{diffusion} \sim 3.0$ kcal/mol have supported this theory (10). Other reports discuss diffusion effects in free radical reactions (11) and diffusion-controlled triplet energy transfer (12).

The mean displacement (x) of a particle in solution or of a molecule in a particle in a defined interval of time t is given by the expression (13) $x = \sqrt{2Dt}$, where D is the diffusion constant of the entity in square

Table I.
Rates of Transfer of Singlet (S) and Triplet Energy (T).

Excited Molecule	Quencher	$k\ (Lmol^{-1}s^{-1})$
Benzene (S)(Solvent a)	Biacetyl	1.9×10^{10a}
Benzene (S)(Solvent b)	Biacetyl	3.3×10^{10a}
Benzene (S)(Solvent c)	1,1'-Binaphthyl	3.5×10^{10b}
	1,2-Benzanthracene	3.3×10^{10b}
p-Xylene (S)(Solvent d)	1,1'-Binaphthyl	4.8×10^{10b}
	1,2-Benzanthracene	3.6×10^{10b}
p-Xylene (T)(Solvent d)	p-Terphenyl	1.1×10^{10b}
	tert-Butene	1.2×10^{8b}
	Anthracene	1.4×10^{10c}
	Biacetyl	5.9×10^{10c}
Dimethylaniline (T)(Solvent e)	Pyrene	9.0×10^{9c}
	Naphthalene	4.8×10^{9c}
	Piperlyene	1.1×10^{10c}
Anthracene (S)(Solvent d)	Perylene	1.2×10^{11a}
	Rubrene	3.2×10^{10a}
Perylene (S)(Solvent d)	Rubrene	1.3×10^{11a}

NOTE: Key to solvents: a, cyclohexane; b, hexane; c, benzene; d, p-xylene; and e, dimethylaniline.
[a] From Ref. 123.
[b] From Ref. 124.
[c] From Ref. 125.

centimeters per second. If t is given in seconds then x is obtained in centimeters. For example, the diffusion constant of a molecule such as pyrene in a vesicle may be 10^{-7} cm²/s. Hence, this molecule will diffuse 2.8×10^{-7} cm during the excited state lifetime of 400 ns.

ENERGY TRANSFER AT A DISTANCE. The preceding discussion assumes that the reactants must approach each other for reaction to occur. However, energy transfer has been observed in systems where the reactants cannot approach each other, and reaction occurs while the species are quite well separated. For the sake of convenience, noncontact-type energy transfer may be divided into two classes: short-range (10–15 Å) and long-range (distances greater than 15 Å).

Short-Range Energy Transfer. An excited species may be quenched provided the quencher molecule is contained within a limited spherical region of space around the excited species (14). This active sphere model postulates that if the quenching molecule is contained within this space, then the excited species is quenched immediately; if the quenching molecule lies outside the active volume, then quenching of the excited species does not take place. The quenching process is purely static.

If the volume of each active sphere is v (cubic centimeters) and n is the number of quencher molecules per cubic centimeter where $n = [Q]\ 6 \times 10^{20}$ ([Q] is the concentration in moles/liter of quencher), then the probability p that the excited molecule P^*, will lie with the active sphere is given by $p = e^{-nv}/V$ where V is the volume of the solution.

The probability is also the experimentally observed probability of quenching (I/I_o, where I_o and I refer to the yield of P^* in the absence and presence of quencher, respectively). Therefore, $I/I_o = e^{-nv}/V = e^{-a[Q]}$ where a is $6 \times 10^{20}\ (v/V)$. The radius R (angstroms) of the active sphere is given by $R = (3v/4\pi)^{1/3} = 6.5\ [Q]^{1/3}$. This equation predicts that the quenching of P^* by Q in the two competing processes:

$$P^* \xrightarrow{\ k_1\ } P + \text{fluorescence}$$

and
$$P^* + Q \xrightarrow{\ k_2\ } P + Q^*$$

depends exponentially on the concentration of Q as given previously. However, for a diffusional description of the two events the Stern–Volmer equation is applicable:

$$I/I_o = \frac{k_1}{k_1 + k_2\,[Q]}$$

The quenching depends linearly on [Q].

The Perrin description of excited state quenching describes the quenching of naphthalene fluorescence by oxygen, nitric oxide, and xenon in rigid glasses (15), and the quenching of anthracene fluorescence by carbon tetrabromide (16). However, an alternative explanation of these experiments suggests a time-dependent rate constant in the diffusion equation (17). The former experiments (16) were carried out by observing the steady-state yield of anthracene fluorescence; the later experiments were carried out by rapid flash photolysis methods in heptane and acetonitrile. An initial rapid quenching of anthracene fluorescence was observed and was explained as being due to the time-dependent part of the rate constant, followed by a slower quenching process that was diffusion controlled. However, large values of the interaction parameter were required to explain the initial quenching, and alternative quenching may be operative (e.g., electron tunneling). This discussion indicates the many and varied problems encountered when trying to ascertain with certainty the mechanism of quenching processes that are suspected to have some static contribution. Only rapid observation of the quenching events can distinguish between the time dependency of the rate constant, electron tunneling, Perrin quenching, and ground-state complexation. The photochemist encounters the same problems in micellar systems.

Electron Exchange Mechanism. The Perrin treatment does not indicate the mechanisms of static quenching. One successful attempt to provide a picture of these processes was the electron or coulombic exchange interaction (18). In this treatment the rate constant for energy transfer k_1 is given by:

$$k_1 = q \, \exp(-2R/L) \int_0^\infty F(\bar{v})\varepsilon(\bar{v})d\bar{v}$$

where q is a specific orbital interaction, R is the donor–acceptor separation relative to the van der Waals radius, and $\int_0^\infty F(\bar{v})\varepsilon(\bar{v})d\bar{v}$ is the overlap integral of the donor emission spectrum and the acceptor absorption spectrum. Because of the exponential term, the effect falls off rapidly with separation of donor and acceptor and is only applicable at short distances (10–15 Å). The effect applies well to triplet–singlet energy transfer where the alternative dipole–dipole reaction is eliminated.

Long-Range Energy Transfer. Long-range singlet–singlet energy transfer is described well by a dipole–dipole interaction. For random orientation of donor and acceptor, the rate constant for energy transfer in the dipole–dipole interaction is given by (19–21):

$$k = \frac{(9000 \ln 10)}{128 \, \pi^5 n^4 N \gamma^6 \tau} \int_0^\infty F(\bar{v})\varepsilon(\bar{v}) \frac{d\bar{v}}{(\bar{v})^4}$$

where n is the refractive index of the medium, τ is the fluorescence lifetime of the donor in the absence of acceptor, γ is the separation of donor and acceptor, N is Avogadro's number, and the integral is the overlap of the donor fluorescence and acceptor absorption spectra. If γ_0 is a critical transfer distance such that the rate of energy transfer to donor is equal to the energy dissipation by fluorescence and radiationless transition, then

$$ k = \frac{1}{\tau} \left(\frac{\gamma_0}{\gamma} \right)^6 $$

The term γ_0 depends only on the acceptor oscillator strength and spectral overlap and not on the donor oscillator strength or τ. Several energy-transfer experiments have been explained successfully by these equations: energy transfer from naphthalene and anthracene separated by 20 Å on a rigid steroid (22), and energy transfer from naphthalene to anilinonaphthalenesulfonate located at various separations from 12 to 46 Å on poly(L-proline). All data indicate that the efficiency of energy transfer varies inversely as the sixth power of the separation of the two chromophores (23).

The donor and acceptor may also be separated by using the monolayer technique (24, 25). These experiments, which will be described later, also show the γ^{-6} dependence of separation on efficiency of energy transfer.

Intersystem Crossing. One mode of decay of the excited singlet state is via intersystem crossing to the excited triplet state. The quantum yield ϕ of this process—molecules transformed per quantum absorbed—varies from close to unity to close to zero. An increased yield of intersystem crossing would gain significant yields of excited triplets and would aid in measuring the quantum yield of triplet formation ϕ_T. The rate of intersystem crossing is increased conveniently by the addition of heavy ions or molecules to the system. The following compounds increase the rate of intersystem crossing: both alkali and alkyl bromides and iodides (26); krypton and xenon (26); oxygen (27); and heavy ions such as Tl^+ (28), Cs^+ (29), and Ag^+ (30). These compounds also increase the rate of intersystem crossing from triplet to singlet ground state. To a rough degree of approximation, the rate constant for the interaction of the quencher with the excited singlet ground state to produce triplet crossing varies with the molecular weight of the atom

$$ S_1 + Q \xrightarrow{\ k_{IS}\ } T + Q $$

For example, k_{IS} is about $10^8 \ M^{-1}s^{-1}$ for Br^- and Cs^+; the corresponding second-order rate constant for I^- is 10^9 $M^{-1}s^{-1}$ and that for Tl^+ is $\sim$ 10^{10} $M^{-1}s^{-1}$. The uncertainty in the numbers shows that the effectiveness of this reaction depends on the nature of S_1. This dependence indicates that a complex between S_1 and Q may be an intermediate in the process.

Oxygen, in spite of its low molecular mass, promotes intersystem crossing very effectively, $(k_{IS} > 10^{10}$ $M^{-1}s^{-1})$. The subsequent reaction of the triplet with O_2 is an order of magnitude slower and often produces singlet oxygen $\Delta'O_2$ $(T + O_2 \rightarrow S + \Delta'O_2)$. The effectiveness of O_2 in inducing intersystem crossing is due to the triplet character of ground state O_2. Free radicals should also induce similar products. This possible reaction should be remembered in radiation chemistry where high concentrations of free radicals are produced together with excited states. The high concentration of free radicals leads to much shortened lifetimes for the excited states.

Excited State–Excited State Interaction. Two excited triplet states will interact to produce an excited singlet state (31): $T + T \rightarrow S_1 + S_o \rightarrow S_o + h\nu$. This reaction gives rise to the well-known p-type fluorescence that is observed for extended periods of time in excess of the normal singlet lifetime. A similar process occurs when two excited singlet states interact. Experimental observations indicate that excited triplet states (T) are formed (32, 33).

$$S_1 + S_1 \rightarrow T + S_o$$

In both processes the energy of one excited state is added to that of the other so that the resultant excited singlet state has an energy of $2E_t$. With T–T annihilation, singlet states higher than the first excited singlet are initially formed (34).

$$T + T \rightarrow S_n + S \rightarrow S_1 \rightarrow S_o + h\nu$$

In excited singlet states, a similar process should lead to ionization because the higher excited singlet state formed has an energy that is greater than that required for photoionization. However, no photoionization is observed! The lack of photoionization could be due to a very rapid recombination of the ions leading to excited singlet and triplet states:

$$S_1 + S_1 \rightarrow S_n \rightarrow S^+ + e^-$$
$$S^+ + e^- \rightarrow S_1 \text{ and } T$$

Future studies should elucidate the exact mechanism for this process.

Mirror Effect on Fluorescence Lifetime

The foregoing processes affect both the excited state lifetime and the luminescence yield. In monolayers, the luminescence lifetime can be changed without any change in the yield (35, 36). This unique situation arises if an excited state S_1 is placed in close proximity to a reflecting surface M. The oscillator, S_1, engulfs the electric field of M. If M does not absorb the radiation, then it only affects S because the electric field takes a finite time to reach M and to then be reflected back to S. A phase shift can result between the oscillation of S and the field from the induced oscillator M. The phase shift can have an accelerating or retarding effect on the lifetime of S_1. The luminescence lifetime of a complex decreased by 30% when the luminescent molecule was located up to 1000 Å from a reflecting silver surface. The film technique used in these experiments will be described later.

Electron-Transfer Reactions

Another fundamental process carried out by excited states is the induction of electron-transfer reactions. These reactions fall into three broad classes.

1. Photoionization, or electron transfer from the excited state to the medium: $D \rightarrow S^+ + e^-$ or e_s^-. In the gas phase the electron is free, but in solution the free electron interacts with the medium to produce a solvated electron, e_s^-.

2. The direct transfer of an electron from an excited molecule D to an acceptor molecule A to form the cation of D and the anion of A ($A + D \rightarrow A^- + D^+$).

3. The direct transfer of an electron from a donor molecule to an excited state to form the anion of the excited state and the cation of the other molecule: $D + A \rightarrow D^- + A^+$.

The reactants can be neutral molecules or anions and cations.

Photoionization. Photoionization is the process whereby an electron is removed to an infinite distance from an atom or molecule. The process occurs in two stages: the delocalization of the electron from the molecule followed by its removal to infinity. The energy required for the whole process is the ionization potential of the species. The work W required for the second stage is given by $W = e^2/D\gamma$, where e is the electronic charge, D is the dielectric constant, and γ is the effective distance at which the electron is delocalized from the molecule. In the gas phase $D = 1$ and γ is 3 Å; thus, W is 5 eV. In the condensed phase, for example, a liquid such as hexane or benzene, $D \sim 2$ and W is 2.5 eV. Therefore, less energy probably will be required to photoionize a molecule in solution compared to the gas phase. This assumption is indeed found to be the case (6, 38, 39) where the ionization potential in solution is lowered more than 2 eV below that of the gas phase.

A more quantitative consideration (40) of the photoionization process states that the ionization potential in solution I_S is related to that in the gas phase I_g by $I_s = I_g + P_+ + V_0$, where P_+ is the polarization energy of the cation in the medium, and V_0 is the energy of the electron in the medium. Both P_+ and V_0 are usually negative. The values of V_0 and P_+ for several liquids are given in Table II. Table II shows that $I_g - I_s$ is > 2. The value of P_+ may also be calculated from the Born charging equation

$$P_+ = \frac{e^2}{2\gamma} \left[1 - \frac{1}{\varepsilon} \right]$$

where ε is the fast dielectric constant of the medium or the square of the medium refractive index.

Table III lists several systems where photoionization occurs either as a one-photon or two-photon process. In all cases the ionization in the condensed phase is much smaller than that in the gas phase. It will be shown later that anionic micelles, such as sodium lauryl sulfate, show an increased photoionization yield compared to other systems. However, this yield is partly due to micellar inhibition of rapid ion neutralization of the photoproduced ion pair (38), which helps the escape of the ions from the ion pair. The observed yield of photoionization is only that of the escaped ions. The fraction of ions that escape increases

Table II.
Gas Phase (I_g) and Solution Phase (I_s) Ionization Potentials of Arene; and Polarization Energies P_+ of Arene in Various Liquids of Known V_0.

Arene	Liquid	$I_g(eV)$	$I_s(eV)$	$-P_+(eV)$	$V_0(eV)$
TMPD	n-Pentane	6.75	3.88[a]	—	—
		—	4.89[b]	−0.01	0.13
TMPD	n-Hexane	—	4.99[b]	1.31	0.10
TMPD	2,2,4-Trimethylpentane	—	4.66[b]	1.37	−0.17
TMPD	Neopentane	—	4.52	1.30	−0.38
TMPD	Tetramethylsilane	—	4.34	1.27	−0.59
Pyrene	n-Pentane	7.41	4.80[a]	—	—
Fluoranthene	n-Pentane	7.72	4.50[a]	—	—

[a] From Ref. 126.
[b] From Ref. 127.

Table III.

Various Systems That Photoionize Readily in the Condensed Phase

System	Wavelength of Light Used (Å)
$I^- \rightarrow I + e_s^-$	2537
$Fe(CN)_6^{4-} \rightarrow Fe(CN)_6^{3-} + e_s^-$	2537
$TMPD \rightarrow TMPD^+ + e_s^-$	<3000
Tetramethylbenzidene $\rightarrow$ cation + e_s^-	3471
Phenothiazine $\rightarrow$ cation + e_s^-	3471
Chloropromazene $\rightarrow$ cation + e_s^-	3471
Pyrene $\rightarrow$ cation + e_s^-	2 quanta 3471
Perylene $\rightarrow$ cation + e_s^-	2 quanta 3471
Aminopyrene $\rightarrow$ cation + e_s^-	3471
Aminoperylene $\rightarrow$ cation + e_s^-	5300

with solvent polarity, reminiscent of the yield of photoionization. The ions that recombine rapidly form excited states P*

$$P \nearrow^{P^*}_{\searrow P^+ + e^-}$$

$$P^+ + e^- \rightarrow P^*$$

The inclusion in the system of electron scavengers, such as $CHCl_3$, reduces the yield of P* but does not increase its decay rate. The reaction of $CHCl_3$ with the excited states studied was very inefficient. The studies indicated that $CHCl_3$ removed electrons from the ion-neutralization reaction and reduced the excited state yield:

$$e^- + CHCl_3 \rightarrow CHCl_3^- \rightarrow CHCl_2^{\cdot} + Cl^-$$

Similar experiments with N,N,N',N'-tetramethyl-p-phenylenediamine (TMPD) in alkanes, have been interpreted in terms of a superexcited state [(TMPD)**] that autoionizes (39):

$$TMPD \rightarrow (TMPD)^{**} \rightarrow TMPD^+ + e^-$$
$$\downarrow$$
$$(TMPD)^*$$
first excited state

Additives such as perfluorohydrocarbons, which are used in this system as specific scavengers for electrons, reduced the yield of the first excited state. The kinetics established by these studies are more in accord with the initial formation of a superexcited state rather than an ion pair. Hence, it is suggested that the perfluoro-electron scavenger must also react efficiently with the superexcited state.

A great deal of understanding is now available on many aspects of photoionization in condensed phases. In particular, the lowering of the ionization potential can be explained in terms of other thermodynamic parameters of the system. However, a more detailed description of the ionization potential in the condensed state is required.

Ions Produced in Radiation Chemistry. Ions are always produced by the action of ionizing radiation on materials. The process is not as specific as in photoionization, and excited states and or radicals are also produced. The yield of ions per 100 eV of energy absorbed (G value) depends on several conditions, such as solvent type, the phase used (i.e., gas, liquid, or solid), and the amount of time following the initial radiation. Some of these points were discussed earlier. The effect of phase is pertinent to the yield of a particular anion or cation, which can be produced in many solvents, both in the liquid state and in the low temperature rigid glassy state. In the liquid state the ions are short-lived (microsecond to millisecond lifetime), and pulse radiolysis techniques have to be used to observe them; this will be discussed later. However, conventional steady-state spectroscopic methods can be used to identify ions trapped in low temperature glasses (41–44). Known quantities of ions of interest can be produced conveniently in many frozen media by the radiolysis method. Thus, the spectroscopic properties of the ion can be well established for subsequent use in other experiments (e.g., photochemistry).

Electron Transfer Between Neutral Molecules. Many molecules will form charge-transfer complexes in suitable solvents. This phenomenon also occurs between two suitable molecules in an inert solvent (45, 46). Suitable examples are I_2 complexes with many solvents and complexes of amines with quinones or nitro compounds in polar solvents. Generally, one partner of the complex should have a low ionization potential, for example, an amine, and the other should have a strong electron affinity, for example, a quinone or nitro compound. Analysis of the absorption spectra of the charge-transfer band shows that the energy of the charge-transfer transition peak E_{CT} is given by:

$$E_{CT} = I_D - A_A - (\Delta E_E - \Delta E_G)$$

where I_D is the ionization potential of the donor D, A_A is the electron affinity of the acceptor A, and ΔE_E and ΔE_G are the energies of formation of the complex DA in the excited state and group state, respectively.

Many molecules will not form charge-transfer complexes in the ground state because the overall free energy change ΔG of the process is unfavorable. However, the additional ΔG necessary to drive the process can become available if one of the molecules is excited. A good example is the system pyrene–dimethylaniline. Figure 1 illustrates the various spectra observed in the laser flash photolysis of pyrene and pyrene–dimethylaniline in several solvents. Flash photolysis of pyrene, in both polar and nonpolar solvents, gives spectra that are indicative of the pyrene excited singlet state P_s^* with absorption maxima at λ_{max} = 4700, 3700, and 2650 Å, and of the triplet excited state P_T^* with absorption maxima at λ_{max} = 4150 and 5250 Å. High laser intensities lead to photolysis of the excited triplet and singlet state to give rise to the pyrene cation P^+ with λ_{max} = 4500 Å, or the solvated electron with λ_{max} = 7200 Å

$$P_s \rightarrow P_s^* \rightarrow P_T^*$$
$$P_s^* \searrow$$
$$\qquad \nearrow P^+ + e_s^-$$
$$P_T^* \nearrow$$

The presence of dimethylaniline (DMA) in the system gives absorption spectra that are quite solvent dependent. In nonpolar solvents, such as alkanes or arenes, the forementioned absorption spectra are replaced by a single spectrum with λ_{max} = 4800 Å associated with the exciplex of pyrene and dimethylaniline, P_s^* + DMA $\rightleftharpoons$ (P, DMA)*, and to the triplet spectrum of pyrene. In polar media, such as acetonitrile or alcohols, the excited state spectra are removed and replaced by the spectra of the pyrene anion P^- and the amine cation DMA^+, P_s^* + DMA $\rightarrow$ P^- + DMA^+. The absorption spectra of P^- and DMA^+ are at λ_{max} = 4900 and 4550 Å, respectively. The spectra in Figure 1 show the marked distinction between the different species. This distinction is a feature that is used in micellar studies. Lifetime measurements are a further aid to identifying species in these studies. Many examples of similar photosystems exist (6, 7).

Electron Transfer from Triplet States. The previous electron-transfer processes involved the excited singlet state of one of the reactants. The lifetimes of such states are shorter than 1 μs and high concentrations of the other reactant have to be used to promote efficient reaction. Several molecules of low ionization potential also promote electron transfer from the triplet excited state. These systems are attractive because the triplet states have long lifetimes; therefore, efficient electron transfer can be obtained at low counter reactant concentrations.

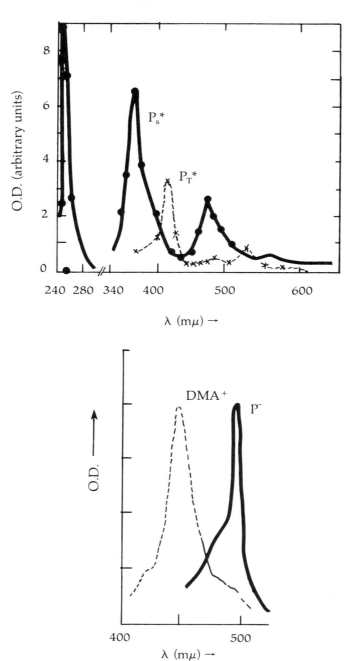

Figure 1. Absorption spectra of excited states and ions of pyrene.

Table IV.
Electron-Transfer Reactions of Excited States

System	Rate Constant $(M^{-1}s^{-1})$
RuII* + Cu^{2+} → RuIII + Cu^{+}	5.7×10^7
RuII* + Fe(CN)$_6^{3-}$ → RuIII + Fe(CN)$_6^{4-}$	3.8×10^{10}
RuII* + MV^{2+} → RuIII + MV^{+}	7.4×10^8
Phenothiazine triplet + Cu^{2+} → cation + Cu^{+}	6.0×10^9
Phenothiazine triplet + Eu^{3+} → cation + Eu^{2+}	4.7×10^9
Tetramethylbenzidene triplet + Eu^{3+} → TMB^{+} + Eu^{2+}	6.9×10^9 (reverse 2.0×10^9)
Tetramethylbenzidene triplet + duroquinone → TMB^{+} + (duroquinone)$^{-}$	2.3×10^{10} (reverse 1.4×10^7)
TMPD triplet + SF$_6$ → TMPD $^{+}$ + SF$_6^{-}$	$\sim 10^{10}$
CO$_3^{=}$ + duroquinone triplet → CO$_3^{-}$ + (duroquinone)$^{-}$	2.0×10^{10}
Tetraphenylporphinezinc triplet + MV^{2+} → cation + MV^{+}	10^9 (reversed micelle)

Table IV lists several systems that have long-lived excited states ($\tau > 1$ μs) that readily undergo electron transfer to suitable electron acceptors upon photoexcitation. All systems have also been studied to advantage in micellar systems.

Electron Transfer from Free Radicals. Several molecules that are strong electron acceptors react with free radicals to produce anions of the acceptor. Presumably, the cations of the radical species are also formed. One example is the reaction of organic free radicals with quinones. The anion of the quinone Q^{-} is observed readily in pulsed studies via its strong visible absorption spectrum (49).

$$R^{\cdot} + Q \rightarrow (R^{+}) + Q^{-}$$

The rate constants (k) for this reaction vary with the nature of R$^{\cdot}$ and Q; however, k values as large as 10^9 LM^{-1}s^{-1} have been observed when Q is benzoquinone and R$^{\cdot}$ is (CH$_3$)$_2$C$^{\cdot}$OH or (CH$_3$)$_2$CO^{-}. Several free radicals [e.g., $^{\cdot}$CH$_2$OH, $^{\cdot}$CH$_2$CHO, and (CH$_3$)$_2$C$^{\cdot}$OH] and the ionized forms of these radicals (e.g., CH$_2$O^{-}) react rapidly with nitrobenzene C$_6$H$_5$NO$_2^{-}$ via electron transfer. The more substituted alcohols react more rapidly than less substituted alcohols, and the ionized radicals are more reactive than the unionized ones (50).

$$^{\cdot}CH_2O^{-} + C_6H_5NO_2 \rightarrow CH_2O + C_6H_5NO_2^{-}$$

Similarly, electron transfer between similar species, such as quinones, may take place. One example is 9,10-anthraquinone-2-sulfonate and duroquinone (51). Micelles can often profoundly influence the direction of such reactions (discussed later) (52).

Electron transfer between arenes, anions, and other solutes has been observed in pulse radiolysis experiments. Pulse radiolysis of alkane solutions of biphenyl quinone (Q_2) leads to the anion Q_2^-, which subsequently transfers electrons to pyrene and other solutes at diffusion-controlled rates. Micelles can either retard or promote such reactions as will be seen later. Table V lists several electron-transfer processes of interest.

Electron Transfer from Cations. Several excited cations readily transfer electrons to solutes (53). One cation of particular interest to micellar photochemistry is tris(bipyridyl)ruthenium dichloride, which has a double positive charge and is denoted by RuII. Excitation of this molecule leads to a long-lived ($\tau > 0.5$ μs) excited state that transfers electrons to several solutes (e.g., Cu^{2+}, ferricyanide ion, and methyl viologen) (54). Methyl viologen, which has a double positive charge (MV^{2+}) is reduced to a single positive charge MV^+ by electron transfer from excited RuII [(RuII)*]

$$(RuII)^* + MV^{2+} \rightarrow RuIII + MV^+$$

The reverse reaction between the triply charged RuIII and MV^+ to give

Table V.
Electron-Transfer Reactions in Nonviscous Solvents (Viscosity ~1.0 cps)

System	Rate Constant $(M^{-1}s^{-1})$
Pyrene anion (P^-) + $CO_2 \rightarrow$ Pyrene (P) + CO_2^-	~10^7
$P^- + MV^{2+} \rightarrow P + MV^+$	2.6×10^{10}
$P^- + O_2 \rightarrow P + O_2^-$	2.0×10^{10}
$P^- + Eu^{3+} \rightarrow P + Eu^{2+}$	2.6×10^9
$P^- +$ cetylpyridinium chloride $\rightarrow$ radical + P	2.6×10^{10}
Biphenyl anion (ϕ_2^-) + P $\rightarrow \phi_2 + P^-$	3.2×10^{10}
$\phi_2^- + SF_6 \rightarrow \phi_2 + SF_6^-$	0.75×10^{10}
$\phi_2^- +$ benzyl chloride $\rightarrow \phi_2 + B_3^{\cdot} + Cl^-$	1.0×10^{10}
$\phi_2^- + O_2 \rightarrow \phi_2 + O_2^-$	2.3×10^{10}
$\phi_2^- + CO_2 \rightarrow \phi_2 + CO_2^-$	<10^4
$CH_3CHO^- + P \rightarrow P^- + CH_3CHO$	1.7×10^8
$(CH_3)_2C=O^- + P \rightarrow P^- + (CH_3)_2C=O$	2.3×10^9

the starting reactants occurs sufficiently slowly for repair of RuIII to RuII to be initiated first. The repair is achieved by carrying out the process in the presence of electron donors (55, 56). Two typical donors are ethylenediaminetetraacetate (EDTA^{2-}) and triethanolamine.

$$RuIII + EDTA^{2-} \rightarrow RuII + EDTA^{-}$$

The radical EDTA^{-} is inert in the system and degrades to innocuous products. The net result is the formation of a permanently reduced methyl viologen MV^{+} and EDTA products; the RuII merely acts as a photocatalyst. This reaction is a form of energy storage because MV^{+} is reactive and can be used to subsequently initiate useful chemistry.

Methyl viologen is a strong electron acceptor. The excited state obtained in alcoholic solution abstracts an electron from the alcohol solvent (57). Similar reactions are reported for excited tetranitromethane in alcoholic solvents (58). The excited states of both molecules are very short-lived, and intimate contact with the alcohol is necessary for reaction to occur. Contact is achieved by using pure alcohols, or by solubilizing the molecules in alcohol clusters in hydrocarbons. In the case of $C(NO_2)_4$ the product is the stable nitroform ion

$$C(NO_2)_4{}^* + ROH \rightarrow C(NO_2)_3{}^- + NO_2 + (ROH)^+$$

Intramolecular Electron Transfer. Electron-transfer processes involving coordination complexes of transition metals may occur by two mechanisms that are commonly called inner or outer sphere mechanisms. In the outer sphere mechanism the electron is transferred to or from the metal ion in the complex without any change in the coordination sphere of the complex. The reactions of excited RuII fall into this category. Some reactions proceed by an inner sphere mechanism whereby the coordination sphere of the complex is dramatically altered. An example is the reaction of Cr^{2+} and $Co(NH_3)_5Cl^{2+}$ in water:

$$Co(NH_3)_5\ Cl^{2+} + Cr^{2+} \rightarrow Co^{2+} + Cr(H_2O)_5\ Cl^{2+}$$

An intermediate complex of the form $(NH_3)_5CoClCr(H_2O)_5{}^{4+}$ is probably formed, and an intramolecular transfer of an electron from divalent chromium to trivalent cobalt takes place. The electron is suggested to proceed from one metal to the other via the ligand. Pulse radiolysis studies have confirmed such a mechanism for some complexes (59). In particular, the complex $Co(NH_3)_5\ O_2C_6H_4NO_2{}^{2+}$ is reduced by anionic radicals R^{-} to produce an intermediate that shows an intense UV absorption characteristic of the anionic $C_6H_4NO_2^-$ species. The first stage of the reaction may be represented as:

$$CoIII(NH_3)_5 O_2^+C_6H_4NO_2^2C + R^- \rightarrow CoIII(NH_3)_5O_2CC_6H_4NO_2^+$$

The product or intermediate subsequently decays via a first-order process to product divalent Co^{2+}

$$CoIII(NH_3)_5O_2CC_6H_4NO_2^+ \rightarrow CoII(NH_3)_5O_2CC_6H_4NO_2^+$$

This reaction is a direct demonstration of a relay of the electron from the donor R^- to the metal ion via the ligand $C_6H_4NO_2$. An excellent review of these processes is available (60).

Reactions of Ions in Unusual Valency States

Many unique and unstable valency states of metal ions can be produced via free radical reactions. The radicals, in turn, are produced in pulse radiolysis or flash photolysis experiments. An immediate example is the hydrated electron e_{aq}^-, which can readily produce several lower and unstable valency states of metal cations:

$$e_{aq}^- + Zn^{2+} \rightarrow Zn^+$$
$$e_{aq}^- + Tl^{3+} \rightarrow Tl^{2+}$$

The OH radical can also produce unstable valency states:

$$OH + Ag^+ \rightarrow Ag^{2+} + OH^-$$
$$OH + Tl^+ \rightarrow Tl^{2+} + OH^-$$

These unusual valency states are observed by fast spectrophotometric methods; consequently, the kinetics of the reactions and the reaction rate constants can be determined. The new valency states formed tend to undergo further electron-transfer reactions, which either revert back to the original valency form or proceed to another stable valency state. These types of reactions are often encountered in micellar chemistry where the stable metal ions may occur as counterions of the micelle. The different valency forms and reactions that have been generated and the reaction rate constants have been summarized (61).

Reactions of the Radicals

Earlier sections have dealt with various radical ions and some of their reactions. This section deals with noncharged radicals or species that have one unpaired electron. Typical examples would include an H atom, an OH radical, and a methyl radical. These species are frequently produced in radiation chemistry or photochemistry. In particular, this section emphasizes aqueous systems or alcohol solutions because they are most pertinent to the later colloidal systems.

Radiolysis of water produces e_{aq}^-, H, and OH radicals; however, information is sparse on the nature of free radical events in the radiolysis of other liquids. The hydrated electron e_{aq}^- will be considered in the next section. Many different free radicals can be produced in photochemistry, for example, photoionization can give e_{aq}^- and bond dissociation can give a variety of alkyl and aromatic free radicals by photolysis of suitable molecules:

$$RX \rightarrow R^{\cdot} + X^{\cdot}$$

where $R^{\cdot}$ is an organic radical, and $X^{\cdot}$ may be another radical or halide atom. Excitation of ketones, such as benzophenone (BQ) leads to excited triplet states BQ^T, which react rapidly with many solvents via H atom abstraction to produce radicals

$$BQ^T + CH_3OH \rightarrow BQH^{\cdot} + {}^{\cdot}CH_2OH$$

This H atom abstraction reaction can occur internally to produce a biradical, which can rearrange to give products. The reaction is called a Norrish type II reaction. These radicals can also react further to produce other radicals, and some of the more important reactions are given as follows.

OH Radicals. Hydroxyl radicals (OH) are formed in many processes. One of the earliest examples of OH is the Fenton reaction, which is the reaction of Fe^{2+} ions and hydrogen peroxide (62): $Fe^{2+} + H_2O_2 \rightarrow Fe^{3+} + OH + OH^-$. Hydroxyl radicals are also produced in high yield ($\phi = 0.5$) in the photolysis of H_2O_2 (63–65): $H_2O_2 \rightarrow OH + OH$. However, the most convenient method of producing OH radicals for subsequent study is via the radiolysis of water. At neutral pH, solvated electrons are also produced, but these species are readily converted to OH by reaction with nitrous oxide.

$$e_{aq}^- + N_2O \xrightarrow{\text{H}^+} OH + N_2$$

A small yield of H atoms is also produced, but if they interfere with the OH reactions they can be removed by selective reactions with a scavenger, such as O_2, thereby producing an inert radical, $H + O_2 \rightarrow HO_2\cdot$.

The rates of reactions of OH radicals with several solutes dissolved in water have been measured by pulse radiolysis and comparative techniques (66). The reactions can be divided into three classes; electron transfer, H atom abstraction, and OH addition. The reaction rates between the three classes can vary from close to the diffusion-controlled limit of $\sim 10^{10}$ L mol^{-1}s^{-1} to about 10^7–10^6 L mol^{-1}s^{-1}. Little is known of the actual mechanisms for most reactions apart from the correlation with Hammett constants, which provide an indication of the electrophilicity of these reactions (67) by comparison with standard thermal reactions.

Electron transfer tends to dominate in the reactions of inorganic materials. In general, these reactions can be written as:

$$OH + M^{(n-1)+} \rightarrow M^{n+} + OH^-$$

A classic example is the oxidation of Fe^{2+} to Fe^{3+} in the Fricke dosimeter ($k = 3.0 \times 10^8$ M^{-1}s^{-1}). Some intriguing examples of electron transfer are found in organic systems. For example, the reaction of OH radicals with polyarenes in water normally gives the OH radical adduct; but in acidic solution, electron transfer can occur and an intermediate polyarene cation is formed and subsequently hydrolyzes to give the OH adduct (68).

$$OH + RH \begin{cases} \nearrow RH(OH)^\cdot \\ \searrow_{H^+} RH^+ + OH^- \end{cases}$$

Transitory intermediates are difficult to observe in OH radical reactions, but they may be more prevalent than was originally thought.

The absolute rate constants for the reactions of OH radicals with benzene (69) and benzene derivatives were among the first to be measured. The product in the case of benzene was identified as the hydroxycyclohexadienyl radical:

Hydroxyl radicals add readily to unsaturated bonds (e.g., ethylenic and acetylenic bonds, and arene compounds), but intermediate electron transfer is a possibility (68).

Hydrogen atom abstraction by OH radicals from organic compounds ($OH + RH \rightarrow R^{\cdot} + H_2O$) has received some detailed attention. Aliphatic alcohols are quite reactive, with $k > 10^9 \ M^{-1}s^{-1}$, and the reactivity increases with increasing chain length (67). Conversely, both carboxylic acids, such as acetic or formic acid, and sulfate esters, such as methyl and ethyl hydrogen sulfate, are quite unreactive, with $k \leqslant 10^8 \ M^{-1}s^{-1}$. The electron withdrawing power of the acid groups reduces the electron density on the α and β carbons of the acids and thus, reduces the reactivity toward the electrophilic OH radical. Increasing the alkane chain increases the reactivity; therefore, C_{12} arenes are quite reactive. The site of H atom abstraction in a large molecule is not known with certainty, but in a long C_{12} chain the probability of OH attack is similar at all positions except those adjacent to $-COOH$ or $-SO_4H$ groups (70).

Reactions of Hydrogen Atoms. Several methods can generate H atoms. Photolysis of aqueous solutions of H_2S produces H atoms and $SH^{\cdot}$ radicals (71), $H_2S \rightarrow H + SH^{\cdot}$. A similar process also occurs with HI although hot H atoms are produced initially (72). The simplest photolytic method produces solvated electrons by photoionization, and subsequently reacts these species with protons

$$e_s^- + H^+ \rightarrow H$$

This method tends to confine the studies to acidic solutions. Once again, radiolysis is the most convenient technique for producing H atoms in water. In neutral aqueous solution the H atom yield is well below the e_{aq}^- and OH yields. However, in acidic solution all the e_{aq}^- are converted to H atoms, and the OH radicals can be converted to an alternative inert radical that does not compete with the H atom chemistry. Because it reacts rapidly with $OH^{\cdot}$ radicals and quite slowly with H atoms, *tert*-butyl alcohol is used widely.

$$\begin{array}{c} CH_3 \\ CH_3 - C-OH + {}^{\cdot}OH \rightarrow \\ CH_3 \end{array} \begin{array}{c} {}^{\cdot}CH_2 \\ CH_3 - C-OH + H_2O \\ CH_3 \end{array}$$

Hydrogen atoms may abstract H atoms from organic molecules, add to sites of unsaturation, or enter into redox reactions. A good example of a redox reaction is the reduction of cupric to cuprous ions, $H + Cu^{2+} \rightarrow H^+ + Cu^+$. Possibly the initial reaction is the formation of a hydride that subsequently decomposes to products. Addition of H atoms to sites of unsaturation is usually quite rapid and approaches diffusion-controlled rates. The reaction of H atoms with oxygen to produce the hydroperoxy radical is very efficient and has a rate constant k of 2×10^{10} $M^{-1}s^{-1}$.

Hydrogen atom abstraction reactions are slower than the corresponding OH radical reactions.

$$H + RH \xrightarrow{\ k_1\ } R^{\cdot} + H_2$$

$$OH + RH \xrightarrow{\ k_2\ } R^{\cdot} + H_2O$$

This feature may be understood from the Polyani relationship, which relates the activation energy of a reaction E to the heat of reaction ΔH: $E = \text{constant} - \alpha\Delta H$, where α is often 0.25. If the Arrhenius factors A for H atom and OH radical reactions are similar, then

$$\frac{k_2}{k_1} = e^{-\frac{(E_2 - E)}{RT}} = e^{\frac{(\Delta H_1 - \Delta H_2)}{RT}}$$

The quantity $\Delta H_1 - \Delta H_2$ is $\sim$ 11 kcal/mol, which gives $k_2/k_1 \sim$ 100 at room temperature, as found experimentally.

Most H atom reactions are reducing in nature. However, a few examples of oxidation reactions have been reported in acidic solution:

$$H + H^+ + Fe^{2+} \rightarrow H_2 + Fe^{3+}$$
$$H + H^+ + I^- \rightarrow H_2 + I$$

Table VI lists several OH and H atom reactions of interest. An excellent and extensive survey is available (66, 67, 73).

OH Radical Reactions with Polymers. The radiolysis of aqueous solutions of polymers follows two patterns: cross-linking is observed with polymers such as polyvinylpyrrolidone or polyvinyl alcohol (74), and degradation occurs with polymers such as polymethacrylic acid (75). The chemistry in both instances is controlled by the OH radical: it attacks the polymers to give organic radicals, which then undergo the aforementioned reactions.

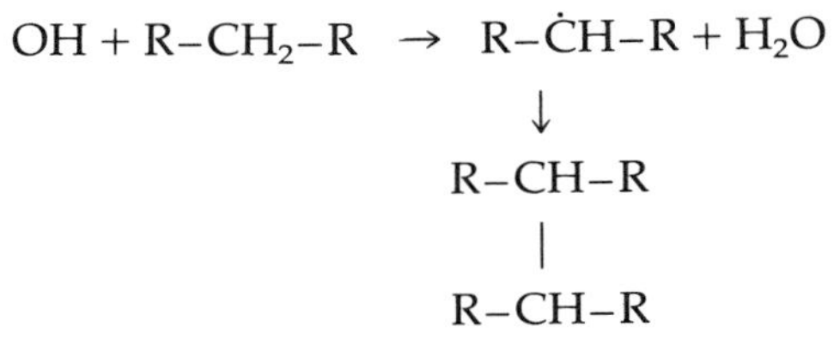

$$OH + R-CH_2-R \rightarrow R-\dot{C}H-R + H_2O$$

$$\downarrow$$

$$R-CH-R$$
$$|$$
$$R-CH-R$$

Cross-linking

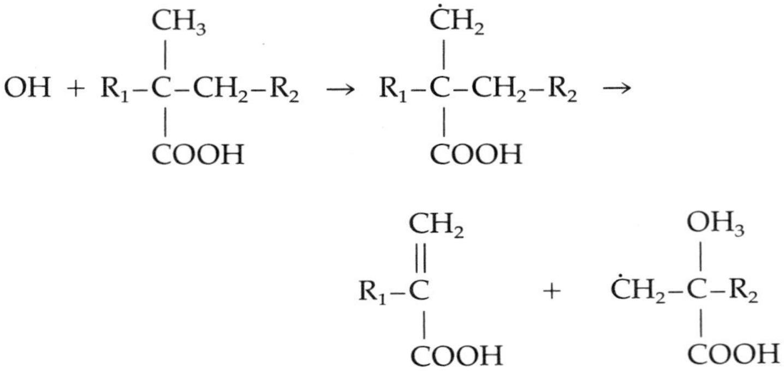

$$\begin{array}{ccc} CH_3 & & \dot{C}H_2 \\ | & & | \\ OH + R_1-C-CH_2-R_2 & \rightarrow & R_1-C-CH_2-R_2 \rightarrow \\ | & & | \\ COOH & & COOH \end{array}$$

$$\begin{array}{ccc} CH_2 & & OH_3 \\ || & & | \\ R_1-C & + & \dot{C}H_2-C-R_2 \\ | & & | \\ COOH & & COOH \end{array}$$

Degradation

Both processes are impaired by O_2, which reacts with the intermediate radical. Thioalcohols RSH repair the intermediate radical R_1 by the reaction $R_1 + RSH \rightarrow R_1H + RS^{\cdot}$ (RS$^{\cdot}$ is inert) (76). Similarly, ferricyanide reacts with the radicals and prevents cross-linking or degradation by oxidizing the intermediate radical to an alcohol. The latter process has been used to measure the rate of decay of the intermediate polymethacrylic acid radical to give the degradation products. The rate of this process at 25 °C is 10^4 s^{-1}.

Reactions of Secondary Radicals. The reactions of H, e_{aq}, and OH are rapid and tend to be nonselective. Many attempts have been made to produce secondary radicals that have more selectivity or to react the organic radicals formed in the events described previously with reactive solutes. Some of these systems are discussed here (79).

The sulfate radical SO_4^- is produced by the reaction of e_{aq}^- with persulfate:

$$e_{aq}^- + S_2O_8^{2-} \rightarrow SO_4^{2-} + SO_4^-$$

This radical tends to react via electron transfer and forms the cation of several solutes. Hydrogen atom abstraction and addition are not important reactions with this species; hence, unique electron-transfer re-

Table VI.
Reactions of H Atom and OH Radicals

System	Rate Constant $(M^{-1}s^{-1})$
$H + H \rightarrow H_2$	1.4×10^{10}
$H + O_2 \rightarrow HO_2$	2.4×10^{10}
$H + H_2O_2 \rightarrow H_2O + OH$	1.2×10^8
$H + CH_3OH \rightarrow \cdot CH_2OH + H_2$	1.9×10^6
$H + benzene \rightarrow$ (cyclohexadienyl radical, $C_6H_7\cdot$)	1.1×10^9
$H + isopropyl\ alcohol \rightarrow (CH_3)_2 C\cdot -OH + H_2$	5×10^7
$OH + OH \rightarrow H_2O_2$	5.3×10^9
$OH + H \rightarrow H_2O$	1.2×10^{10}
$OH + H_2O_2 \rightarrow H_2O + HO_2$	4.5×10^7
$OH + CH_3OH \rightarrow \cdot CH_2OH + H_2O$	4.7×10^8
$OH + benzene \rightarrow$ (hydroxycyclohexadienyl radical)	5.0×10^9
$I + OH^- \xrightarrow{\ I^-\ } I_2^-$	1.1×10^{10}
$OH + isopropyl\ alcohol \rightarrow (CH_3)_2 C\cdot -OH + H_2O$	1.84×10^9

actions can be initiated by SO_4^-. A similar reaction of e_{aq}^- with per-
oxodiphosphate ion leads to the three acid–base forms of the phosphate
radical: H_2PO_4, HPO_4^-, and PO_4^{2-}. These radicals abstract H atoms
from aliphatic compounds quite slowly ($k \sim 10^5$ for acetic acid), and
addition to a double bond is also slow. The reactivities of the phosphate
radicals and SO_4^- are quite similar. Therefore, these secondary radicals
may give greater reaction selectivity than the more reactive OH radical.

Other radicals of the X_2^- type, such as Br_2^-, formed by the reactions

$$OH + Br \rightarrow Br + OH^- \xrightarrow{\ Br^-\ } Br_2^-$$

show unique selectivity. For example, at a pH value of 7.8, Br_2^- reacts only with tryptophan ($k = 8 \times 10^8$ M^{-1}s^{-1}), tyrosine ($k = 2 \times 10^7$ M^{-1}s^{-1}), and histidine ($k = 1.5 \times 10^7$ M^{-1}s^{-1}) via electron transfer, and not with other common amino acids (79). This property of Br$_2^-$ has been used to investigate the degree of exposure of these amino acids in surfaces of erythrocyte membranes in situ and when treated with the surfactant sodium lauryl sulfate (NaLS) (80). The reaction of Br$_2^-$ with these amino acids in the membrane gives an amino acid cation spectrum. The yield of this species is intensified by addition of NaLS because of the breakup of the membrane.

The organic radicals formed by the reaction of these entities (OH, H, e_{aq}^-, etc.) with organic solutes often react via electron transfer with ferricyanide or benzoquinone. The reaction may be followed by observing the bleaching of ferricyanide:

$$R^\cdot + Fe(CN)_6^{3-} \rightarrow R^+ \overset{H_2O}{\nearrow} \begin{matrix} ROH + H^+ \\ \\ Fe(CN)_6^{4-} \end{matrix}$$

or the formation of the quinone anion:

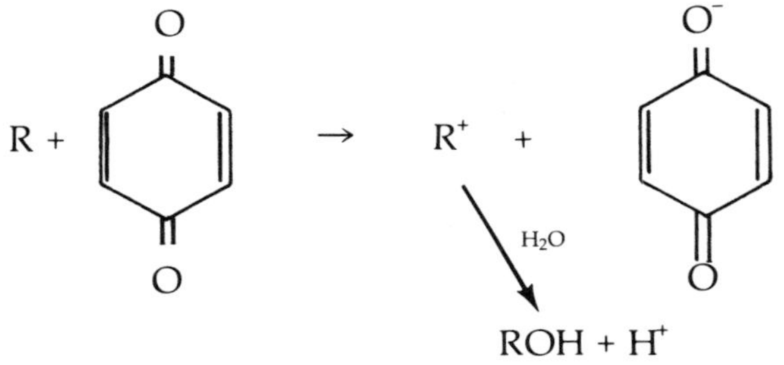

The nature of the radical R$^\cdot$ controls the rate of the reaction. For example, $^\cdot$CH$_2$OH is very reactive with Fe(CN)$_6^{3-}$ ($k > 10^9$ M^{-1}s^{-1}); however, $^\cdot$CH$_2$CH$_2$OH is quite unreactive. Such reactions are useful in determining the nature of the radicals formed in the radiolysis of an aqueous system.

Other selective reactions of radicals use nitrobenzene (81) and N,N,N',N'-tetramethyl-p-phenylenediamine. These systems are useful in probing important processes in more complex systems, such as the sites of inactivation of enzymes by free radicals (76), and they are useful as guidelines in determining the nature of the radicals formed in irradiated systems.

Limited space permits the mention of only a few of the free radicals that occur and that have been studied in chemistry. Later discussions illustrate how organized assemblies affect such well-established reactions, and how these reactions contribute useful information on the nature of the assemblies themselves.

Reactions of Ions

Photochemical excitation may produce ions by direct photoionization or by electron transfer between two reactants, one of which is excited. The quantum yield of ions observed, either by product analysis in stationary state techniques or by probe techniques, may deviate considerably from the ideal quantity yield of unity. Rapid recombination of the photoproduced geminate ion pair in the solvent cage is a major factor that decreases quantum yields.

Radiolytic formation of ions is, as stated previously, primarily from direct ionization of the solvent. Electron transfer between excited reactants may also occur, but direct ionization of the solvent is the most sought after process in radiation chemistry. The meaningful parameter here is the G value of ions produced (i.e., ions produced per 100 eV of energy absorbed). In nonpolar media this yield is low because of ion recombination in the solvent cage or "spurs." This behavior is quite similar to the events in photochemistry. Much attention has been directed to geminate ion recombination in radiation chemistry.

Geminate Ion Recombination

In the early days of radiation chemistry the initial deposition of energy was believed to lead to isolated regions of high ion and free radical density. The chemical kinetics that develop from this type of energy deposition can be demonstrated dramatically by the observation of ion yields as a function of time in the radiolysis of hydrocarbons. Figure 2 shows the recombination kinetics of biphenyl anions and cations in the pulse radiolysis of biphenyl in cyclohexane (82). The initial radiolytic act $[C_6H_{12} \rightarrow (C_6H_{12})^+ + e_s^-]$ produces solvated electrons e_s^- and solvent cations. The biphenyl solute then reacts rapidly with both ions

$$(C_6H_{12})^+ + (C_6H_5)_2 \rightarrow [(C_6H_5)_2]^+ + C_6H_{12}$$
$$(C_6H_5)_2 + e_s^- \rightarrow [(C_6H_5)_2]^-$$

The experimental data in Figure 2 show the variation in (biphenyl)$^-$ yield observed at $\lambda = 6100$ Å versus time. A rapid initial decay gradually becomes slower with time. These kinetics are indicative of geminate ion recombination, where pairs of geminate ions, which are closely spaced, react rapidly; ions that are more widely separated react more slowly or diffuse into the bulk of the solution. The solid line in Figure 2 gives a calculation based on the above mechanism of geminate ion recombination with a distribution of ion pair separation distances (82–84). Reasonable estimates of several parameters are required for the

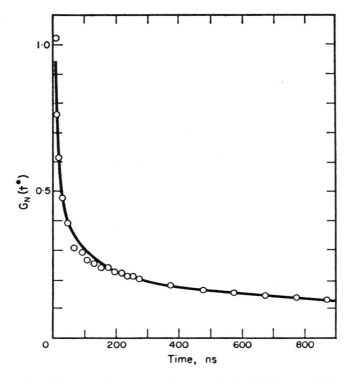

Figure 2. Comparison of the observed yield with the theoretical calculation of diphenylide anion present after a 10-ns pulse irradiation of a cyclohexane solution (0.1 mol/L) in diphenyl (82).

calculation; however, the general concept of geminate recombination and an adequate description of these kinetics seem to be borne out by agreement of experiment and theory.

The products of the neutralization process are singlet and triplet excited states.

$$(C_6H_5)_2{}^+ + (C_6H_5)_2{}^- \rightarrow (C_6H_5)_2{}^* + (C_6H_5)_2$$

escape to bulk

The rate of growth of the excited states matches the decay of the solute ions exactly (85). Several empirical equations have been suggested to rationalize the data obtained for the scavenging of the geminate ions by added solutes. In particular, the scavenging of $e_s{}^-$ by added solutes has received much attention (84). Figure 3 shows typical data for the yields of $CH_3^{\cdot}$ and $C_2H_5^{\cdot}$ from the radiolysis of CH_3Cl, C_2H_5Br, and CH_3Br in C_6H_{12} solution. The reactions involved are:

$$C_6H_{12} \rightarrow C_6H_{12}^{+} + e_s^{-} \rightarrow C_6H_{12}^{*} \text{ (geminate)}$$

$$e_s^{-} + RX \rightarrow R^{\cdot} + X^{-}$$

The experimental data either fit the equation:

$$G(R^{\cdot}) = G_{fi} + \frac{G_{gi}}{1 + \dfrac{1}{(\alpha[S])^{1/2}}}$$

or the revised form,

$$\frac{1}{G(R^{\cdot}) - G_{fi}} = \frac{1}{G_{gi}} + \frac{1}{G_{gi}\alpha^{1/2}} - \frac{1}{[S]^{1/2}}$$

where G_{fi} is the yield (molecules per 100 eV) of the ions that escape geminate recombination (typically $G_{fi} \sim 0.1$ for C_6H_{12}), $G(R^{\cdot})$ is the yield of alkyl radicals, G_{gi} is the yield of ions undergoing geminate recombination ($G_{gi} = 3.9$ for Figure 3), [S] is the solute concentration, and α

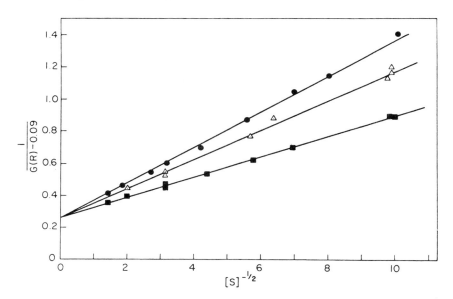

Figure 3. Plot of alkyl radical yields from CH_3Cl ($\bullet$), C_2H_5Br ($\triangle$), and CH_3Br ($\blacksquare$) solutions in C_6H_{12} according to the scavenging equation. ($G_{geminate\ ions} = 3.9$.) (86)

is a constant related to the reactivity of e_s^- with RX, and is proportional to the rate constant for the reaction of e_s^- and RX. Data for the radiolysis of many systems fit this empirical expression, which has some justification in theory (86).

Reactions of Solvated Electrons

Solvated electrons, which are produced readily in photolysis and radiolysis, occupy a unique place among anion species, perhaps reminiscent of the position held by protons among cations. However, unlike protons and many cations, solvated electrons are extremely reactive species and their properties are quite solvent dependent. Usually, electrons in polar proteated media are quite well-behaved, and most of their kinetic properties are understood. The most studied medium for solvated electrons is water; therefore, the hydrated electron will be considered in some detail, and it will be assumed that solvated electrons in alcohols behave similarly.

Hydrated Electrons. Hydrated electrons e_{aq}^- react readily with many cations to produce a lower valency state (87),

$$e_{aq}^- + Cu^{2+} \rightarrow Cu^+$$
$$e_{aq}^- + Ag^+ \rightarrow Ag$$
$$e_{aq}^- + H^+ \rightarrow H$$

Generally, the reactions are very rapid and are diffusion-controlled. The rate constant for a diffusion-controlled reaction between two ions, A and B, is described by the Debye modification (4) of the Smoluckowski reaction (2),

$$k = \frac{4\pi\gamma\ DN}{1000} \left[\frac{z_A z_B e^2}{\gamma EkT} \right] \Big/ \left(\exp \frac{[z_A z_B e^2]}{[\gamma EkT]} - 1 \right)$$

where γ is the encounter distance, z_A and z_B are the charges on the ions, E is the dielectric constant of the medium, and γ is the interaction distance. This equation provides rate data that agree well with those measured experimentally for the reactions given here (5):

$$e_{aq}^- + Cu^{2+} \rightarrow Cu^+$$
$$e_{aq}^- + O_2 \rightarrow O_2^-$$
$$e_{aq}^- + NO_3^- \rightarrow NO_3^{2-}$$
$$e_{aq}^- + e_{aq}^- \rightarrow H_2 + 2OH^-$$

The values of γ and D were obtained independently by other physical techniques.

The species formed by reduction of a compound by e_{aq}^- may be stable (e.g., Cu^+ or Fe^+), but for the most part the intermediates are reactive in their own right (e.g., O_2^-, CO_2^-, or Zn^+). In some instances the e_{aq}^- solute adduct is very short-lived and rapidly breaks apart; organic halocompounds often exhibit this behavior. Good examples are provided by chloroacetic acid, methyl iodide, and carbon tetrachloride (88).

$$e_{aq}^- + ClCH_2COOH \rightarrow Cl^- + {\cdot}CH_2COOH$$

$$e_{aq}^- + CCl_4 \rightarrow {\cdot}CCl_3 + Cl^-$$

$$e_{aq}^- + CH_3I \rightarrow CH_3{\cdot} + I^-$$

Thus, e_{aq}^- in many such processes readily produces several radicals of interest.

Addition of e_{aq}^- to stable organic compounds often produces anions that are of immediate interest to both radiation chemists and photochemists:

$$e_{aq}^- + benzophenone \rightarrow (benzophenone)^-$$

$$e_{aq}^- + CH_3-\overset{\overset{\textstyle O}{\|}}{C}-CH_3 \rightarrow CH_3-\underset{\cdot}{\overset{\overset{\textstyle O^-}{|}}{C}}-CH_3$$

In the latter reaction, the acetone anion can also be produced by ${\cdot}OH$ attack on isopropyl alcohol,

followed by ionization.

Several hundred rate constants for the reaction of e_{aq}^- with a variety of compounds are available (87).

Electrons in proteated solvent may react before thermalization is complete and prior to solvation. Pulse radiolysis studies of e_{aq}^- in the presence of large concentrations of acetone (~1.0 M) show that the yield of e_{aq}^- is decreased before significant reaction of e_{aq}^- with acetone can take place. This result is interpreted as a reaction of acetone with presolvated electrons, sometimes called *dry electrons* (90, 91). Similar effects are observed in monohydric alcohols (90, 92) and glycerol (93). A variety of solutes behave in this way. In particular, benzene at 1 M concentration sharply decreases the yield of e_s^- without any effect on the e_s^- lifetime because e_s^- is unreactive toward benzene. Benzene reacts with the precursor of e_s^-, that is, the dry electron. An alternative explanation of these data is that some of the scavenger solutes are in close proximity to the site of production of e^-, or close to the point of solvation of e^- when forming e_s^-, and that rapid reaction occurs in these instances. Other sites of e^- or e_s^- do not possess scavenger molecules in close proximity; hence, solvated electrons form normally and exhibit conventional kinetic behavior.

Electrons in Nonpolar Media. Solvated electrons in nonpolar media, such as hydrocarbons, exhibit properties quite different to those exhibited in polar media. The spectra of electrons in alkanes are far shifted to the red part of the spectrum with maxima at $\lambda > 18,000$ Å (94–97). This shift indicates a shallow trapping of the electrons in these liquids, a feature that also manifests itself in the kinetic properties of e_s^-. Pulse conductivity studies determined that the mobilities of electrons in nonpolar media were much larger than those of electrons in polar media (98–100). In many cases, the mobilities of e_s^- in alkanes were 100–1000-fold those of e_s^- in alcohols or water. The more symmetrical branched alkanes, such as isooctane and neopentane, have larger mobilities than the straight chain alkanes, such as hexane. For e_s^- in nonpolar liquids a shallowly trapped electron is believed to diffuse very rapidly throughout the system.

The first direct observations of the reaction rates of e_s^- with solutes in various hydrocarbons revealed that the rate constants were indeed very large with $k > 10^{14}$ M^{-1}s^{-1}. These data were, to some extent, in agreement with the high mobilities of e_s^- in these liquids. However, the agreement was not quantitative and the rate constants tended to vary with the square root of the e_s^- mobility. This unusual result, not predicted by the Smoluckowski diffusion theory where k varies linearly with mobility, suggests that electrons in nonpolar liquids do not behave as classical particles. Later experiments confirmed these data and indicated even more unusual behavior for e_s^- reactions in alkanes. Features such as negative activation energies, which have no place in simple kinetic theory, were found (104, 105). The simplest explanation

3. CHEMISTRY OF INTERMEDIATES

for the kinetic behavior of e_s^- in alkanes invokes a comparison of the rates of reaction of e_s^- with solutes in liquid alkanes to the corresponding reactions in the gas phase. The rate data are discussed in terms of V_0, which is the energy of the lowest extended state of the electron in the liquid relative to that in vacuum. The rate constant for reactions of e_s^- with a solute in the liquid depends on V_0 in much the same way that the rate of the reaction varies with electron kinetic energy in the gas phase. The rate of reaction of e_s^- with a solute is written as an equilibrium

$$e_s^- + \text{solute} \rightleftharpoons (\text{solute})^-$$

The forward rate is dominant if V_0 is large, as in hexane, and the back reaction is poorest if V_0 is low, as in neopentane.

In conclusion, the kinetics of e_s^- in liquids are well-established and can be related to other parameters of the system under investigation. This point is important in discussing e_s^- reactions in micellar systems.

Electron Tunneling. In the early days of radiation chemistry (106) it was noted that the radiolysis of rigid glassy alkanes at 77 K produced trapped electrons and cations. Similarly, the photolysis of several organic compounds in the same rigid media at low temperature led to photoionization and to the trapping of the photolytic ions by the matrix (108). Solutes included in the irradiated frozen matrix apparently underwent electron transfer processes to produce solute cations and anions. Reliable spectra of the cations and anions produced are known (107). These data show that ion migration occurs in rigid media, and they may indicate an electron tunneling or hopping mechanism. Pulse radiolysis studies showed that some of the solute ions were formed rapidly, in less than 1 μs, and that the rest were formed slowly, over seconds (109). Similar studies in aromatic glasses at 77 K showed that rapid energy transfer takes place in these systems rather than ion transfer (110). Electron transfer from trapped electrons or solute anions takes place over considerable periods of time, and the kinetics for a hopping model (112) may be explained by an empirical equation (111, 112):

$$\frac{d[e^-]}{[e_t^-]dt} = C\, t^{(\alpha-1)}\, (1 - P)$$

where $P = \exp(-2.52S\gamma^3)$, and is the probability that e^- remains outside the reaction sphere of radius γ centered on S; α and C are time-independent constants. The rate constant $k(R)$ (s^{-1}) in the tunneling model (111) is given by

$$k(R) = \nu_0 \, F \exp \, - \, [2(2mB)^{1/2}R]$$

where ν_0 is approximately $10^{15} \, s^{-1}$, F is the Franck–Condon factor, B is the binding energy of e^-, and R is the separation of donor and acceptor. The rate of reaction decreases by about an order of magnitude for each 2-Å increase in the separation of reactants; this decrease explains the unusual kinetics discussed (i.e., the continually decreasing rate of electron transfer with time, due to reactants at increased separation). Some controversy exists regarding the exact mechanism for the migration of charge (111–113). However, the important feature for this discussion is that migration of charge over significant distances (10–20 Å) has been observed experimentally. Excited reactants are often partially immobilized on organized assemblies at such separations, and electron tunneling mechanisms could be operative in micelles and other particles, a point to be discussed in more detail later.

Subsequent ion recombination in low temperature rigid glassy systems often leads to the formation of excited states. Excellent sources summarize this work (114–117).

Reactions of Cations

Several electron-transfer reactions of cations have been observed in the radiolysis of low temperature systems (107). A matrix of high ionization potential is used so that the electron-transfer reaction of interest, (matrix)$^+$ + solute → (solute)$^+$ + matrix, is exothermic. In these studies, radiolytic techniques have many advantages over photolytic methods because of the higher energy of the quanta in the former case. Thus a matrix of high ionization potential (e.g., CCl_4 or butyl chloride) can be ionized readily to produce a cation, which subsequently extracts an electron from an added solute.

Similar experiments may be carried out in fluid solution where fast response pulse radiolysis methods are necessary to observe such short-lived species. Radiolysis of liquid CCl_4 (118) and $C_2H_2Cl_2$ (119, 120) containing solutes of interest produces a variety of solute cations:

$$RCl \rightarrow (RCl)^+ + e^- \rightarrow (RCl)^- \rightarrow Cl^- + R^{\cdot}$$
$$(RCl)^+ + solute \rightarrow (solute)^+ + RCl$$

The electrons produced are scavenged by the halide matrix producing Cl^-, which does not interfere with the cation reactions. Large rate constants, $k > 10^{10} \, M^{-1}s^{-1}$ are observed (see Table VII). Cations of some

Table VII.
Reactions of Cations in 1,2-Dichloroethane at
24 °C.

System		Rate Constant $(M^{-1}s^{-1})$
$PhCH_2^+$	$+ I^-$	4.9×10^{10}
	$+ Br^-$	5.2×10^{10}
	$+ Cl^-$	9.3×10^{10}
Ph_2CH^+	$+ I^-$	5.2×10^{10}
	$+ Br^-$	7.0×10^{10}
	$+ Cl^-$	9.1×10^{10}
Ph_3C^+	$+ I^-$	3.6×10^{10}
	$+ Br^-$	5.6×10^{10}
	$+ Cl^-$	8.0×10^{10}
$PhCH_2^+$	$+$ Isobutylene	1.9×10^7
	$+$ 1,3-Butadiene	8.7×10^5
Ph_2CH^+	$+$ Isobutylene	9.5×10^6
	$+$ 1,3-Butadiene	$<10^5$

solvents, for example, C_6H_{12}, are very mobile, and the electron exchange reaction with added solutes is more rapid than a conventional diffusion-controlled reaction (121).

Cations (e.g., pyrene cation) produced in various photoinduced processes can also undergo rapid electron transfer with solutes of low ionization potential (such as iodide ion) and amines (such as triethylamine). The much lower electron affinity of the pyrene cation and other photoproduced ions, compared to the alkyl halide cations, excludes the general use of photolysis to produce radical cations. However, the redox processes of photoproduced radical cations are very important in solar storage of energy. Here, it is important to convert the cation back to the parent compound; hence, several electron donor systems, such as ascorbic acid, triethanolamine, and cysteine, have been developed which donate electrons back to the cation to repair the solute (124).

Literature Cited

1. Rehm, D.; Weller, A. *Isr. J. Chem.* **1970**, *8*, 259.
2. Von Smoluckowski, M. *Z. Phys. Chem.* **1917**, *92*, 129.
3. Noyes, R. M. *Progress in Reaction Kinetics*, "Solvated Electron"; **1961**, *1*, 129.
4. Debye, P. *Trans. Electrochem. Soc.* **1942**, *82*, 265.
5. Matheson, M. In ADVANCES IN CHEMISTRY SERIES No. 50, American Chemical Society: Washington, D.C., 1965, p. 45.
6. Birks, J. B. "Photophysics of Aromatic Molecules"; London, 1970.

7. Turro, N. "Modern Molecular Photochemistry"; Benjamin, New York, 1978.
8. Lamola, A. A. Energy Transfer & Organic Chemistry Interscience, New York, 1969.
9. Wilkinson, F. *Adv. Photochem.* **1964**, *3*, 241.
10. Anbar, M.; Hart, E. J. "The Solvated Electron"; Wiley: New York, 1970.
11. Schick, H.; Fischer, H. *Hel. Chim. Acta.* **1978**, *61*, 2130.
12. Saltiel, J.; Shannon, P. J.; Zafision, O. C.; Urate, A. K. *J. Am. Chem. Soc.* **1980**, *102*, 6799.
13. Adamson, A. "A Textbook of Physical Chemistry"; Acad. Press.: New York, 1973, p. 67.
14. Perrin, F. *C. R. Acad. Sci. Paris* **1924**, *178*, 1978.
15. Siegel, S.; Judickis, H. S. *J. Chem. Phys.* **1968**, *48*, 1613.
16. Bowen, E. J.; Metcalf, W. S. *Proc. Roy. Soc.* **1951**, *A206*, 937.
17. Ware, W. R.; Novros, T. S. *J. Phys. Chem.* **1966**, *70*, 3246.
18. Dexter, D. L. *J. Chem. Phys.* **1953**, *21*, 836.
19. Förster, T. H. *NaturwissSchaften*, **1946**, *33*, 166.
20. Förster, T. H. *Discuss. Faraday Soc.* **1959**, *27*, 7.
21. Förster, T. H. *Naturforschung* **1949**, *49*, 321.
22. Latt, S. A.; Chung, H. T.; Blont, E. R. *J. Am. Chem. Soc.* **1965**, *87*, 995.
23. Stryer, L. S.; Haugland, R. P. *Proc. Nat. Acad.* **1967**, *58*, 720.
24. Kuhn, H.; Mobius, D.; Bucher, H. *Phys. Methods Org. Chem.* **1972**.
25. Mobius, D. *Ber. Bunsenges. Phys. Chem.* **1978**, *82*, 849.
26. Medinger, T.; Wilkinson, F. *Trans. Faraday Soc.* **1965**, *61*, 620.
27. Richards, J. T.; West, G.; Thomas, J. K. *J. Phys. Chem.* **1970**, *74*, 4137.
28. Almgren, M.; Greiser, F.; Thomas, J. K. *J. Am. Chem. Soc.* **1979**, *101*, 279.
29. Paterson, L.; Rzad, S. *Chem. Phys. Letts.* **1975**, *31*, 254.
30. Humphrey-Baker, R.; Moroi, Y.; Grätzel, M. *Chem. Phys. Letts.* **1978**, *58*, 207.
31. Parker, C. A.; Hatchard, C. G. *Proc. Roy. Soc.* **1962**, *A269*, 574.
32. Thomas, J. K. Unpublished data.
33. Kira, A.; Thomas, J. K. Unpublished data.
34. Nickel, B. *Chem. Phys. Letts.* **1974**, *27*, 84.
35. Drexhage, K. H.; Fleck, M.; Kuhn, H.; Schäfer, F. P.; Sperling, W. *Ber. Bunsenges. Phys. Chem.* **1966**, *70*, 1179.
36. *Ibid* **1968**, *72*, 329.
37. Holroyd, R. A.; Russell, R. L. *J. Phys. Chem.* **1974**, *78*, 2128.
38. Thomas, J. K.; Piciulo, P. In ADVANCES IN CHEMISTRY SERIES No. 184; American Chemical Society: Washington, D.C., 1980, p. 97.
39. Wu, K. C.; Lipsky, S. *J. Chem. Phys.* **1977**, *66*, 5614.
40. Raz, B.; Jortner, J. *Chem. Phys. Letts.* **1969**, *4*, 155.
41. Hamill, W. H. In "Radical Ions"; Kaiser, E. T.; Kevan, L. Eds.; Interscience: New York, 1968, p. 321.
42. Claridge, R. F. C.; Willard, J. E. *J. Am. Chem. Soc.* **1965**, *87*, 4992.
43. Kevan, L. "Advances in Radiation Chemistry"; Burton, M.; Magee, L. J. Eds.; Wiley: New York, 1974, Vol. 4, p. 181.
44. Miller, J. *J. Chem. Phys.* **1972**, *56*, 5173.
45. Mulliken, R. S. *J. Am. Chem. Soc.* **1951**, *19*, 514; *23*, 397, 1955.
46. Briegleb, G. "Eleckronen-Donator-Acceptor Komplexe"', Springer-Verlag: Berlin, 1961.
47. Razem, B.; Wong, M.; Thomas, J. K. *J. Am. Chem. Soc.* **1978**, *100*, 1679.
48. Richards, J.; West, G.; Thomas, J. K. *J. Phys. Chem.* **1970**, *74*, 4137.
49. Chen, T., Ph.D. Thesis, University of Notre Dame 1978.
50. Wardman, P. *Curr. Top. Radiat. Res. Q.* **1977**, *11*, 347
51. Meisel, D.; Neta, P. *J. Am. Chem. Soc.* **1975**, *97*, 5198.
52. Almgren, M.; Greiser, F.; Thomas, J. K. *J. Phys. Chem.* **1979**, *83*, 3232.
53. Kalyanasundaran, K.; Grätzel, M. *J. Chem. Soc. Chem. Commun.* **1979**, 1137.
54. Whitten, D. G. *Acc. Chem. Res.* **1980**, *13*, 83.

55. Willner, I.; Ford, W. E.; Otvos, J. W.; Calvin, M. *Nature* **1979**, *280*, 823.
56. Grätzel, M. *Acc. Chem. Res.* **1981**, *14*, 376.
57. Rodgers, M. A. J. *Photochem. Photobiol.* **1979**, *29*, 1031.
58. Frank, A. J.; Grätzel, M.; Henglin, A. *Ber. Bunsenges. Phys. Chem.* **1976**, *80*, 593.
59. Neta, P.; Simie, M. G.; Hoffman, M. Z. *J. Phys. Chem.* **1976**, *80*, 2018.
60. Hoffman, M. Z.; Whitbum, K. D. *J. Chem. Educ.* **1981**, *58*, 119.
61. Sellers, R. M. *J. Chem. Educ.* **1981**, *58*, 114.
62. Stein, G.; Weiss, T. *J. Chem. Soc.* **1951**, 3265.
63. Baxendale, J. H.; Wilson, R. *Trans. Faraday Soc.* **1957**, *53*, 344.
64. Baxendale, J. H.; Thomas, J. K. *Trans. Faraday Soc.* **1958**, *54*, 1515.
65. Hunt, R.; Taube, H. *J. Am. Chem. Soc.* **1952**, *74*, 5999.
66. Dorfman, L. M.; Adams, G. E. *N.S. RDS. NBS.* **1973**, *46*.
67. Anbar, M.; Farkatazis; Ross, A. B. *N.S. RDS. NBS.* **1975**, *51*.
68. Schested, K.; Corfitzen, H.; Christenson, H. C.; Hart, E. J. *J. Phys. Chem.* **1975**, *79*, 310.
69. Dorfman, L. M.; Tasch, I. A.; Buhler, R. E. *J. Chem. Phys.* **1962**, *36*, 3051.
70. Thomas, J. K. "Advances in Radiation Research"; Magee, J. L.; Burton, M., Eds.; Wiley: New York, 1970; Vol. I, p. 103.
71. Darwent, B. de B.; Roberts, R. *Proc. Roy. Soc.* **1953**, *A216*, 344.
72. Martin, R. M.; Willard, J. E. *J. Chem. Phys.* **1964**, *40*, 2999.
73. Farhataziz; Ross, A. B. *N.S. RDS. NBS.* **1977**, 59.
74. Henglein, A.; Schnable. Ein Führung in die Strahlenchemie Weinhelm/ Bergst. Velag. Chemie 1969.
75. Baxendale, J. H.; Thomas, J. K. *Chem. Ind. London* **1956**, 377.
76. Adams, G. E.; McNaughton, G. S.; Michael, B. D. *Trans. Faraday Soc.* **1968**, *64*, 902.
77. Adams, G. E.; Wardman, P. "Free Radicals in Biology"; Acad. Press: New York, 1977; Vol. III, p. 53.
78. Neta, P. *Adv. Phys. Org. Chem.* **1976**, *12*, 223.
79. Adams, G. E.; Redpath, J. L.; Bisby, R. H.; Cundall, R. B. *Isr. J. Chem.* **1972**, *10*, 1073.
80. Bisby, R. H.; Cundall, R. B.; Wardman, P. *Biochim. Biophys. Acta.* **1975**, *389*, 137.
81. Asmuss, K.-D.; Wigger, A.; Henglein, A. *Ber. Bunsenges. Phys. Chem.* **1966**, *70*, 862.
82. Thomas, J. K.; Johnson, K.; Klippert, K.; Lowers, R. *J. Chem. Phys.* **1968**, *48*, 1608.
83. Rjad, S. J.; Infelta, P. P.; Warman, J. M.; Schuler, R. H. *J. Chem. Phys.* **1970**, *52*, 3971.
84. Hummel, A. *J. Chem. Phys.* **1968**, *48*, 3268; **1968**, *49*, 4840.
85. Gangwer, T. E.; Thomas, J. K. *Int. J. Radiat. Phys. Chem.* **1975**, *7*, 305.
86. Warman, J. R.; Asmus, K. D.; Schuler, R. H. "Radiation Chemistry II"; ADVANCES IN CHEMISTRY SERIES No. 82; American Chemical Society: Washington, D.C., 1968, Vol. II, p. 25.
87. Hart, M. "Solvated Electron"; ADVANCES IN CHEMISTRY SERIES No. 50, American Chemical Society, 1965.
88. Thomas, J. K. *J. Phys. Chem.* **1967**, *71*, 1919.
89. Anbar, M.; Bambenek, M.; Ross, A. B. *Nat. Bur. Stand. Rej. Data. Ser., N.B.S.* **1973**, *43*, 67.
90. Hamill, W. H. *J. Phys. Chem.* **1969**, *73*, 1341.
91. Wolff, R. K.; Bronskill, M. J.; Alrich, J. E.; Hunt, J. W. *J. Phys. Chem.* **1973**, *77*, 1350.
92. Bromberg, A.; Thomas, J. K. *J. Chem. Phys.* **1975**, *63*, 2124.
93. Kajiwara, T.; Thomas, J. K. *J. Phys. Chem.* **1972**, *76*, 1700.
94. Hamill, W. H.; Gallivan, J. B. *J. Chem. Phys.* **1966**, *44*, 1279.
95. Richards, J. T.; Thomas, J. K. *J. Chem. Phys.* **1970**, *53*, 218.
96. Fuochi, P. G.; Freeman, G. R. *J. Chem. Phys.* **1972**, *56*, 2333.

97. Baxendale, J. H.; Bell, C.; Wardman, P. *Chem. Phys. Letts.* **1971**, *12*, 347.
98. Schmidt, W. F.; Allen, A. O. *J. Chem. Phys.* **1970**, *52*, 4788.
99. Davis, H. T.; Schmidt, L. D.; Minday, R. M. *Chem. Phys. Letts.* **1972**, *13*, 413; *J. Phys. Chem.* **1972**, *76*, 442.
100. Allen, A. O. *N.S. RDS. NBS.* **1976**, 58.
101. Thomas, J. K.; Beck, G. *Chem. Phys. Letts.* **1972**, *13*, 295.
102. Beck, G.; Thomas, J. K. *J. Chem. Phys.* **1972**, *57*, 3649.
103. Beck, G.; Thomas, J. K. *J. Chem. Phys.* **1974**, *60*, 1705.
104. Holroyd, R. A.; Allen, A. O. *J. Chem. Phys.* **1971**, *54*, 5014.
105. Allen, A. O.; Holroyd, R. A.; Gangwar, T. E. *J. Phys. Chem.* **1975**, *79*, 25.
106. Holroyd, R. A.; In "Radiation Research"; Nygaard, O.; Adler, H.; Sinclair, W. Eds.; Acad. Press: New York, 1975; p. 378.
107. Skelly, D. W.; Hayes, R. G.; Hamill, W. H. *J. Chem. Phys.* **1965**, *43*, 2795.
108. Albrecht, A. *Acc. Chem. Res.* **1970**, *3*, 328.
109. Thomas, J. K.; Richards, J. T. *Chem. Phys. Letts.* **1971**, *8*, 13.
110. Thomas, J. K.; Richards, J. T. *J. Chem. Phys.* **1971**, *55*, 3636.
111. Millar, J. R.; Beitz, J. V. In "Radiation Research"; Okada, S.; Smamura, M.; Tersima, T.; Yamugnichi, H. Eds.; 1979; p. 301.
112. Burton, G. V. P. 317, in Ref. 111.
113. Hamill, W. H.; Funabashi, P. *Phys. Rev.* **1977**, *B16*, 5523.
114. Brocklehurst, B.; Porter, G.; Savadatti, M. I. *Trans. Faraday Soc.* **1964**, *60*, 2017.
115. Kieffer, F.; Lapersonne, C.; Minigaut, J. *Int. J. Radiat. Chem. Phys.* **1974**, *6*, 79.
116. Cordier, P.; Kieffer, F.; Lapersonne, C.; Regeut, J. *Radiat. Res.* **1975**, 426.
117. Bernas, A. P. p. 325 in Ref. 106.
118. Cooper, R.; Thomas, J. K. In Radiation Chemistry II, ADVANCES IN CHEMISTRY SERIES No. 82, American Chemical Society, 1968; p. 351.
119. DePalma, V. M.; Yang, Y.; Dorfman, L. M. *J. Am. Chem. Soc.* **1978**, *100*, 5416.
120. Wang, Y.; Dorfman, L. M. *Macromolecules* **1980**, *13*, 63.
121. Zador, Z.; Warman, J. M.; Hummell, A. *Chem. Phys. Letts.* **1973**, *23*, 363.
122. Krasna, A. I. *Photochem. Photobiol.* **1980**, *31*, 75.
123. Birks, J. "Photophysics of Aromatic Molecules"; Macmillan: London, 1970.
124. Gangwer, T.; Thomas, J. K. *Rad. Res.* **1973**, *54*, 192.
125. Land, E. J.; Richards, J. T.; Thomas, J. K. *J. Phys. Chem.* **1972**, *76*, 3805.
126. Holroyd, R. A.; Russell, R. L. *J. Phys. Chem.* **1974**, *78*, 2128.
127. Siomos, K.; Christopheron, L. G. In "Proceedings of the 7th International Conference in Conduition and Breakdown"; Scheidt, W. F., Ed.; Berlin, 1981; p. 84.
128. Thomas, J. K. NSRDS-NBS46 In "Advances in Radiation Chem."; 1969, Vol. I, p. 103.
129. Dorfman, L. M. In "Fast Reactions in Energetic Systems"; Capellos, C.; Walker, R. F., Eds.; D. Reidel; Boston, 1981; p. 95.

4

Experimental Techniques

A THOROUGH STUDY OF THE RADIATION-INDUCED CHEMISTRY of a system in or on an organized assembly requires many experimental techniques. For example, the nature of the assembly should be understood, and parameters such as size, aggregation number, and charge have to be determined. Conventional techniques, such as light scattering, laser, Raman and IR spectroscopy, sedimentation studies, electron microscopy, conduction, viscosity, and NMR and electron spin resonance spectroscopy may all contribute to an understanding of the assembly.

The photosystem itself will probably experience an unusual environment on the assembly. Hence, emission and absorption spectral techniques are required to give some measure of the photosystem's surroundings. Finally, the actual excitation technique has to be specified, that is, steady-state or pulse irradiation, and photolysis or radiolysis. This great variety of observations often makes the work very appealing because a detailed but satisfying picture of a complex system can often be produced.

Environment of the Photosystem

The location of a molecule in an assembly must always be known with some precision. A first approach uses whatever knowledge of the system is available to construct a molecule that should be located in a desired position of the assembly. Often, unexpected factors overrule the initial expectations of this molecule; however, a start has to be made somewhere. A good example is to locate the pyrene chromophore close to a positively or negatively charged surface. The former condition is achieved by using sodium pyrenesulfonate, where the sulfonate interacts strongly with the positive charge of the surface. The latter condition is achieved by synthesizing a quaternary ammonium salt of pyrene, for example, pyrene–methylene–trimethylammonium bromide, or pyrene–butyltrimethylammonium bromide. Often, a simple polar group will encourage the chromophore to attend the surface of an assembly system. In a large microemulsion system, pyrene tends to locate in the oil interior and to a much lesser extent in the surface. However, amino-pyrene, pyrene methanol, or pyrenesulfonate locate primarily in the surface region and close to the lipid–water interface.

0065-7719/84/0181-0071$06.75/1

Once a system has been selected for a precise purpose, the location must be confirmed by other techniques. Emission and absorption spectroscopy are often very useful in this regard.

Emission Spectroscopy. Steady-state emission spectroscopy is a well-established technique that is now carried out exclusively on commercial equipment. The experimental details are controlled by the equipment manufacturer; therefore, they will not be described here (1).

Many molecules have quite different solvent interactions in the excited and ground states that lead to different absorption and emission spectra. To be useful for probe studies in assemblies, the solvent shifts have to be significant and readily interpretable. Several molecules have been investigated for use as probes in biological and chemical assemblies. The high sensitivity of emission studies makes this type of measurement very desirable in probe studies because only small amounts of probe material are needed and little, if any, distortion of the system results.

FINE STRUCTURE CHANGES. Symmetrical molecules such as benzene, pyrene, and coronene show marked effects of solvents on the O–O band of absorption and emission. Pyrene, because of its usefulness in other photophysical studies, has been studied in detail, and pyrene fluorescence shows quite a strong solvent dependence (2, 3). Figure 1 shows pyrene fluorescence in several solvents of different polarities. For convenience the peaks are labeled I to V, and the variation in the peak ratio III/I is noted. Figure 1 and other studies (3) show that the III/I ratio is quite solvent dependent. The ratio decreases with increasing solvent polarity and may be correlated with medium polarity. Other effects may also influence the III/I ratio; for example, pyrene forms weak complexes with alcohols (4) and quaternary ammonium compounds (5). The III/I ratio, although affected by such chemistry, is also influenced by temperature and high pressure. Nevertheless, if applied with care, the pyrene III/I ratio can provide very useful data on the probe environment in an assembly.

The first singlet absorption of pyrene, $^1Ag \rightarrow {}^1B_{3u}$, is very weak and symmetry forbidden because the $^1B_{3u}$ state is short-axis polarized. The fluorescence of the first excited state shows mixed polarization, probably due to a mixing of the excited state $^1B_{3u}$ (short-axis polarized) with the second excited state $^1B_{2u}$ (long-axis polarized). The fluorescence spectrum of pyrene shows vibronic bands corresponding to allowed vibrations ($^1B_{1g}$) and forbidden vibrations (^1Ag), including the O–O band. Solvents markedly affect the intensity of forbidden vibronic bands in weak electronic transitions. With pyrene, Peak III is strong and allowed, and it shows little variation with solvent. However, Peak I

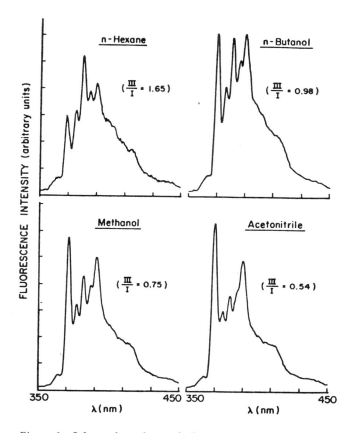

*Figure 1. Solvent dependence of vibronic band intensities in py-
rene monomer fluorescence. Conditions: [pyrene], 2 μM; and λ_{excit},
310 nm (3).*

(O–O) is forbidden and weak in poor, nonpolar solvents. The ratio
III/I can, thus, serve as a measure of the solvent environment surrounding pyrene.

Substitution on the pyrene ring to form derivatives, such as meth-
ylpyrene and pyrenebutyric acid, destroys the symmetry of the molecules and makes the O–O Peak I allowed. The derivatives of pyrene
usually exhibit a III/I ratio less than 1. However, although these molecules do not exhibit solvent-dependent III/I changes, they do show
small fluorescence red shifts with increasing solvent polarity, which are
useful in investigating these molecules in many assemblies.

MOLECULES WITH TWO CHROMOPHORES. Molecules with two chromo-
phoric groups, such as pyrenecarboxaldehyde (6), N-phenylnaph-

thylamine (NPN) (7), and anilinonaphthalenesulfonic acid (ANS) (8), often show marked spectral changes both in the fluorescence maximum and fluorescence quantum yield with changing solvent polarity. These molecules and others have been used extensively in biological systems (8). The observed effects can be explained in terms of the photophysics of pyrenecarboxaldehyde (PCHO).

The fluorescence of PCHO shows a marked red shift with increasing solvent polarity. Figure 2 shows the fluorescence spectra of PCHO in several media, and Figure 3 shows a plot of the wavelength maximum λ_{max} versus the dielectric constant ε of the medium; a linear portion appears when $\varepsilon > 10$. This behavior is explained by the interplay of two chromophores of the system, the $\eta-\pi^*$ CHO chromophore and the $\pi-\pi^*$ aromatic pyrene. In nonpolar media the fluorescence maximum lies at $\lambda_{max} \sim 4000$ Å and is due to the $\eta-\pi^*$ transition. When the polarity of the medium is increased, the $\pi-\pi^*$ level, which lies close to the $\eta-\pi^*$ level, is brought below that of the $\eta-\pi^*$ level by solvent interactions with the excited state which lead to $\lambda_{max} > 4000$ Å.

Figure 3 may be used to measure the local ε in the vicinity of the micelle surface. Several other molecules, such as trisbipyridylruthenium (9), and indole derivatives (10) show solvent-dependent emissions used to probe assemblies.

Fluorescence Polarization. The degree of polarization of fluorescence is a very useful method for determining the restrictions to motion that

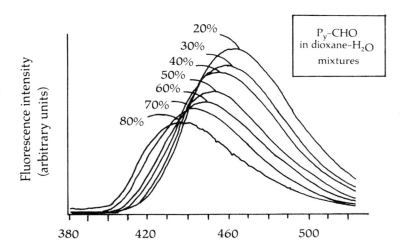

Figure 2. Solvent dependence of pyrene-3-carboxaldehyde (PCHO) in monomer fluorescence in various dioxane–water mixtures (percent dioxane is given). Conditions: [PCHO], $10^{-5}M$; and λ_{excit} 356 nm (6).

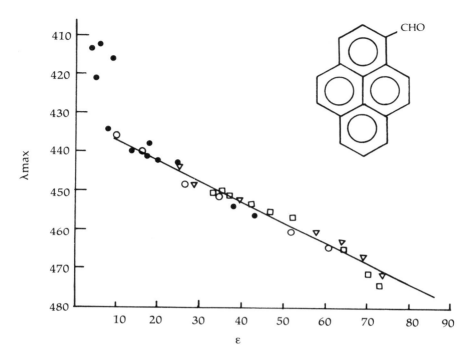

Figure 3. Variation of λ_{max}^{fluor} for pyrene-3-carboxaldehyde with solvent dielectric constant ε.

Key: □, methanol–water mixtures; ▽, ethanol–water mixtures; ●, dioxane–water mixtures; and ○, pure solvents. Key to ε(fluorescence maximum): chloroform, 4.8 (421); chlorobenzene, 5.6 (412); dichlorobenzene, 8.9 (416); 1-hexanol, 13.5 (440); 2-propanol, 18 (438); 1-butanol, 17.5 (441); 1-propanol, 20.3 (442); ethylene glycol, 38 (456); and glycerol 43 (456) (6).

a molecule experiences in its particular environment. The experimental technique is difficult to carry out with any degree of accuracy with standard spectrophotometers; however, two commercial machines provide reliable fluorescence polarization data (Elscint Corporation, Israel; S.L.M. Instruments, Urbana, Illinois).

The fluorescence polarization technique was originated by Weber (11) from suggestions by Perrin (12). A probe molecule is located in an assembly where it is excited with polarized light. The degree of polarization of the fluorescence P is equal to $(I_{11} - I_1)/(I_{11} - I_1)$, where I_{11} and I_1 refer to the fluorescence intensities with either both polarizers parallel or the polarizers crossed in the excitation and emission channels, respectively. P is related to the viscosity η of the probe's environment by

$$\frac{(1/P) - (1/3)}{(1/P_o) - (1/3)} = 1 + \frac{\lambda RT}{V_o\, \eta}$$

where P_o is the degree of polarization in an extremely viscous or rigid medium, V_o is the effective volume of the molecules, T is the temperature, R is the gas constant, and λ is the lifetime of the fluorescence. The fluorescence lifetime acts as an internal clock to time the tumbling motion of the molecule. This technique may be used to estimate the local microviscosity of the probes' environment (13). However, the exact motion of the probe leading to fluorescence depolarization is not known. If the probe is chemically bound to a large molecule, such as a protein, then in a solvent of known η V_o can be measured, and hence, the size of the protein can be measured (11, 14).

Phosphorescence. The phosphorescence of aromatic molecules is more prevalent in assemblies than in homogeneous solution. However, to date phosphorescence data have not provided information on the assemblies themselves or on the environment of the probes. The greater occurrence of phosphorescence in assemblies enables the photochemistry of the triplet state at room temperature to be studied much more readily. This measurement was confined to low temperatures in the past. Phosphorescence in micelles has been an aid in analytical studies (15, 16).

Steady-State Irradiation. Simple techniques are used for steady-state irradiation of materials in assemblies. In photolysis studies, mercury resonance lamps are used for UV irradiation at 2537 Å, and quartz iodine lamps, or simple 500–1000 W photographic projector lamps are used for irradiation with visible light. Detailed aspects of steady-state irradiations are available (1, 17). The technique for steady-state radiolysis of a sample is more complex only because of the biological shielding of personnel that is necessary; details of various steady-state high-energy irradiation techniques are given elsewhere (18).

Pulsed Irradiation. Pulse techniques of irradiation are used primarily to investigate reactive short-lived intermediates. The basic concept in both pulse radiolysis and flash photolysis is to rapidly produce a high yield of short-lived intermediates and to subsequently observe their properties, both spectral and kinetic, by a variety of methods. In pulse radiolysis, high-energy linear accelerators, or van de Graaff generators, produce short (10^{-11}–10^{-6} s) pulses of high-energy electrons (3–30 MeV) that are used for exciting the sample. In flash photolysis, flash tubes can be used to produce intense short (10^{-9}–10^{-3} s) pulses of white light. Lasers that produce light of selected wavelengths in short pulses (10^{-11}–10^{-6} s) are ideal for situations where high intensities are required (e.g., in absorption studies). The detection systems are similar in both techniques, even down to the very short 10^{-12}-s laser pulses, where special techniques are used to utilize the full time resolution capability of the laser equipment.

Flash Photolysis. Figure 4 shows a typical laser flash photolysis arrangement for excitation and observation in the 10^{-9}-s time region. The basic design is that used in 1950 (*19*), with modifications to attain greater time resolution. The laser pulse, or light flash, from a flash tube impinges on the sample contained in a suitable rectangular pyrex or suprasil cell that is freed of oxygen by bubbling with nitrogen or by the well-known repeated freeze–thaw vacuum technique. Short-lived species in the sample are monitored by absorption or emission spectroscopy (*20, 21*). For absorption spectroscopy, a beam of light from a xenon arc lamp is directed through the sample cell and onto the slits of a high intensity monochromator that selects the spectral region of interest. For emission spectroscopy the luminescence of the sample is guided by lenses to the slits of the monochromator. The light output of the monochromator is monitored by a suitable photomultiplier or diode. The output is displayed on an oscilloscope, a transient capture device (e.g.,

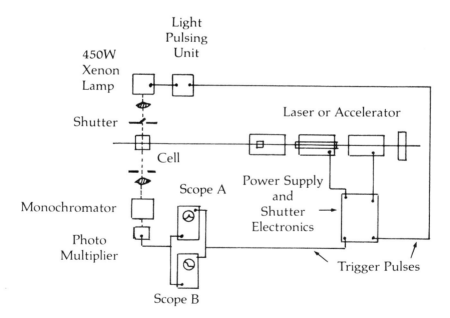

Scope A for Transient Measurement
Scope B for I_o Measurement

Figure 4. Diagrammatic layout of laser flash photolysis equipment.

a Tektronix 7912), or a biomation transient unit. The data stored in these devices are conveyed to a computer for subsequent analysis. Photomultipliers may be used conveniently for high speed (1 ns) detection over the spectral range from 2000 to 10,000 Å. Special high current circuits are needed for high speed absorption studies (22, 23). The time resolution of the apparatus can be increased to about 50 ps with fast photodiodes, [e.g., ITT4014 (24)], and the spectral response can be increased to $\lambda > 20,000$ Å with photodiodes (the time response now >3 ns).

Additional techniques are available for emission spectroscopy, such as single photon counting, which, with short pulses of excitation light (~1 ns), can give a time resolution of less than 0.5 ns (25). This technique has the advantages of (1) it requires only small light pulses, and (2) the photomultiplier tube does not have to have a fast response that precisely follows the pulse. In the direct time observations described previously, the diodes and photomultipliers must have high fidelity and fast rise and fall time response characteristics. The main disadvantage of single photon counting is that the technique cannot be used for absorption studies, and the pulse repetition rate has to be high. The development of streak cameras also increases the response of the recording equipment for absorption and emission studies (26, 27).

Picosecond Laser Flash Photolysis

The development of mode-locked lasers has heralded in the art of picosecond (28–30) and subpicosecond (30) laser flash photolysis. These techniques require time responses in the observation techniques that are superior to those time responses available with diodes and photomultiplier tubes. Optical methods have been developed to observe the physical and chemical phenomena generated in these systems. Streak cameras with responses in the picosecond range have also been used in such experiments with good results. The main disadvantage of the technique is the cost of the mode-locked lasers, streak cameras, and peripheral equipment; however, unique data have been obtained.

Pulse Radiolysis

The pulse radiolysis technique is similar to the laser flash photolysis technique; however, an accelerator produces short pulses of energy rather than a laser. The detection techniques are similar, although the pulse radiolysis technique requires that the observer be some distance from the irradiated sample and shielded from the high-energy radiation. Therefore, long cables are required to carry the photomultiplier signals from the irradiation rooms. In addition, the analyzing light beam, or emission beam, has to be transported over a large distance to a safe radiation-free environment, where it may be monitored. Quartz sample cells must be used because pyrex vessels rapidly develop a deep brown discoloration because of the high-energy radiation. The maximum time resolution is from 10 to 20 ps, at least an order of magnitude less than that attained in picosecond laser studies. This loss is mainly due to the long irradiation pulses that are used in pulse radiolysis.

Data handling in pulse radiolysis is similar to that in flash photolysis. The full spectrum of a species can be taken with a single pulse by using a streak camera, an image converter tube, and a television monitoring system. This spectrum is displayed as a function of time (31). The spectrum of a transitory species is developed by repeated pulsing of the sample, which must be flowed to replenish the spent solvent or solute; each pulse produces a datum point for one wavelength. A typical scan, which could represent a pulse radiolysis or laser photolysis experiment, is shown in Figure 5, and this scan shows the rate of for-

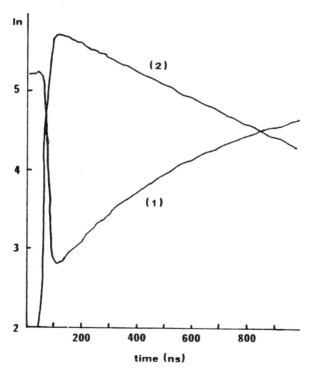

Figure 5. Fluorescence decay of [Ru(II)]* in water where λ_{em} is 610 nm, the rate is 1.71×10^6 s^{-1}, and the half-life is 414 ns. Key: 1, fluorescence vs. time in arbitrary units; and 2, natural log of fluorescence vs. time.

mation and decay of trisbipyridylruthenium. The sample was excited at 3000 Å and the excited state of the organometallic ion was monitored at 6100 Å via its characteristic luminescence. The lower curve shows these data; the upper curve is the logarithmic plot of the fluorescence versus time. This latter plot is linear, thus, it agrees with the first-order decay kinetics of the excited state back to the ground state

$$RuII \rightarrow (RuII)^* \xrightarrow{k_1} RuII + h\nu$$

$$\ln \frac{(RuII)_o^*}{(RuII)^*} = k_1t \text{ or } \ln (RuII)^* \propto - k_1t$$

The slope of the $\ln[C]$ versus t gives the half-life for decay $t_{1/2} = 4.4$ ns, and the rate constant $k_1 = 1.71 \times 10^6 \text{ s}^{-1}$. A series of such curves at different wavelengths can be used to generate the spectrum of the emission (or absorption if the apparatus is used in the absorption mode) at various time intervals after the pulse for $t = 0$ to $t = \infty$.

Form of the Kinetics

The kinetics observed in pulsed experiments can be quite complex; however, in the deliberate measurement of a rate constant for reaction, the system is designed so that the kinetics are as simple as possible, and so that the required data are obtained readily and with maximum accuracy.

If the rate constant for reaction of a species (S) with a solute is required, the solute is added at high concentration so that [solute] $>>$ [species], and so that the rate of decay of S is increased greatly over that in the absence of S. Under these conditions the following processes apply

$$S \xrightarrow{k_0} \text{natural decay}$$
$$S + \text{solute} \xrightarrow{k_1} \text{product}$$

For excited states the decay of S is usually first order, $\ln [S_0]/[S] = k_0t$. The rate increases in the presence of solute and if $\ln [S_0]/[S]$ is plotted against t, then an overall rate constant k is measured, and is given by $k = k_0 + k_1 [\text{solute}]$. If k is plotted versus [solute] the slope of the line gives k_1 and the ordinate intercept gives k_0. Some problems result if the decay of S is not first order, but is a mixture of first and second order (i.e., a rapid initial decay followed by a slower one). This problem is overcome by making the decay in the presence of solute much faster than the actual decay. Accurate data on second-order reactions are difficult to obtain by pulsed irradiation techniques because of the inhomogeneity of the radiation act, which also leads to inhomogeneity in the species concentration throughout the sample. Under ideal conditions the data fit the expression $(1/[S]) - (1/[S_0]) = k_2 t$ for the reaction, $S + S \rightarrow S_2$ (k_2).

Often, two first-order rates of decay of S with different rate constants can appear to fit a second-order plot. The validity of the fit of data with a first second-order expression can be investigated by changing the intensity of the irradiation pulse that produces S, thus varying its concentration. If S decays via second-order processes, then the rate of decay will vary with pulse intensity; first-order processes are not affected by pulse intensity.

These expressions apply to decay kinetics. If the growth of a species is observed, that is, the observation of a product [S + solute $\rightarrow$ product (k_1)] then the same expressions can be used provided that [S] is derived from the relationship, $[S] = [product]_a - [product] = [S_0] - [product]$, where $[product]_a$ and [product] refer to the product concentrations when all S has decayed (i.e., completion of product growth) and at time t, respectively.

Reaction kinetics in assemblies are often more complicated than those for simple homogeneous system. Several examples will be given later.

Stroboscopic Technique

As discussed previously, ultrafast time resolution in mode-locked laser studies is achieved by optical rather than electronic detection methods. In this technique a portion of the irradiation pulse is delayed in a precise manner with respect to the excitation pulse and can then be used to probe the events caused by excitation. A similar technique has been developed in pulse radiolysis studies (32–34). Figures 6a and 6b illustrate this method as a diagrammatical layout. The irradiation source in the pulse radiolysis method is a linear accelerator that produces a row of short, fine-structure electron pulses, the duration of which is 15 ps with a pulse separation of 360 ps. The beam of pulses is passed through a quartz cell containing xenon gas where it produces pulses of Cerenkov radiation, which are in direct phase with the electron beam. Immediately following the Cerenkov cell a thin mirror deflects the visible and UV Cerenkov light around an optical pathway, while the energetic electron beam passes through the thin mirror and into the sample cell. After a suitable delay, which is defined by its light path and the velocity of light, the Cerenkov beam is redirected by another mirror into the electron beam path and follows the electron beam through the sample cell. The intensity of the Cerenkov light is monitored by a photomultiplier tube after passing through the sample cell. The variation in the

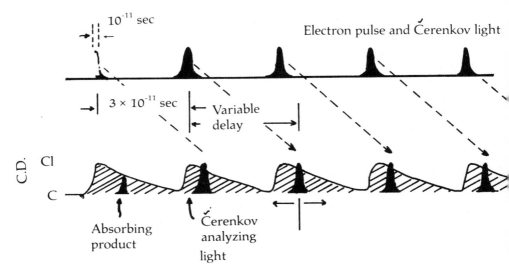

Figure 6a. Basic stroboscopic pulse-radiolysis technique using linac beam fine-structure and Cerenkov light to detect short-lived absorption signals.

The top line shows the electron fine-structure and simultaneous Cerenkov light pulses, separated by 0.3 ns, the period of the microwave power. The bottom line shows the production and decay of an absorbing product. The Cerenkov light (short, dark pulse) is suitably delayed and used to measure the concentration of the absorbing product (35).

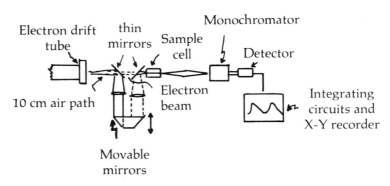

Figure 6b. Components of a stroboscopic pulse-radiolysis system.

The electron beam (--) emerges from the electron drift tube, passes through 10 cm of air and two thin mirrors and finally irradiates the sample. The analyzing Cerenkov light (—) is produced in the air path of the electron beam, is transmitted over a variable length optical path, and is focused to pass through the irradiated sample. This Cerenkov light then is focused through a monochromator and detected by a photomultiplier. Special integrating circuits give a DC signal that may be displayed on a recorder (33, 34).

intensity of the Cerenkov light pulse, as it is delayed behind the excitation pulse, effectively monitors the time profile of the absorbing species produced in the cell by the electron beam. A time resolution of about 10–20 ps is attained by this technique. The concepts of this experiment, and those of picosecond laser photolysis, are unlike the more conventional nanosecond and microsecond electronic experiments, and they add a great deal of invention and artistry to pulsed methods.

Conduction

The techniques of pulse radiolysis and laser flash photolysis have been successfully combined with detection methods other than those of emission or absorption spectroscopy. The detection of species by conduction methods adds another dimension to the detection of short-lived species, especially when combined with optical methods. The first pulsed conductivity data reported a rapidly decaying current spike on irradiation of liquid neopentane with X-rays (35). These data were later confirmed and attributed to radiation-produced electrons of high mobility (36). A mobility (μ) of 70 $cm^2 V^{-1} s^{-1}$ was measured in neopentane; a value that is four orders of magnitude larger than that of an ordinary ion in this liquid. The magnitudes of electron mobilities in different liquids can be correlated with the symmetry or sphericity of the solvent molecules. Thus, the electron mobility in liquid neopentane is three orders of magnitude larger than that in n-pentane, and liquid methane is even larger with $\mu = 450$ $V^{-1} s^{-1}$ (37). The mechanism that can explain such high mobilities is still debated. A popular model envisages the electron as being quasi-free in the liquid and occasionally located at sites of low energy, which are erected by suitable arrangement of the solvent molecules. In polar media the electron is caught in deep solvent traps, and the mobility of the solvated electron is similar to or even a little larger than that of a simple ion (38). In nonpolar liquids the electron moves from site to site in the liquid; here, the shallow traps lead to rapid electron migration, and, hence, to a large mobility to the kinetics. The measurement of electron mobility is not as important as the measurement of the electron rate of reaction.

   The rate of reaction is conveniently measured by two allied conductivity methods, the AC and the DC methods. The DC method works well for nonconducting liquids; a typical experimental arrangement is shown in Figure 7 (39, 40). The irradiation cell is constructed of pyrex or quartz and contains two platinum electrodes. The laser beam, or X-ray pulse, is directly between the electrodes, where it produces the ions of interest. These ions cause a current to flow if a potential is applied to the two electrodes. The current or voltage change on a load

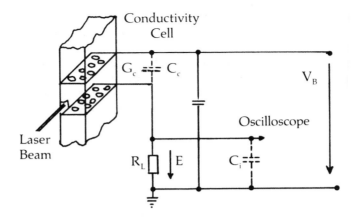

Figure 7. Schematic diagram of the photoconductivity system. Key: G_c, cell conductance; C_c, cell capacitance; R_L, load resistor; C_i, input capacitance of the oscilloscope; E, observed signal; and V_B, supply voltage (41).

resistor may be displayed on a fast oscilloscope or a transient capture device. The voltage E at the oscilloscope is related to the ion concentration by

$$E = (V_B R_L F / 10^3 k_c) \sum_j c_j |z_j| \mu_j$$

where C_j is the concentration of charged species, z_j is the charge per ion, μ_j is the mobility of the ion, F is the Faraday constant, and k_c is the cell constant. All ions, both negative and positive, give rise to the signal, and a prior knowledge of the system (from optical measurements) is useful in interpreting the data. In conduction studies a display of voltage versus time is equivalent to a display of [ion] versus time in optical studies. Therefore, the conduction data can be treated in the conventional manner to obtain rate constants. Comparison of the radiation-induced voltage change in the unknown sample with a standard ion sample can be used as a measure of the mobility of the unknown species with respect to the standard. The fine structure pulses of a linear accelerator now allow this technique to be used in the picosecond region (41). In colloid systems, the kinetics and rates of electron capture by water pools in reversed micelles can be measured by this technique (42). This technique is difficult to apply in conducting samples as electrolysis of the sample takes place. The DC voltage may be switched rapidly to overcome this problem (43), or the AC method may be used.

An example for an AC conductivity circuit (44) is shown in Figure 8. Two cells of identical geometry are made, two arms of a wheatstone

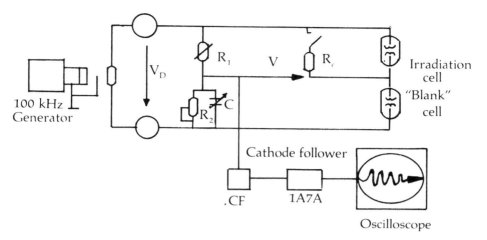

Figure 8. Block diagram of the AC-conductivity setup (50).

bridge are powered by a DC voltage V_o from a high frequency generator, and the solution to be investigated is flowed through the two cells. Adjustment to zero signal voltage V before irradiation is achieved by varying resistor R_2 and capacitance C. Formation or consumption of charge carriers in the irradiation cell results in a change of V that is recorded oscillographically.

Many chemical systems, in which little or no changes occur in optical density upon irradiation, have been investigated by the conductivity method (45, 46). One example from the field of inorganic chemistry is the investigation of the reactions of nitrogen dioxide formed in the pulse radiolysis of aqueous nitrate and nitrite solutions. A combination of optical and conductivity techniques revealed that NO_2 disappears via the reactions

$$2NO_2 \rightleftarrows N_2O_4 \xrightarrow[\text{(H}_2\text{O)}]{k} NO_2^- + NO_3^- + 2H^+$$

The value of the rate constant ($k = 10^3$ $M^{-1}s^{-1}$) was determined by the conductivity technique because two hydrogen ions, which have a large equivalent conductance, are formed during the hydrolysis of a single N_2O_4 molecule. Many other examples of the application of pulsed conductivity to determining the mechanism of reaction are given in References 4–6.

In general, the advantages of the conductivity techique are the high sensitivity of the method, the easy identification of ions involved in photolysis and radiolysis processes, and the information provided on the mobility of transient charge carriers in aqueous and nonaqueous solvents.

Fast Polarography in Pulse Radiolysis and Photolysis

Fast pulse-polarography detects short-lived intermediates formed by a pulse irradiating a sample solution, on the basis of their redox behavior at a mercury electrode. The combination of polarography and pulse radiolysis and flash photolysis has mainly four different aspects:

1. It provides a further independent technique to determine rate constants for fast radical reactions in solutions.

2. Short-time polarograms of radicals may be used for their identification, as in ESR spectra.

3. The redox behavior of short-lived transients that have biological importance may be investigated.

4. The electrokinetic analysis of the polarographic current–time curves yields values for the rate of the electron transfer from the electrode to the electroactive species in solution and the transfer coefficient. These two parameters characterize the rate by which the transient is reduced or oxidized at the electrode.

Figure 9 shows the layout of the Notre Dame pulse-polarographic layout. The layout is a version of the original Berlin apparatus (49). The irradiation cell consists of a three-electrode 1-cm³ suprasil compartment

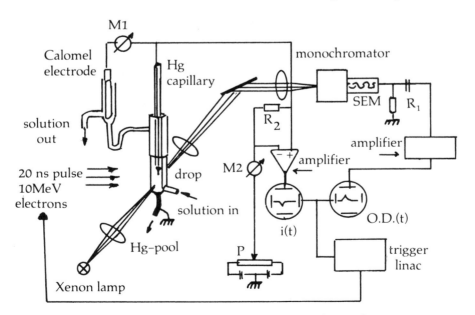

Figure 9. Measuring device for short-time polarography.

consisting of a hanging drop electrode, a mercury pool, a counterelec-
trode, and a saturated calomel element (SCE) reference electrode. The
solution flows through the cell and is in electrical contact with the
hanging drop mercury and the SCE electrodes. To make the solution
conduct, Na_2SO_4 is used as an inert electrolyte. The radicals and ions
are formed around the mercury drop by irradiation with a pulsed electron beam or laser pulse; these species may also be monitored by optical
spectroscopy. If the transitory species are reduced or oxidized at the
mercury drop upon application of a potential, a faraday current will
flow across the double layer separating the drop and the electrolyte.
The variation in current with time is recorded on an oscilloscope or
transient capture device. Repetition of the experiment at different potentials (calculated against the SCE electrode) produces a cathode current versus applied potential curve as shown in Figure 10a. The time
resolution of the device is about 3 μs.

The determination of the rate constants for the disappearance of a
short-lived intermediate by the polarographic technique is done at an
electrode potential that lies in the limiting current region of this transient. In this region, the charge transfer at the electrode surface occurs
rapidly, and, therefore, has no influence on the shape of the current–

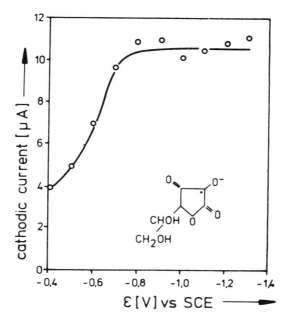

*Figure 10a. Current (8 μs after the pulse) vs. the
potential for the reduction of the ascorbic acid radical
(52).*

time curves. Hence, only the diffusion of the electroactive transient to the electrode and its bulk disappearance have to be considered in the derivation of an expression for the current–time function. If the short-lived depolarizers disappear with second-order kinetics, then, for times smaller than the first half-life of the bulk reaction, the approximate expression, $1/[i(t)^{1/2}] = (A/c_0) + k_2 At$, applies, where A is a constant, c_0 is the concentration of depolarizer immediately after the pulse, and k_2 is the bimolecular rate constant for the bulk reaction of the depolarizer. The validity of this equation is checked by recording current–time curves for different short-lived radicals with a great variety of half-lives (50). A satisfactory agreement of the polarographically determined rate constants to within 20% of the values determined by the optical method was found. Short-time polarograms are typical for certain classes of radicals and may be used in their identification (49). For instance, α alcohol radicals have reducing properties and undergo oxidation at a drop potential of -0.7 V (vs. SCE) according to:

$$\text{Ç-OH} \rightarrow \;\; {\searrow}{\atop{\diagup}}\text{C=O} + \text{H}^+ + e^-$$

However, α keto radicals are reduced at this potential:

$$ {\searrow}{\atop{\diagup}}\overset{\displaystyle O}{\overset{\displaystyle \|}{\text{Ç-C-}}} + e^- + \text{H}^+ \rightarrow {\searrow}{\atop{\diagup}}\underset{\displaystyle H}{\overset{\displaystyle O}{\overset{\displaystyle \|}{\text{C-C-}}}} $$

The possibility of distinguishing between these two radical types by the polarographic method helped to elucidate the radiation chemical reactions of alcohols with vicinal OH groups (51).

Because current values for short-time polarograms are taken only a few microseconds after the formation of the electroactive transient the polarographic waves obtained are mostly irreversible. Therefore, the current at that time is influenced by the transfer reaction of potentials that lie outside of the limiting current region. An example of an irreversible wave is shown in Figure 10b, which describes the reduction wave of the ascorbic acid radical, an intermediate in the enzymatic oxidation of ascorbic acid. Current–time curves, from which this polarogram was obtained, are shown in Figure 10a. Because the ascorbic acid radical is long-lived with respect to the measuring time, these curves are determined only by diffusion and transfer kinetics. A transfer coefficient α of 0.23 was obtained in the case of the reduction of the ascorbic acid radical. Similarly, transfer coefficients and reduction rate constants were determined for other radicals.

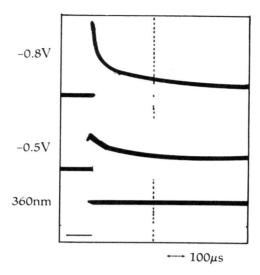

-0.8V

-0.5V

360nm

⟵ 100μs

Figure 10b. Current vs. time curves and an optical absorption vs. time curve for the reduction of the ascorbic acid radical (52).

Fast polarography has been applied to the study of the free radicals formed by the reaction of OH radicals with the surfactant cetyltrimethylammonium bromide (*47, 48*). Analysis of the data gave a rate constant of 10^8 $M^{-1}s^{-1}$ for the recombination of the surfactant radicals.

Light Scatter

Light scattering, which has been used to measure particle size, can also be studied in laser flash photolysis and pulse radiolysis with a response of about 1 μs (*52, 53*). The basic concept of the technique is that of conventional steady-state light scattering. An intense steady beam of analyzing laser light is directed through the sample cell, which can be irradiated with a pulse of exciting laser light or high-energy electrons. Macromolecules contained in the sample cell scatter a portion of the analyzing light, which is conveyed away from the cell and to a photomultiplier tube by means of a light guide. Radiation-induced molecular weight changes are reflected by changes in the intensity of the scattered analyzing light. The intensity of the light scatter by the solution I_s is related to the weight-average molecular weight M by the expression,

$$\frac{KcI}{I_s - I_o} = \frac{1}{PM} + 2BC$$

where C is the polymer concentration in grams per milliliter, I is the incident light intensity, I_o is the light scatter by the solvent, P is the particle scattering factor, K is a constant, and B is the second virial coefficient. If the polymer breakdown giving rise to the light scatter changes is assumed to occur from radiation-produced centers that follow first-order decay kinetics, then an equation that relates the light changes observed by the phototube is

$$\ln \frac{(U_\infty - U_L)^{-1}(U_t - U_L)^{-1}}{(U_\infty - U_L)^{-1}(U_o - U_L)^{-1}} = kt \qquad (1)$$

where k is the rate constant for decay; t is the time; U_o, U_t, and U_∞ represent the phototube signals (proportional to light scatter) at $t = 0$, $t = t$, and $t = \infty$, respectively; and U_L is the scattering signal from the pure solvent.

The degradation decay kinetics of poly(methacrylate) (PMA) and poly(phenyl vinyl ketone) (PUK) fit Equation 1. For PUK, one decay, independent of O_2, was observed, and this decay was probably due to disentanglement of the two polymer radicals. The technique holds much promise in the investigation of rapid molecular weight changes in macro systems.

Light-scattering techniques involving resonance Raman scatter have been developed (54, 55). Again, conventional steady-state laser Raman spectroscopy is applied to short-lived species. The species are produced by intense pulses of excitation radiation and are then irradiated into their absorption band by intense monitoring laser pulses. This latter process produces resonance Raman scatter useful in interpreting the structure of the short-lived species. This analysis technique is not as convenient as the methods for kinetic analysis; however, it does give a more precise picture of the free radical or ionic species studied, a feature that is absent in the other techniques. The possibility of achieving more precise details of the radiation-produced species, rather than kinetic rate data alone, is a highly desirable feature of pulse work studies. Electron spin resonance investigations of the intermediates produced also provide additional data.

Electron Spin Resonance Studies

Electron spin resonance (ESR) techniques are very specialized and can detect radicals formed in radiolysis (56–59) and photolysis (60, 61). The signals observed are quite specific for the radicals produced; consequently, a detailed structural picture of the free radical can be constructed. This method transcends previous analytical techniques, but it suffers from several problems, the most immediate being lack of time resolution (the response time is confined to about 1 μs).

The technique used irradiates a sample in an ESR cavity by a steady or pulsed radiation, either light or high-energy radiation. In the steady-state method a low stationary concentration of free radicals is produced in the cavity. The free radicals are monitored by the conventional ESR method. With pulse irradiation the ESR machine has to be modified to monitor the ESR spectrum at set intervals of time following irradiation. The data consist of a series of ESR spectra at various time intervals; in the steady-state method one spectrum is produced. The variation in the intensity of the spectrum with time can be used to measure rate constants for decay of short-lived species. The more pertinent measurement is the ESR spectrum, which defines the structure of the species.

The free electron in the free radical can possess two energy states in a magnetic field H because of concentrations of its magnetic vector with respect to the applied field. In addition to the applied magnetic field, the free electron also experiences magnetic fields from the nuclei in its vicinity. This field lends much structural information in these studies. The ESR spectrum of the free electron comments on the various nuclei surrounding it and on their interaction or orientations in the molecule. (An example of an ESR spectrum will be given when steady-state ESR is discussed in micellar systems.)

The sensitivity of the free electron to its environment rapidly becomes more of a menace than a blessing because of the large number of possible transitions that the free electron can make in a more complex radical. For example, the radical C_6H_5CHOH, with seven protons, has 128 lines of equal intensity. This total reduces the intensity of an individual line; hence, the sensitivity of the measurement is reduced, and the interpretation of the spectrum is complicated. Restriction of rotation of the free radical, such as in large biomolecules or polymers, gives rise to line broadening, which also reduces the sensitivity of the technique. In spite of these drawbacks, pulsed ESR studies of transitory species nearly always give useful and important additional data to complement other optical and conduction studies.

Nuclear magnetic resonance (NMR) and chemically induced dynamic nuclear polarization (CIDNP) have been measured following radiolysis of a liquid sample (62). It is too early to comment on the usefulness of these techniques in radiolysis, although they have considerable utility in photochemical studies.

Literature Cited

1. Parker, C. A. Photoluminescence of Solutions. Elsever 1968; also, E. Wehny Series on Modern Fluorescence Spectroscopy.

2. Nakajima, A. *Bull. Chem. Soc. Jpn.* **1971**, *44*, 3272.
3. Kalyanasundaran, K.; Thomas, J. K. *J. Am. Chem. Soc.* **1977**, *99*, 2039.
4. Lianos, P.; Georghiou, S. *Photochem. Photobiol.* **1979**, *30*, 355.
5. Almgren, M.; Grieser, F.; Thomas, J. K. *J. Am. Chem. Soc.* **1979**, *101*, 279.
6. Kalyanasundaram, K.; Thomas, J. K. *J. Phys. Chem.* **1977**, *81*, 2176.
7. Overath, P.; Träuble, H. *Biochemistry*, *12*, 2625 **1973**; *B.B.A.* **1973**, *302*, 491.
8. Dodd, G. H.; Radda, G. K. *Biochem. J.*, **1969**, *114*, 407.
9. Meisels, P.; Matheson, M. S.; Rabani, J. *J. Am. Chem. Soc.* **1978**, *100*, 117.
10. Schore, N. E.; Turro, N. J. *J. Am. Chem. Soc.* **1974**, *96*, 306.
11. Weber, G. *Annu. Rev. Biophys, Bioeng.* **1972**, *1*, 553.
12. Perrin, F. *J. Phys. Radium* **1936**, *7*, 1.
13. Grätzel, M.; Thomas, J. K. In "Modern Fluorescence Spectroscopy"; Wehry, E., Ed.; **1976**, p. 169.
14. Thomas, J. K.; Castellino, F. J.; Brockway, W. J.; Tiao, J.-T.; Rawitch, A. B. *Biochem. J.* **1973**, *12*, 2787.
15. Cline-Love, L. J. *Aal. Chem.* **1980**, *52*, 754.
16. Cline-Love, L. J.; Skvilec, M. *Am. Lab.* **1981**, *13*, 103.
17. Calvert, T. G.; Pitts, J. N. "Photochemistry"; J. Wiley: New York, **1966**.
18. Spinks, J. W. T.; Woods, R. J. "An Introduction to Radiation Chemistry"; J. Wiley: New York, 1975.
19. Porter, G. *Proc. R. Soc.* **1950**, *A200*, 284.
20. McNeil, R.; Richards, J. T.; Thomas, J. K. *J. Phys. Chem.* **1970**, *74*, 2290.
21. Kajiwara, T.; Thomas, J. K. *J. Phys. Chem.* **1972**, *76*, 1700.
22. Hunt, J. W.; Thomas, J. K. *Radiat. Res.* **1967**, *32*, 149.
23. Beck, G. *Int. J. Radiat. Phys. Chem.* **1969**, *1*, 361.
24. Beck, G.; Thomas J. K. *J. Phys. Chem.* **1972**, *76*, 3856.
25. Ware, W. R. In "Creation and Detection of the Excited State"; Lamola, A., Ed.; Dekker: New York, 1971; Vol. I, p. 213.
26. Bradley, D. J.; Bryant, S. F.; Sibbett, W. *Rev. Sci. Instrum.* **1980**, *51*, 824.
27. Wang, Y.; Crawford, M. K.; McAuliffe, M. J.; Eisenthal, K. B. *Chem. Phys. Lett.* **1980**, *74*, 160.
28. Eisenthal, K. *Acc. Chem. Res.* **1975**, *8*, 118; Kaufman, K. J.; Rentzepis, P. M. *Acc. Chem. Res.* **1975**, *8*, 407.
29. Porter, G.; West, M. A. In "Investigations of Rates and Mechanisms of Reactions, 3rd Ed"; Hammas, G. C., Ed.; Wiley: New York, 1974; p. 367.
30. Ippen, E. P.; Shank, C. V. *Phys. Today* **1978**, *31*, 41.
31. Gordon, S.; Schmidt, K. H.; Martin, J. E. *Rev. Sci. Instrum.* **1974**, *45*, 552.
32. Bronskill, M. J.; Wolff, R. K.; Hunt, J. W. *J. Chem. Phys.* **1970**, *53*, 4201.
33. Wolff, R. K.; Bronskill, M. J.; Hunt, J. W. *J. Chem. Phys.* **1970**, *33*, 4211.
34. Aldrich, J. E.; Foldary, P.; Hunt, J. W.; Taylor, W. B.; Wolf, R. V. *Rev. Sci. Instrum.* **1972**, *43*, 991.
35. Tewari, P. H.; Freeman, G. R.; *J. Chem. Phys.* **1968**, *49*, 4394.
36. Schmidt, W. F.; Allen, A. O. *J. Chem. Phys.* **1971**, *52*, 4788.
37. Bakale, G.; Schmidt, W. F. *Chem. Phys. Lett.* **1972**, *17*, 617.
38. Munday, R. M.; Schmidt, L. D.; Davis, H. T. *J. Chem. Phys.* **1969**, *50*, 1473.
39. Schmidt, K. H.; Buck, W. L. *Science* **1966**, *151*, 70.
40. Beck, G.; Thomas, J. K. *J. Chem. Phys.* **1972**, *57*, 3649.
41. Beck, G. *Rev. Sci. Instrum.* **1979**, *50*, 1147.
42. Beck, G. p. 279 in *Rad. Res.*, Okada, I.; Imamura, M.; Tarashemia, T.; Yamaguchi, H., Eds.; Tokyo 1979.
43. Bakale, G.; Beck, G.; Thomas, J. K. *J. Phys. Chem.* **1981**, *85*, 1062.
44. Lillie, J.; Fessenden, R. W. *Mellen Institute Rgns, R.R.S.* **1972**, *5238*, 410.
45. Schmidt, K. H. *Int. J. Radiat. Phys. Chem.* **1972**, *4*, 439.
46. Asmus, K. D. In "Fast Processes in Radiation Chemistry and Biology"; Adams, G.; Fielden, M.; Michael, B., Eds.; Wiley: London, 1974, p. 40.
47. Chen, T. S. Ph.D. Thesis, Notre Dame, Ind. 1978.
48. Grätzel, M.; Bansal, K. M.; Henglein, A. p. 493 in Rad. Res., Ed. Nygaard, O. F.; Adler, H. I.; Sinclair, W. K. Acad. Press, N.Y. 1975.

4. EXPERIMENTAL TECHNIQUES

4.   EXPERIMENTAL TECHNIQUES   93

49. Lillie, J.; Beck, G.; Henglein, A. *Ber. Bunsenges Phys. Chem.* **1971**, *75*, 458.
50. Schöneshöfer, M.; Beck, G.; Henglein, A. *Ber. Bunsenges Phys. Chem.* **1970**, *74*, 1011.
51. Bansal, K. M.; Grätzel, M.; Henglein, A; Janata, E. *J. Phys. Chem.* **1973**, *77*, 16.
52. Beck, G.; Kiwi, J.; Lindenau, P.; Schnabel, W. *Eur. Polymer J.* **1974**, *10*, 1069.
53. Beck, G.; Lindenau, D.; Schnabel, W. *Macromolecules.* **1977**, *10*, 135.
54. Pagsberg, P.; Wilbrandt, R.; Hansen, K. B.; Weisberg, K. V. *Chem. Phys. Lett.* **1976**, *39*, 538.
55. Dallinger, R. F.; Guanci, J. L.; Woodruff, W. H.; Rodgers, M. A. J. *J. Am. Chem. Soc.* **1979**, *101*, 1355.
56. Fessenden, R. W.; Schuler, R. H. *J. Chem. Phys.* **1963**, *39*, 2147.
57. Avery, E. C.; Remko, J. R.; Smaller, B. *J. Chem. Phys.* **1968**, *49*, 951.
58. Smaller, B.; Remko, J. R.; Avery, E. C. *J. Chem. Phys.* **1968**, *48*, 5174.
59. Fessenden, R. W. p. 60 in Reference 46.
60. Atkins, P. W.; Gard, R. C.; Simpson, A. F. *Chem. Commun.* **1970**, 513.
61. Ayscough, P. B.; English, T. H.; Tong, D. A. p. 76, reference 46.
62. Trifunac, A. D.; Johnson, K. W.; Lowers, R. H. *J. Am. Chem. Soc.* **1976**, *98*, 6067.
63. Closs, G. L. In "Chemically Induced Magnetic Polarisation"; Lepley, A. R.; Closs, G. L., Eds.; Acad. Press: New York, 1973; p. 95.

5

Colloidal Systems

A COLLOID MAY BE DEFINED as a suspension of a substance in a very fine state of subdivision. The dimensions of the particles are larger than molecular sizes, but the particles are still invisible to the naked eye. Our concern is with suspensions of these small particles in fluid systems.

The many different types of colloidal systems, for the sake of convenience, will be divided into two classes: those formed primarily from organic constituents and those formed primarily from inorganic components. The first group contains the various systems formed from surfactant molecules, such as micelles, reversed micelles, emulsions, microemulsions, liquid crystals, and vesicles. The second group contains such entities as silver and gold sols, colloidal metal oxides, sulfides, silicas, and clays. These systems have points of similarity as well as quite stark differences. Such properties are often useful from a photochemical viewpoint, both as a method of gaining insight into mechanisms of radiation-induced reactions in a system, and also as a means of promoting a specialized feature of a reaction. Micelles and organic systems are probably the most studied of all colloidal systems, at least photochemically, and a brief description of these systems is immediately constructive and useful for later work.

Organic Assemblies—Micelles

The term micelle is usually reserved for the assembly formed in water by the aggregation of surfactant or soap molecules (1–6). Several surfactant and soap molecules (*see* structures) all have a common feature, namely, that the molecule consists of two distinct moieties: a hydrophobic region, usually a long hydrocarbon chain, and a hydrophilic region such as an anionic group or an ethylene oxide residue. For this reason they are called amphipathic molecules. Beyond a defined concentration, called the critical micelle concentration (CMC), these molecules tend to aggregate together to form micelles. This behavior is most common in water, but similar behavior does occur for certain organic solvents and with selected surfactants, in particular disodium diisooctyl sulfosuccinate, AOT. Further details of this process will be

0065-7719/84/0181-0095$07.75/1
© 1984 American Chemical Society

$CH_3(CH_2)_{11}-OSO_3^-Na^+$
Sodium lauryl sulfate
NaLS

$CH_3-(CH_2)_{15}-\overset{+}{N}(CH_3)_3Br^-$
Cetyl trimethyl
    ammonium bromide
    CTAB

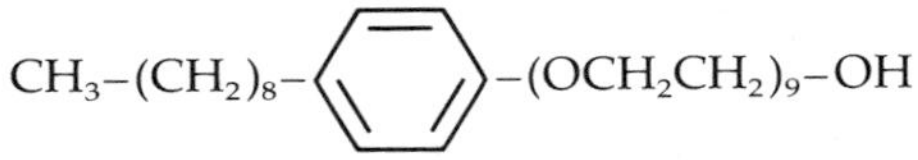

$CH_3-(CH_2)_8-$⟨benzene ring⟩$-(OCH_2CH_2)_9-OH$
Igepal 60-630

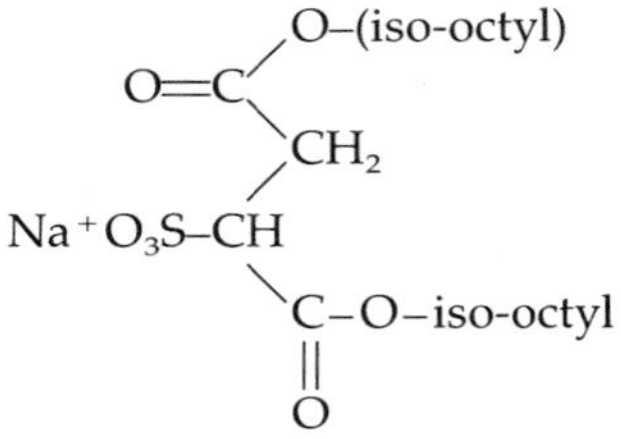

$$O=C\begin{cases}O-\text{(iso-octyl)}\\CH_2\end{cases}$$
Na^+O_3S-CH
    $\diagdown C-O-\text{iso-octyl}$
        $\overset{||}{O}$
Bis(2-ethylhexyl) sodium
    sulfosuccinate
    Aerosol OT

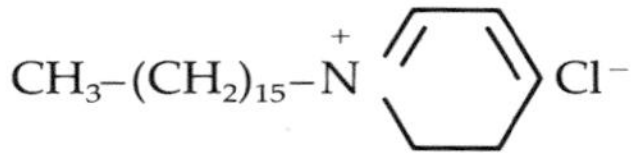

$CH_3-(CH_2)_{15}-\overset{+}{N}$⟨pyridinium ring⟩$Cl^-$
Cetyl pyridinium chloride

$(CH_3-(CH_2)_{17})_2(CH_3)_2\overset{+}{N}Br^-$
Dioctadecyldimethyl
    ammonium bromide

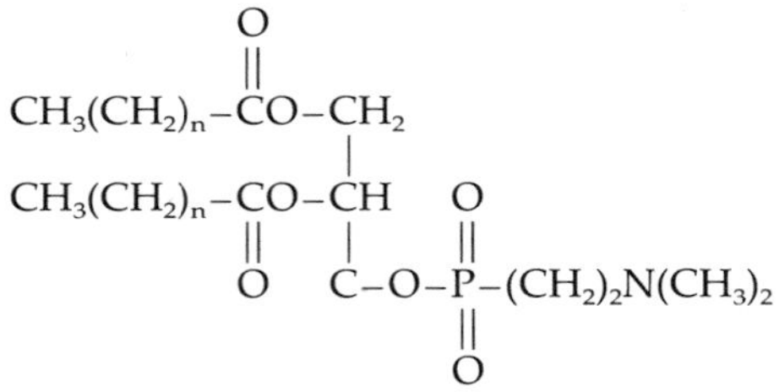

$$CH_3(CH_2)_n-\overset{\overset{O}{||}}{C}O-CH_2$$
$$CH_3(CH_2)_n-\underset{\underset{O}{||}}{C}O-CH\quad O$$
$$\qquad\qquad C-O-\overset{}{\underset{\underset{O}{||}}{P}}-(CH_2)_2N(CH_3)_2$$
Lecithin

given later. The exact structure of a micelle is unknown, although several intelligent guesses have been put forth, many of which agree with the observed data connected with micelles. Figure 1 shows a typical micelle, which is roughly spherical with a sequestering of the hydrophobic hydrocarbon chains away from the aqueous phase to form a

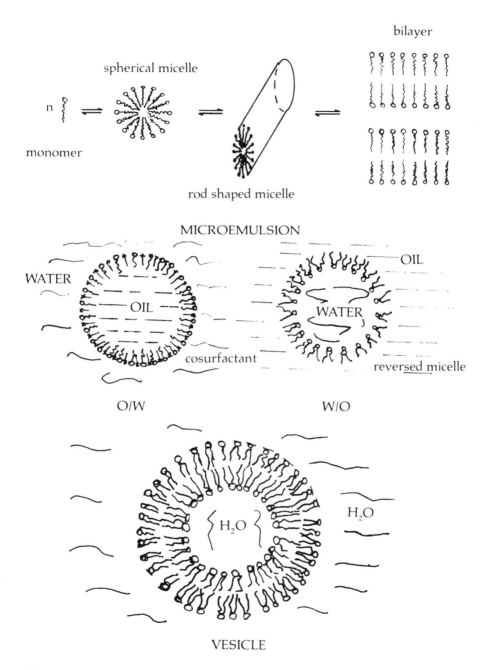

Figure 1. Aggregated structures: spherical micelle, cylindrical micelle, bilayer, oil-in-water microemulsion, water-in-oil microemulsion, reversed micelle, swollen micelle, and vesicle.

micelle core. The polar groups are located at the hydrocarbon–water interface. Water is associated with micelles most certainly with the polar head group or Stern layer region, and probably penetrating a little into the hydrocarbon core. Some controversy exists with regard to the degree of water penetration; most support is given to the limited penetration indicated earlier, but some support is given to extensive water penetration. For an anionic micelle such as NaLS or CTAB most of the inorganic counterions (e.g., Na^+ or Br^-) are situated in the Stern layer. However, a sizable fraction (10–20%), which varies with the type of surfactant used, is situated in the aqueous bulk or in the immediate surroundings of the micelle, or Gouy–Chapman layer. This arrangement leads to a large net charge in a micelle; if the aggregation number is 70–100, then the net charge can be 10 to 20 negative or positive groups, depending on the micelle.

Micellar Shape and Size

A micelle, as defined, is not a permanent entity but exists in kinetic equilibrium with its monomer. However, for the purposes of present interests, where the micelle–monomer equilibrium is always operative, a micelle can be considered as a discrete entity that can be examined by conventional methods. The micelle diagram in Figure 1 shows a roughly spherical entity with a low degree of order. The diagram also hints that the radius of a micelle should not be larger than the length of a hydrocarbon chain (e.g., C_{12} or C_{16} as the case may be). Larger micelles, formed from nonionic surfactants, may be disclike in shape as indicated by viscosity measurements (3). At this stage, the micelle should be defined as follows:

1. Its shape is roughly spherical or disclike.

2. The core consists of mainly hydrocarbon chains and may be looked upon as an oil drop.

3. The polar or ionic regions of the surfactant tend to locate at the hydrocarbon–water interface, thus presenting a unique environment. If the micelle is ionic, then this region (or Stern layer) has a high charge.

Many micelles become larger on addition of an inert salt such as NaCl or on increasing surfactant concentration. The addition of inert salt tends to decrease the degree of ionization of the micelle surface and allows the head groups to locate in closer proximity to each other and thus to maintain a structure with a smaller curvature. In some cases the spherical micelles coalesce and form long rod-shaped micelles. These micelles have a radius close to that of a small micelle but a length considerably larger. Under suitable conditions these entities may form hexagonal close-packed systems, and eventually, at high surfactant concentration, they may form bilayer systems. A typical representation of these structures is shown in Figure 1.

Addition of long chain alcohols, such as pentanol or hexanol, to an aqueous micellar solution leads to incorporation of the alcohol or cosurfactant into the micelle. The OH group of the alcohol is located in the Stern layer, and the alkyl chain extends into the micellar interior. This alignment leads to an increased degree of dissociation of the micelle, but a small micelle size is still maintained. Long alcohols, such as dodecanol, also associate with micelles, and large bilayer structures are formed (8, 9). In some cases vesicles are formed from such structures (10, 11). The small micellar structures of surfactants and short chain alcohols are called swollen micelles. These structures will solubilize large quantities of alkanes or arenes to form microemulsions. Microemulsions are spherical and have a hydrocarbon or oil core and a surface consisting of surfactant and cosurfactant in contact with the aqueous phase. Many and varied structures are formed by surfactant mixtures, and because all the systems described are optically clear, they are ideally suited to investigate the role of environment on radiation-induced reactions.

The introduction of two long alkyl chains onto a surfactant (e.g., didodecyldimethylammonium bromide) produces vesicular structures (12–14). These species are of the same family as micelles, as shown in Figure 1, and really consist of a spherical bilayer. The vesicle has the unique property of enclosing a volume of the solution in its interior. These structures have great biological interest because they are similar to biological cell membranes, and indeed are exactly analogous to vesicles formed by sonication of lecithin, a component of cell membranes. The advent of surfactant vesicles will contribute a great deal to the well-established field of lecithin vesicles and liposomes (15). The various lecithins are costly materials that are often difficult to obtain in a pure form, while double chain surfactants that form vesicles are relatively inexpensive and are readily available in a pure form. Many changes can be made in the surfactant structure to investigate structural details in the vesicles. The surfactant vesicles possess many of the properties of vesicles constructed from natural components. More details will be given later.

All the systems discussed so far have contained water as the major constituent. However, a complimentary micellar-type chemistry also exists in nonaqueous solvents (4, 16, 17). The most studied of these systems is the Aerosol OT–alkane–water system. AOT readily dissolves in most hydrocarbons to form reversed micelles. These structures can incorporate quite large quantities of water in the micelle core, and they are called reversed micelles, although in some areas they are categorized as microemulsions of the water-in-oil type, w/o. Similar structures, which are genuine w/o microemulsions, can be constructed from the same components that formed oil-in-water, o/w, microemulsions, that is, surfactant and cosurfactant with oil and water. All of these structures, apart from the vesicles, are thermodynamically stable and are constructed by simply mixing together the components of the system; the aggregates form spontaneously. Vesicles are usually formed by sonication of the surfactant or lipid in water, although some techniques involve precipitation (15). Gross oil-and-water emulsions that are dispersed by a single simple surfactant are not thermodynamically stable and require mechanical energy of mixing for their formation. The special group of surfactants, such as AOT, that do disperse oil and water to give theromodynamically stable reversed micelles have special structural details such as two long or branched side chains, which are reminiscent of a grouping involving a surfactant and cosurfactant. Thus, as will be seen later, the geometry required to form a stable structure is contained in one surfactant.

Inorganic Aggregates

Colloidal inorganic materials can be both a blessing and a curse to the chemist: they often provide the chemist with first class catalysts, while they plague the analytical chemist who finds great difficulty in filtering these solutions. This paradox gives a clue as to the nature of these systems. They are composed of small particles of material formed by repeated fracture of larger ones or are insoluble materials that fail to crystallize into large filterable solutions. The common class of these materials is of the silicate variety.

Colloidal Silica. Much is known about colloidal silica because of its many everyday uses (18). It is usually formed by neutralizing a water-soluble silicate when insoluble silicic acid is formed,

$$SiO_3^= + 2H^+ \rightleftharpoons H_2SiO_3$$

The silicic acid tends to aggregate and form small clusters that lose water to give an aggregate that is mainly polymerized SiO_2

$$\eta H_2SiO_3 \rightarrow \eta H_2O + \eta SiO_2 \rightarrow \text{polymerized } SiO_2$$

These particles can be of various sizes and can be as small as a fraction of a micrometer. The surfaces of the particles contain silanol groups, $-SiOH$, which can be neutralized with alkali to give a particle that is negatively charged with $-SiO^-$ groups and a counterion of Na^+. In alkaline solution the particles adsorb OH^-, and this adsorption leads to a further increase in their negative charge and then to further stabilization of the particles. The particles repel one another strongly, and hence particle growth, which leads to coagulation or flocculation, is prevented. The particles have a much diminished stability to coagulation at pH 7.0, but the stability increases again at acidic pH because of adsorption of ions from the solution. The addition of cationic material can lead to rapid coagulation because the particle negative charges are neutralized. This neutralization leads to a hydrophobic hydrocarbonlike surface which has an affinity for other particles with similar surfaces. Figure 2 shows schematically the action of this process.

Clay Particles. A very common higher form of silica is a class of compounds known as aluminum silicates or clays (19–21). These materials (*see* structures) are all naturally occurring and play a major role in our everyday lives from ceramics to brickwork. The colloidal form is achieved by crushing the clay into small particles that can be suspended in water.

$$(OH)_4Si_8(Al_{3.34} \cdot Mg_{0.66})O_{20}$$
$$\downarrow$$
$$Na_{0.66}$$

Montmorillonite

$$(OH)_4Si_8(Mg_{5.34} \cdot Li_{0.66})O_{20}$$
$$\downarrow$$
$$Na_{0.66}$$

Hectorite

$$Al_2Si_2O_5(OH)_4$$

Kaolin

In particular, the bentonite clays form excellent colloidal systems for reasons apparent in the structure of a clay as shown in Figure 3. The theoretical formula for such a clay may be $(OH)_4Si_8Al_4O_{20} \cdot nH_2O$, where the nH_2O is the interlayer water. Clays have layered structures composed of units made up of two silica tetrahedral sheets with a central alumina octahedral sheet. Figure 3 shows a typical structure with the tips of the tetrahedrons pointing toward the center of the unit; both tetrahedral and octahedral sheets are arranged so that the tetrahedron tips of each silica sheet and one of the octahedral sheet hydroxyl layers form a common layer. The individual sheets take up molecular species where these guest molecules are intercalated between the clay sheets. Substitution of Si^{4+} by Al^{3+} at the tetrahedral sites gives rise to a residual negative charge on the layer, and the structure absorbs Na^+ or Ca^{2+} ions to achieve electrical neutrality. The Na^+ or Ca^{2+} cations are interchangeable with other cations in a cationic exchangelike process. The incorporation of large cationic surfactant molecules (e.g., octadecyl ammonium) at the Na^+ sites leads to separation of the clay

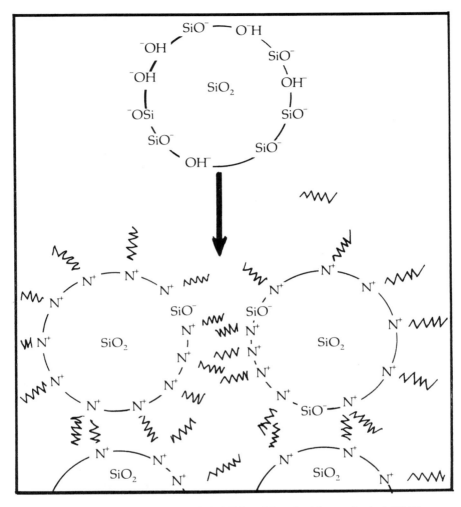

Figure 2. Representation of colloidal SiO₂ with and without adsorbed CTAB.

layers, as has been observed by X-ray techniques (*28, 32–36*). This property of clays to adsorb organic molecules and enclose them in unique inorganic environments is of utility in photochemical experiments. Clays promote certain specific reactions of chemical interest (*32–41*) and may be looked upon as catalysts. A logical extrapolation of this knowledge would indicate that clays might also catalyze desirable features of photochemical reactions. Crushing the clay leads to small particles that, on suspension in water, give rise to negatively charged particles with countercations (e.g., Na^+). The layered structure of the clay is maintained, and molecules may be adsorbed in the layers, at the broken edges of the particle, or on the external surface. All three regions provide different environments for the guest molecules. Certain clays (e.g., kaolins) do not solubilize molecules in internal layers.

Solubilization in Organic Assemblies

The picture of a micelle given in Figure 1 suggests that a solute might be located in three possible positions in ionic micelles and two possible positions in nonionic micelles. In a micelle a solute might be located in the micelle interior, in the micelle surface region, or, if ionic, it might

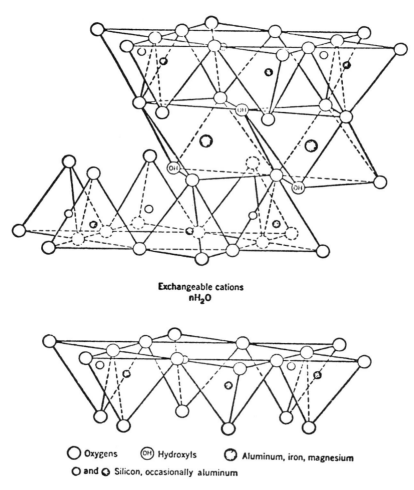

Exchangeable cations
nH_2O

○ Oxygens   (OH) Hydroxyls   ◐ Aluminum, iron, magnesium
○ and ◓ Silicon, occasionally aluminum

Figure 3. Representation of clay structure (59).

be incorporated as a counterion. The last site presents no difficulty; a solute (e.g., Tl^+ in NaLS, or I^- in CTAB micelles) will locate in the Stern layer. The two remaining sites, micelle interior or surface, are not easily resolved. In many ways a nonionic micelle presents the same difficulty where a micelle hydrocarbonlike interior is available, such as a polar organic surface or mantle region. This mantle region is much larger than the Stern layers of an ionic micelle. The volume of the region where a guest molecule can lie will play an important role in the site of solubilization because, if no other factors enter, then the percentage of probe located at a particular region will be proportional to the volume of that region with respect to the total micelle volume. The inescapable conclusion for such an argument is that solutes should locate toward the peripheral regions of a micelle. However, the head group regions could maintain some rigidity because of the ionic binding of the whole surface, and thus lead to solute exclusion. Thus, ideas on the possible sites of solubilization of guest molecules may be presented, but a conclusion awaits experimental evidence.

Both micelles and the solubilization process are dynamic in nature. The solute enters and exits a micelle at rates that locate its position primarily on the micelle. The micelle surface is in a continual state of flux as monomer units enter and leave the micelle and also rapidly oscillate back and forth on the micelle surface (22). A solubilization site is thus an average environment that the molecule experiences; experimental data provide information on such sites.

Alkanes. Solubility studies indicate that alkanes such as ethane are solubilized in the micelle interior along with the alkane chains of the surfactant (23).

Aromatic Compounds. The location of nonpolar aromatic compounds in micelles and other assemblies can be ascertained by spectroscopy. The absorption spectra of aromatic compounds are quite solvent dependent and may be used as a gauge of the polarity of the compounds' environment (4, 24, 25). A comparison of the absorption spectrum of an aromatic probe molecule in a micelle with those measured in standard solvents then provides information on the probes' environment in the micelle. Benzene and other simple molecules situate themselves in the micelle surface region; penetration further into the micelle may occur at high benzene or solute concentrations.

Table I lists some of the many probe molecules that have been used and their particular useful function. The probe molecule pyrene is frequently used in photochemical experiments, and a knowledge of its micellar environment is essential for a meaningful description of its micellar photochemistry. As indicated earlier the pyrene fluorescence,

Table I.
Common Fluorescent Probes for Structural Investigation

Probe	Information
Pyrene	Polarity of environment Degree of lateral movement
Pyrene carboxaldehyde	Polarity of environment
Dipyrene Methylanthracene Diphenylhexatriene Perylene	} Local rigidity or microviscosity
Anilinonaphthylamine	Local polarity of environment and local microviscosity
Indole derivatives	Local environment polarity
Pyridinium iodide	Interface polarity

or the ratio of fluorescence peaks III/I provides an ideal tool for environment studies. The pyrene fluorescence III/I ratio has been investigated in many micelles (26), and in all cases spectroscopy indicates an environment somewhere between that of an alkane and water, and thus a surface site of solubilization. With quaternary ammonium micelles such as CTAB, the III/I ratio is low and indicates a very polar environment. Pyrene forms weak complexes with quaternary ammonium groups; thus it is located at the CTAB surface (27). Other micelles show varying III/I ratios. Perfluoromicelles do not solubilize pyrene (28), and this fact shows that pyrene must have some contact with the methylene groups of the surfactant for effective solubilization to take place. The pyrene is visualized as being located on the surface region of the micelle but with sufficient micellar penetration to come in contact with the hydrocarbon chains.

The pyrene fluorescence III/I ratio can provide a vivid illustration of events in a micelle as it is progressively changed to a microemulsion by addition of cosurfactant and oil (29, 30). Figure 4 shows the pyrene fluorescence spectra, with the III/I marked, in water, micellar 0.1 M NaLS, NaLS + pentanol, and NaLS + pentanol + dodecane. The III/I ratio is much larger in NaLS compared to bulk water as the pyrene is solubilized by the micelle. Addition of a cosurfactant, pentanol, to the NaLS solution increases the III/I ratio further. The pentanol interacts with the head group region of the micelle. This interaction loosens the structure, as will be seen shortly, and enables the pyrene to penetrate

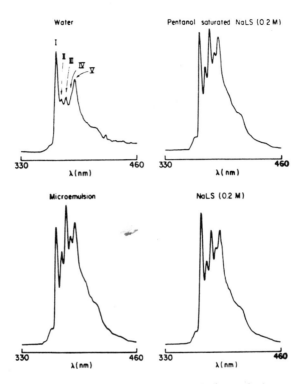

Figure 4. The fluorescence of pyrene in four solutions, showing the variations of the vibronic band structure in various solvents. The concentration of pyrene was about 10^{-6} M in water and 10^{-5} M in the other solutions. The wavelength of excitation was 320 nm in all cases (29).

further into the structure. A more hydrophobic environment is provided for the pyrene, as illustrated by an increase in the III/I ratio. Introduction of an oil (e.g., dodecane) to the swollen micelle structure increases the size of the assembly from a radius $v \sim 15$ Å to $v \sim 150$ Å. The classical picture of such a microemulsion is given in Figure 1 and depicts a fluid core of hydrocarbon surrounded by a film of surfactant and cosurfactant. The increased III/I ratio in this system compared to water, micelle, or swollen micelle indicates that pyrene now resides mainly in the hydrocarbon core. The site of solubilization is, however, dynamic because the microemulsion III/I ratio is not as large as that of pure dodecane; thus, pyrene moves around the microemulsion system over the head group area and also the hydrocarbon core. The picture of a gradual penetration of pyrene from a micelle surface into the core of a microemulsion is well illustrated by the fluorescence probe technique.

The degree of polarization of the probe fluorescence may be used to ascertain the restriction to movement that the probe experiences in the assembly. Probes such as perylene and methylanthracene are similar in structure to pyrene, and their solubilization behaviors are also similar (27). This short lifetime of the fluorescence of these molecules makes it possible to ascertain the microenvironment of the systems described. The probes experience more restriction to movement in a micelle compared to a simple homogeneous solvent where the viscosity η is about unity. Depending on the micelle type a "microviscosity" may be calculated from the Perrin equation, and it is 10 to 30 times that of hexane. Regarding use of the calculated microviscosity in these systems, the exact mode of depolarization of the probe's fluorescence is not known, and the microviscosity is just used for convenience or to promote discussion. It is not an absolute measurement as obtained in bulk fluids. However, relative changes in microviscosity, as measured by the technique, are useful. Addition of a cosurfactant dramatically reduces the degree of fluorescence polarization to a value that is comparable to simple nonviscous solvents. This reduction is due to a disruption of the micellar surface by the long chain alcohol. The degree of polarization of these probes in microemulsions is also approximately that found in the pure oil used to form the system. Thus fluorescence fine structure and fluorescence polarization give a concise picture of the behavior of nonpolar aromatic solutes in micelles and microemulsions. The ability to locate a solute at the assembly water interface is often desirable and, as shown, is not achieved by use of nonpolar molecules. However, a molecule with useful spectroscopic properties can be located at the interface.

Surface Regions of Assemblies

The intriguing properties of the assembly lipid–water interface (i.e., the unusual structural details and the high electric field) suggest the construction of probe molecules that will be located uniquely at the interface. This construction is simply achieved by placing a polar substituent on the chromophore of interest (e.g., pyrenesulfonic acid, pyrenebutyric acid, anilinonaphthalenesulfonic acid, and pyrenecarboxaldehyde). The spectroscopic properties of the chromophore are often changed, but a judicious choice of probe can often provide structural information of considerable value to colloid chemists and photochemists. The merits of the probe pyrenecarboxaldehyde (PCHO) were discussed in an earlier section. Its application to a study of the transition, water → micelle → swollen micelle → microemulsion, is quite fruitful in discussing aggregate surfaces. The PCHO study is similar to the pyrene fluorescence III/I ratio work, which was also studied in this system. Figure 5 shows

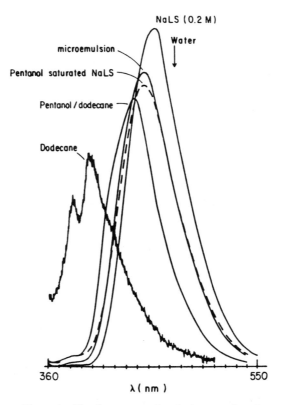

Figure 5. The fluorescence band of pyrene-3-carbox-yaldehyde in various solvents. The emission intensity is in arbitrary units, and the various bands have been shifted in height to facilitate the identification of the maxima. The emission in dodecane is at least a factor of 10 less intense than in the other solvents. The arrow indicates the position of the maximum in water. The excitation wavelength was 365 nm, and the concentration of pyrene-3-carboxaldehyde was 5×10^{-5} M in all solutions (29).

the fluorescence peak of PCHO in NaLS micelles, NaLS–pentanol swollen micelles, NaLS–pentanol–dodecane microemulsions, and in dodecane; the position of the water peak is also given. The PCHO fluorescence spectrum is blue shifted in NaLS micelles compared to water, but the peak in dodecane exhibits an extremely large blue shift. The position of the spectrum in NaLS–pentanol swollen micelle is identical to that in the microemulsion system and quite different to that observed in other environments. The PCHO surface probe shows that the pentanol immediately interacts with the micelle surface and causes a change in the PCHO environments. Further addition of dodecane does not affect the aggregate surface (apart from increasing the curvature), which still remains a mixture of surfactant and cosurfactant. The PCHO probe may also be used to inspect the changes of a micellar surface on addition of inert salt that reduces the charge at the Stern layer (31).

Location of Solutes in Organic Assemblies

The dynamic nature of most organic assemblies makes the precise location of a solute in the assembly a little obtuse. A microemulsion provides a good example of this problem where data indicate that a solute such as pyrene moves throughout the assembly even into the aqueous phase but that its main location is the oil core of the assembly. The nature of the solute also controls the location because polar solutes tend to locate at assembly surfaces. A charged anionic solute such as sodium pyrenesulfonate certainly interacts with the cationic quaternary ammonium region of a CTAB micelle, or CTAB microemulsion. Several considerations indicate that the primary location of a nonpolar solute in a micelle is close to the surface, and that penetration into the micelle interior may occur at high solute concentration. This penetration may in part be due to a cosurfactant effect of the solute on the micelle, as already observed for pentanol and hexanol. Other systems that are similar to micelles, such as vesicles, should show similar solubilization effects. Indeed, the environment of pyrene is quite polar in lecithin vesicles but decreases abruptly at the phase transition (32, 33). The increased fluidity of the hydrocarbon chains at the phase transition allows pyrene to penetrate further into the vesicle, an effect that indicates a nonpolar region, possibly in the choline area of the lipid, prior to the transition temperature. The picture of solubilization in organic assemblies is by no means complete, but for the sake of convenience in discussing radiation-induced reactions in these systems, some cautious definition of solubilization site has to be given.

Solubilization in Inorganic Media

The use of inorganic materials such as aluminium oxide and silica in the chromatography of organic compounds illustrates that many organic compounds are strongly bound to these and other inorganic materials in nonpolar media. The property has been used to study the photochemistry of many systems adsorbed in inorganic materials (34). The situation is quite different in polar media such as those encountered in aqueous colloidal solutions. The silica systems show some of the variation in type of solubilization that is possible.

Colloidal Silica. Aqueous colloidal silica has little affinity for uncharged organic molecules such as pyrene or pyrenecarboxyaldehyde. However, cationic materials exchange with the sodium counterions of the particle and in this way become attached to the particle. Large organic cations such as cetylpyridinum chloride, CTAB, or methyl viologen bind so strongly to the silica surface that the silica particles coagulate. To use aqueous colloidal silica, a cationic group must be introduced into the probe molecule. For example, to bind pyrene to silica, a cationic group must be introduced and form a compound such as pyrenebutyltrimethylammonium bromide. Some luminescent probes are cationic and may be bound directly onto the silica (e.g., tris(bipyridyl)ruthenium or terbium ions).

As expected, the behavior of silica is carried over to clay particles, and aqueous colloidal clays solubilize only cationic materials to any significant extent, although weaker binding of polar organic compounds also occurs. In nonaqueous media large amounts of polar organic material may be inserted between the clay layers. The incorporation of polar organic molecules or of cationic surfactants (e.g., CTAB) between these layers provides hydrophobic sites for the solubilization of nonpolar organic molecules such as pyrene. Complete exchange of the clay counterions by CTAB or similar cationic surfactants leads to an organo-clay that has little affinity for water but can be dispensed in organic solvents.

Other Inorganic Colloids. Other inorganic colloids such as colloidal metals and semiconductors show no pronounced affinity for organic molecules in aqueous solutions apart from a small adsorption of surfactant or other polar molecules that may stabilize the colloid. Some metal oxides (e.g., TiO_2) can adsorb significant quantities of organic material. A water-insoluble material can be precipitated in the presence of an inorganic colloid and produce a coating of the material on the colloid. The net effect is to situate the material at the surface of the colloid in a random fashion. Other factors such as affinity of the precipitate affect the final product. Although such systems are of great use in general catalysis, the lack of definition has precluded them from general use as assemblies in radiation-conducted reactions.

Nuclear Magnetic Resonance Spectroscopy

Numerous NMR studies of micellar systems (4, 35–44) have been conducted because the technique lends itself well to investigating these systems. Some examples to illustrate the utility of the technique are given as follows. By taking the appropriate NMR spectrum of a surfactant molecule in the free form or in the various aggregates mentioned, information regarding the various components of the molecule can be obtained. The chemical shifts, line widths, or spin–lattice relaxation times give detailed information regarding the sequential mobility of the various units of the molecule. The interaction of an aromatic probe molecule solubilized in a micelle with the various micellar sections can be observed, and hence the location of the probe in the assembly can be ascertained.

NMR studies showed that pyrene interacted strongly with protons of the cetyl chain of CTAB and thus that it was in contact with the micelle interior (44). Interaction with the NCH_3 protons was also observed, and this observation is in agreement with pyrene being near the surface of the micelle with some penetration into the interior. The molecules pyrenesulfonic acid and pyrenebutyric acid showed stronger interactions with the NCH_3 protons in accordance with their position in the micelle surface. Such measurements can be used to collaborate suggestions from other techniques (e.g., UV–visible spectroscopy).

Figure 6 presents typical proton NMR spectra of micellar solutions of Igepal CO-630 in D_2O. The assignments for the various proton signals are also indicated. The spectrum shows partially resolved resonances for the alkyl and aryl protons; the ethylene oxide protons appear as a broad band consisting of several overlapping resonances. The observation that the ethylene oxide band consists of several overlapping

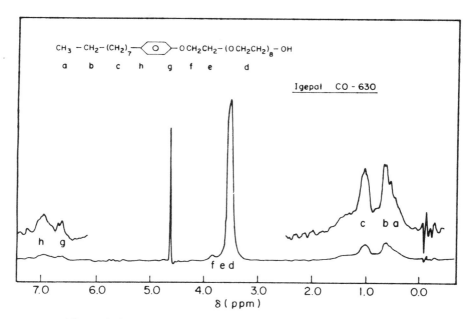

Figure 6. ¹H-NMR spectrum of 5 mM Igepal CO-630 in D₂O (45).

resonances is clearly demonstrated in Figure 7. This figure presents a series of partially relaxed Fourier transform (PRFT) spectra (45) for this band. Although the protons on the high field side relax quite rapidly, the protons on the low field side relax much more slowly (see, for example, the spectrum corresponding to a delay of 250 ns between the 180° and 90° pulses). In fact, under high resolution (100 MHz) as many as six resonance lines can be identified and the T_1 values for these lines can also be measured. This type of behavior has also been observed in Triton X-100 micelles (42, 43).

Chemical shift assignments indicate that the protons on the low field side arise from ethylene oxide units at the end of the chain, while those of the high field side arise from ethylene oxide units adjacent to the phenoxyl unit. Detailed studies with Triton X-100 have clearly established that the micellar core is free of water. Apparently this type of behavior, namely, a micellar core free of water, seems to be a general feature of the nonionic micelles. Measurements of the proton spin–lattice relaxation times, the T_1 values, for various protons provide a direct method of determining the viscosity surrounding a given spin along the chain and thus the segmental mobility of the hydrocarbon chain inside the micelles.

The relative magnitudes of the T_1 values for various protons (alkyl, aryl, and methylenes of the ethylene oxide units) are considerably less than those observed in pure hydrocarbon liquids. This result indicates that the micellar interiors in these nonionic micelles are more rigid than in hydrocarbon liquids. Furthermore, the gradation in T_1 values observed indicates a gradient in the segmental mobility of the various units. Farther down from the phenoxyl unit toward the inner core, the

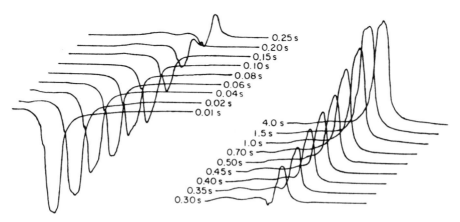

Figure 7. Partially relaxed Fourier-transform spectra for the ethylene oxide band in Igepal CO-630 PMR spectrum (45).

freedom of motion for the alkyl chain increases. This condition is also true for the ethylene oxide units: those units toward the end of the hydrophilic chains possess maximum freedom of motion. This picture of segmental mobility for the Igepal CO-630 micelle is consistent with similar results reported recently for the Triton X-100 micelles.

Carbon-13 NMR Relaxation Studies

As was shown earlier with ionic micelles, a major shortcoming of the proton NMR studies is that the methylene signals are, on average, composed of contributions from different methylene groups of the alkyl chain (34, 35). Carbon-13 NMR makes it possible to detect the motions of the individual carbon atoms forming the backbone of the surfactant molecule. The proton-decoupled ^{13}C-NMR spectra of micellar solutions of a typical nonionic surfactant, Triton X-100, are presented in Figure 8. The carbon-13 spectrum shows more resonances than the proton spectrum. The chemical shift assignments relative to TMS for the various lines are also indicated in the figure. The assignments are based on earlier assignments for pure hydrocarbon liquids (17, 18), ionic micelles (8, 16), and model compounds.

The chemical shifts and spin–lattice relaxation times for various carbon resonances in micellar solutions of the nonionic micelle Triton X-100 are summarized in Table II. The carbon-13 spin–lattice relaxation

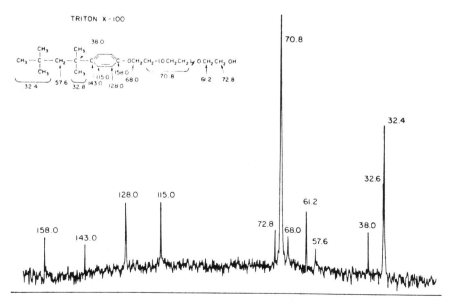

Figure 8. Carbon-13 NMR spectrum of micellar Triton X-100 solutions in D_2O (45).

Table II.
Carbon-13 Spin-lattice Relaxation Times
(T_1 Values) for Nonionic Micellar System
Triton X-100

Peak#	δ(ppm)	T_1(ms)	Peak	δ(ppm)	T_1(ms)
a	32.4	380	g	128.0	480
b	57.6	320	h	158.0	2000
c	32.8	1800	i	68.0	350
d	38.0	1850	j	70.8	380
e	143.0	1900	k	61.2	1100
f	115.0	460	l	72.8	1000

Source: Reproduced from Ref. 31. Copyright 1980 American
Chemical Society.

data provide more detailed information on the segmental mobility of
the fatty acid chains as well as that of ethylene oxide units in the
nonionic micelles. As with proton relaxation times, the carbon-13 T_1
values for alkyl chain carbon atoms are considerably less than those
observed in pure liquids and ionic micelles. This result indicates that
the micellar core in a nonionic micelle is considerably more rigid in the
other systems. A gradient in the segmental mobility of the chain (both
the alkyl as well as the ethylene oxide units) is indicated by the increase
in the T_1 values in the direction of the inner core or the surface with
respect to the phenoxyl unit. UV spectroscopic studies of the phenoxyl
unit confirm these observations.

Electron Spin Resonance

Electron spin resonance (ESR) studies have been used extensively to
study the structure of free radicals. The behavior of a stable free radical
such as nitroxide radical can be used to gain information on its environment. A free radical such as a nitroxide free radical, >N–O, possesses an unpaired electron that can be aligned either parallel or antiparallel to an external magnetic field as applied by an ESR spectrometer.
On also applying an oscillating magnetic field at right angles to the
static field, the free radical may undergo transitions between two energy
levels defined by the conditions of the two magnetic fields in the spectrometer. If the frequency of the oscillating magnetic field is v, then hv
$= G \beta_e H$ where h is Planck's constant, H is the static field, β_e is the
Bohr magnetron, and G (or the G-value) is a property of the free radical.
As v is changed a resonance condition occurs; this condition satisfies
the equation and leads to absorption of energy. Because the nitrogen
nucleus has a spin of 1 and may polarize parallel, perpendicular, or

antiparallel to the static magnetic field, the three resonance conditions are observed for the >NO˙ radical, as shown in Figure 9.

Because the nitroxide radical has two resonance forms, >N–O˙ and >N⁺–O⁻ (which is more prevalent in polar media), the environment of the spin probe can be ascertained. Figure 9 also shows the spin probe in water and in micellar NaLS; two distinct environments of the spin probe are shown at the right. Restriction of the probe in an assembly leads to line broadening, and is a measure of the fluidity of the assembly. Various derivatives of the nitroxide radical have been used to investigate micellar systems (4, 46). A small nitroxide radical such as di-*tert*-butyl nitroxide shows a much broadened ESR spectrum in NaLS and CTAB micelles compared to water (47, 48). Interpretation of the spectrum gives the partition constant of this radical between the micelle and water and the rates of entry and exit of the probe. The effects of various additives on the rigidity of the micelles are also reflected in the degree of broadening of the ESR spectrum in much the same fashion as that found with fluorescence polarization studies. The ESR spectrum of a surfactant with a nitroxide radical chemically attached to its hydrocarbon chains shows spectroscopic properties indicative of a polar environment and such a restricted return of the free radical (49). The nitroxide probe probably lies close to the surface of the micelle. Similar

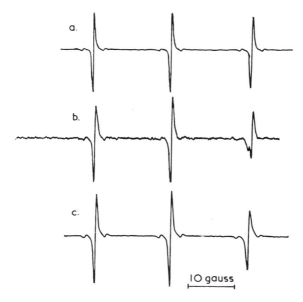

Figure 9. ESR spectra at 296 K: (a) 5 × 10⁻⁴ M DTBN in water; (b) 6 × 10⁻⁵ M DTBN in 1.39 × 10⁻² M NaLS; (c) 5 × 10⁻⁴ M DTBN in 1.75 × 10⁻¹ M NaLS (DTBN is di-tert-butyl nitroxide) (47).

probes have been used extensively in biochemical studies (50, 51) to study the structure of lipid vesicles and various interaction of substrates with such biomimetic systems.

ESR measurements of radicals generated in micelles show marked effects of the micellar environment, mainly viscosity effects, on the reaction kinetics of the radicals (50). Such measurements, although playing a minor role compared to NMR measurements, often convey important data regarding the rigidity, phase changes, and interaction of solutes in and with organic assemblies.

Light Scattering Studies

Many different techniques of light scatter are now used on a conventional basis for the investigation of particle size and structure. The standard and well-known Rayleigh scatter has been used extensively to measure the sizes of all types of layer systems, from polymers to cell membranes. Most micelle size measurements that give aggregation numbers are derived from light scatter measurements. Many textbooks describe the technique in detail (53).

However, the new technique of quasi-light scatter deserves some further description, and as more commercial equipment becomes available, it is finding increasing utility in many colloid laboratories (54, 55). The technique basically measures the diffusion constant D of an assembly or particle in a system, and the radius of the assembly γ is then calculated from the standard Stokes–Einstein equation,

$$D = \frac{kT}{6\pi\gamma\eta}$$

where k is the Boltzmann constant, T is the temperature, and η is the viscosity of the system. The calculation relies on the fact that the assembly is spherical, a feature not often evident in many reports that just list radii of unknown particles.

The basic principles of the technique can be understood from the following simplified discussion (56). A high-intensity continuous-wave laser beam, of wavelength that is not absorbed by the sample (e.g., λ 4880-Å argon ion laser or 6328-Å helium neon laser), is focused into the sample contained in a small cuvette. The scattered light will experience a small Doppler frequency shift because of the motion of the scattering particles. The scattered light will consist of a random Lorentian distribution of frequencies distributed about the fundamental wavelength. If the particles are reasonably monodispensed then the half-width of the Lorentian ion frequency distribution τ_1 is given by τ_1 = DK^2 where K is a scattering factor and D is the diffusion coefficient of the particle. The scattering factor K is readily calculated from the refractive index of the medium n, the laser wavelength λ, and the scattering angle θ by the expression,

$$[K] = \frac{4\pi n}{\lambda} \sin \frac{\theta}{2}$$

The small frequency shifts are measured by optical mixing techniques (57). The technique rapidly measures the particle diffusion constant and enables the particle radius to be calculated from the Stokes–Einstein equation.

Laser Raman

Raman spectroscopy is ideally suited to studies of assemblies in water because, unlike IR spectroscopy, the aqueous environment does not interfere with the measurement. A good introductory text that illustrates applications of laser Raman spectroscopy and some resonance Raman spectroscopy to many systems is given in Reference 1. Basically, the low-energy motions of the molecule interact with the exciting light impinging on the particle and give rise to scattered light that is shifted a set amount from the fundamental scattering frequency. The shift depends on the type of interaction involved. Some examples of the motions involved are CH_2 rock, C–O and C–C stretch, CH_2 twisting, CH_2 wagging, and CH_2 scissoring. Information on such motions in a molecule will reflect on its structure and the nature of its interaction with its surroundings in an assembly. Many applications of the technique to investigate micelle structure are given in Table III. The method is a powerful tool because it does not perturb the structure of the micelle by introducing a probe molecule of any sort. However, it suffers from lack of sensitivity, and fairly large surfactant concentrations in excess of 10^{-2} M must be used to obtain reliable data.

Table III.
Raman Frequencies for the Cationic Surfactant Solids CTAB, DAC, and DeTAB

CTAB	Hexadecane[a]	DAC[b]	Dodecane[a]	DeTAB[c]	Decane[a]	Tentative assignments
128	150	184	194	177	231	Longitudinal accordion modes
		256				
		268				
452		452		452		
				463		
494				472		
504				506		
537				535		
753				759		CH_3 rock from $N^+(CH_3)_3$ group
763				770		
776				806		
800						
888	888	888	892	891	886	CH_3 rock (terminal methyl)
910		910		909		$C-N^+$ stretch
939		948		950		
960				961		CH_2 rock
988				988		
1013		1014		1033		CH_2 twist, crystalline
1048		1045		1055		$C-C$ sym stretch + CH_2 wag
1063	1058	1064	1061	1064	1060	$C-C$ stretch, crystalline
1093		1073		1079		
1101						
1128	1135	1122	1136	1124	1136	$C-C$ asym stretch + CH_2 wag from (TG_mT) with m large CH_2 rock
1151		1163		1151		

1179		1177		1186	1217	CH_2 wag, crystalline
1212						
1229		1224				
1241		1240				
1276						
1296	1295	1298	1297	1296	1295	CH_2 twist
		1331				CH_2 wag
				1356		
1371		1372		1372		$C-CH_3$ sym bending
1398				1399		$C-H$ sym bend from $N(CH_3)_3$ group
1418		1420				$-CH_2$ bend
1441	1442	1442	1441	1441	1447	$-CH_2$ bend
1467	1471	1453	1462	1463	1475	$-CH_2$ bend, crystalline
1480		1470				$-CH_2$ bend
2849	2846	2848	2845	2851	2843	$C-H$ sym stretch of $-CH_2^-$
2857		2857		2858		
2882	2878	2882	2879	2883	2877	$C-H$ sym stretch of CH_3^-
2889		2888				
2904		2902		2906		
2914		2914				
2933	2934	2932	2933	2933	2935	$C-H$ asym stretch of $-CH_2^-$
2944		2950		2943		
2972	2963	2973	2964	2965	2964	$C-H$ asym stretch of CH_3^-
2981		2981		2982		$C-H$ sym stretch of CH_2-N^+ $(CH_3)_3$ group

[a] Source: Ref. 58.
[b] DAC is dodecylammonium chloride
[c] DeTAB is decyltrimethylammonium bromide

Literature Cited

1. McBain, J. W. *Trans. Faraday Soc.* **1913**, *9*, 99.
2. Hartley, G. S. "Aqueous Solutions of Paraffin Chain Salts"; Heman et cit.: Paris, 1936.
3. Tanford, C. "The Hydrophobic Effect"; J. Wiley: New York, 1973.
4. Fendler, J. H., Fendler, E. J. "Catalysis in Micellar and Macromolecular Systems"; Acad. Press: New York, 1975.
5. Shinoda, K., Nakagawa, T.; Tainamushi, B.; Isemura, T. "Colloidal Surfactants"; Acad. Press: New York, 1963.
6. Thomas, J. K. "The Micelle"; Van Nostrand Encyclopedia, 1980.
7. Menger, F. *Acc. Chem. Res.* **1979**, *12*, 111.
8. Winsor, P. A. *Chem. Rev.* **1968**, *68*, 2.
9. Fujiwara, F. Y.; Reeves, L. W., *J. Am. Chem. Soc.* **1976**, *98*, 6790.
10. Hargreaves, W. R.; Deamer, D. W., *Biochemistry* **1978**, *17*, 3759.
11. Other reference to strong structures.
12. Kunitake, T.; Okahata, Y. *J. Am. Chem. Soc.* **1977**, *99*, 3880.
13. Kano, K.; Fendler, J. H. *Chem. Phys. Lipids* **1979**, *23*, 189.
14. McNeil, R.; Thomas, J. K. *J. Colloid and Interface Sci.* **1980**, *73*, 522.
15. Bangham, A. D. In "Progress in Biophysics and Molecular Biology"; Butler, T. A. U., Ed.; Noble: London, 1968; p. 29.
16. Fendler, J. H. *Acc. Chem. Res.* **1976**, *9*, 153.
17. Menger, F. M.; Vitale, A. C. *J. Am. Chem. Soc.* **1973**, *95*, 4931.
18. Iler, R. K. "The Chemistry of Silica"; Wiley: New York, 1979.
19. Grim, R. E. "Clay Mineralogy"; McGrams Hall: New York, 1968.
20. Van Olphen, H. "An Introduction to Clay Colloid Chemistry"; Interscience: New York, 1963.
21. Theng, B. K. G. "Chemistry of Clay-Organic Reactions"; Wiley: New York, 1974.
22. Aniansson, G. E. A. *J. Phys.Chem.* **1978**, *82*, 2805.
23. Matheson, I. B. C.; King, A. D. *J. Colloid Interface Sci.* **1978**, *66*, 464.
24. Mukerjee, P.; Cardinal, J. R.; Degai, N. R. "Micellisation, Solubilisation, Microemulsions"; Mittal, K. L., Ed.; Plenum Press: New York, 1977, Vol. I, p. 171.
25. Klein, A.; Hauser, M. *Acta. Phys. Chem.* **1973**, *19*, 363; *Inst. Physch. Chem.*, **1974**, *90*, 215.
26. Kalyanasundaran, K.; Thomas, J. K. *J. Am. Chem. Soc.* **1977**, *99*, 2039.
27. Almgren, M.; Grieser, F.; Thomas, J. K. *J. Am. Chem. Soc.* **1979**, *101*, 279.
28. Paino, T.; M. S. Thesis, University of Notre Dame, Ind., 1981.
29. Almgren, M.; Grieser, F.; Thomas, J. K. *J. Am. Chem. Soc.* **1980**, *102*, 3188.
30. Gregoritch, S.; Thomas, J. K. *J. Phys. Chem.* **1980**, *84*, 1491.
31. Kalyanasundaran, K.; Thomas, J. K. *J. Phys. Chem.*, **1977**, *81*, 2176.
32. Morris, D. A. N.; Thomas, J. K. Ref. 24 Vol. II, p. 913.
33. Morris, D. A. N.; Castellino, F.; McNeil, R.; Thomas, J. K., *Biochim. Biophys. Acta.*, **1980**, *599*, 380.
34. Nichols, C. H.; Thermakers, P. A. *Advances in Photochemistry* **1971**, *8*, 315.
35. Lindman, B.; Forsen, S. "NMR Basic Principles and Progress"; Springer: Berlin, 1976; Vol. 12.
36. Clifford, J.; Pethcea, A. B. *Trans. Faraday Soc.* **1964**, *60*, 1483.
37. Clifford, J. *Ibid* **1965**, *61*, 182; **1965**, *61*, 1276.
38. Williams, E.; Scans, B.; Allerhand, A.; Cordes, E. H. *J. Am. Chem. Soc.*, **1973**, *95*, 4871.
39. Roberts, R. T.; Chackaty, C. *Chem. Phys. Lett.* **1973**, *22*, 348.
40. Corkill, L. M.; Goodman, J. F.; Wyer, J. *Trans. Faraday Soc.* **1969**, *65*, 9.
41. Clemett, C. J. *J.Chem. Soc.* **1970**, *A*, 2251.
42. Podo, F.; Roy,A.; Nemethy, G. *J. Am. Chem. Soc.*, **1973**, *95*, 6164.
43. Riberio, A. A.; Dennis, E. A. *Chem. Phys. Lipids*, **1975**, *14*, 193; *Biochemistry* **1975**, *14*, 3746.

44. Kalyanasundaram, K.; Grätzel, M.; Thomas, J. K. *J. Am. Chem. Soc.*, **1975**, 97, 3915.
45. Kalyanasundaram, K.; Thomas, J. K. Ref. 24, Vol. II, p. 569.
46. Fox, K. K. *Trans. Faraday Soc.* **1971**, 67, 2802.
47. Atherton, N.; Strach, J. J. *J. Chem. Soc., Faraday Trans, 2*, **1972**, 68, 374.
48. Grätzel, M.; Thomas, J. K. *J. Am. Chem. Soc.* **1973**, 95, 6885.
49. Bakalik, D. P.; Thomas, J. K. *J. Phys. Chem.* **1977**, 81, 1905.
50. Griffith, O. H.; Waggoner, A. S. *Acc. Chem. Res.* **1969**, 2, 17.
51. Levine, Y. K. *Prog. Biophys. Mol. Biol.* **1972**, 24, 1.
52. Trifunac, A. D.; Nelson, D. J.; Mottley, C. *J. Magn. Resson.* **1978**, 30, 263.
53. Tanford, C. "Physical Chemistry of Macromolecules"; Wiley, New York, 1961.
54. Mazer, N. A.; Kwasnick, R. F.; Carey, M. C.; Benadek, G. B. "Mittal Micelles"; Vol. I, p. 383.
55. Holzback, R. T.; Oh, S. Y.; McDonnell, M. E.; Jamieson, A. M. "Phys. Math. for Chem."; Vol. I, p. 403.
56. Clark, N. A.; Lunacek, J. H.; Benedek, G. B. *Am. J. Phys.* **1970**, 8, 575; 3.
57. Chu, B. "Laser Light Scattering"; Academic Press: New York, 1974.
58. Mizushima, S.; Simanooti, T. *J. Am. Chem. Soc.* **1949**, 71, 1320.
59. Krenske, D.; Abdo, S.; Van Damne, H.; Cruz, M.; Friplat, J. J. *J. Phys. Chem.* **1980**, 84, 2447.

6

Micellar Systems

O$_F$ THE MANY COLLOIDAL SYSTEMS that have been studied, probably micelles and systems similar to micelles have received the most consistent attention because these systems are important in industrial projects, and they could serve as models for biological membranes and enzymes (1). Indeed the statement, "micelles are reminiscent of enzymes," is carefully phrased to project the vital interest in these systems without undue claim as to their exact identification with an enzyme. The basic amphiphilic character of a surfactant or soap molecule, which makes use of the basic properties of water in constructing unique structures such as micelles, is a challenging concept that is rooted in modern concepts of biological organization (2). From the point of view of photochemists and radiation chemists, micelles are well-defined physical systems with a vast backlog of well-established chemical and physical properties, which can be utilized to discuss radiation-induced events in these systems. Micelles and similar structures are thus worthy of a more detailed discussion, both from kinetic and thermodynamic points of view.

Thermodynamic Consideration

Much has been written concerning the forces that operate to promote micelle formation (1–9). The general concept is concerned with the interaction of the long methylene chains of the surfactant with the water molecules of the bulk water medium. In the simplest possible picture, the surfactant molecules attempt to eliminate, as much as possible, the water–hydrocarbon chain interactions. Elimination is achieved by clustering the surfactant molecules into a micelle. This structure (discussed previously) is roughly spherical; the interior of the sphere contains hydrocarbon chains that are surrounded by a Stern layer consisting of the polar head groups, which are in contact with water. Other models have been suggested, and they maintain that water readily penetrates into the micelle (10). This idea mainly accommodates observations that solutes solubilized by the micelle experience quite a polar environment. These data can also be explained by assuming that the solutes are

0065-7719/84/0181-0123$07.25/1
© 1984 American Chemical Society

solubilized near the Stern layer of the micelle. However, the low solubility of water in hydrocarbons suggests that water does not penetrate to any great extent into the micelle. Water is believed to be in close contact with the first CH_2 after the polar head group (11).

Size and Shape Considerations

The most commonly used micelle model is the so-called "Hartley micelle" or "roughly" spherical micelle. Other shapes such as rods occur under the correct conditions and are discussed later. The roughly spherical micelle conveys a useful picture of a micelle. Although the "true" picture of a micelle will vary with surfactant type and is not known, the picture can be inferred from various pieces of data and reasoning.

Micelles are probably not truly spherical. The surfactant molecules pack together as closely as possible to eliminate as much water as possible from the structure. The degree of packing is affected by the repulsion of the polar head groups of the structure, which try to maintain as great a separation as possible. A compromise between these two opposing forces, the aggregation of the alkane chains and the repulsion of the head groups, is achieved by forming a structure with a curved surface where the separation of head groups is maximized. Thus roughly spherical or cylindrical surfaces are possible. Furthermore, one dimension or radius of the structure may not exceed the dimension of the alkane chain.

Structures such as a sphere, a disc, an oblate ellipsoid, or a rod can be accommodated within the above restrictions. However, a perfect sphere does not allow a sufficient number of surfactant molecules to be packed into the micelle with a radius (r) that is less than or equal to the length of the alkane chain. A disclike structure, however, could be drawn to cover any degree of micellar aggregation (12). Because the micelle is a dynamic structure with considerable motion in its surface, a rounded-off disc or oblate shape could be a more realistic structure. The head groups should be packed together so that the surface area per head group (A) is as small as possible and eliminates as much hydrocarbon–water interaction as possible. Calculation shows that A is 66.2 Å^2, 63.5 Å^2, 80.0 Å^2, and 74.2 Å^2 for an oblate ellipsoid (12), a flat disc with a semitoroidal edge, a prolate ellipsoid, and a straight cylinder, respectively. The calculations were performed for these structures where the surfactant contained 12 CH_2 groups and had an aggregation number of 100 with a limiting trans-micelle dimension of 24.36 Å. These calculations indicate that a distorted disc structure, or roughly spherical micelle, is a realistic picture for small micelles with aggregation numbers in the vicinity of 100 or so.

Aggregation Number

The number of surfactant molecules per aggregate, or the aggregation number, is controlled by the optimum packing of hydrocarbon chains, to maximize interhydrocarbon interaction, and by the repulsion of the charged head groups. The latter effect produces curvature in the assembly, which acts as a boundary containing a select number of units. Reduction in the head group repulsion leads to a closer packing of these charged groups and to a smaller degree of surface curvature. A bigger assembly forms with a larger aggregation number. Thus, the first picture that controls aggregation number is the length of the surfactant alkyl chain, which controls one dimension (the radius) of a roughly spherical micelle; the larger the radius is, the larger the aggregation number. Surfactant shape plays a most important role because it controls the second feature of packing, which will be discussed later. For example, with two long alkane chains, the surfactant forms vesicles rather than micelles, with a corresponding increase in aggregation number.

The third feature is the charge on the micelle surface, which reflects on the degree of head group packing. Decreasing the effective charge by increasing the surfactant concentration or by deliberate addition of a salt leads to an increase in aggregation number. A roughly spherical micelle then takes on a more oval appearance. A radical change in shape from roughly spherical to rod micelle or to a bilayer structure may take place if the aggregation number increases quite markedly in each case. The addition of a cosurfactant, such as long chain alcohols, aids in the formation of the more crystalline bilayer structures.

Implications of Aggregation Number on Kinetics of Micelle Formation

The formation of an aggregate from monomers (M) can be written as a series of association reactions.

$$M + M \underset{}{\overset{K_1}{\rightleftharpoons}} M_2$$

$$M_2 + M \underset{}{\overset{K_2}{\rightleftharpoons}} M_3$$

$$M_{n-1} + M \underset{\phantom{K_{n-1}}}{\overset{K_{n-1}}{\rightleftharpoons}} M_n$$

Some dye systems show behavior described by these open-ended self association reactions where, $K_1 = K_2 = K_3 = K_n$. The dyes are planar molecules that form stack-type aggregates where an added monomer lies flat on top of a stack containing other monomers (13). No geometric restrictions exist as to packing of the monomers in the aggregate, and the polymerization can continue to quite large aggregation numbers.

The constraints in forming a micelle limit the degree of polymerization to an upper aggregation number and favor this structure over smaller multimer structures. If the equilibrium constant for the various stages of monomer addition to a multimer is defined as K_n, then K_n increases with n up to moderate values of n, that is, it becomes easier to add a monomer surfactant unit to an existing multimer structure with increasing n up to a limiting n value. This effect is a cooperative effort, which is not present in the dye stacking structures. Beyond a certain value of n, K_n decreases as addition of other monomer units to the structure becomes increasingly more difficult. This difficulty is connected with the physical constraints of packing more units into an existing structure; a major issue is repulsion of the polar surfactant head groups. Thus, micelles tend to be fairly discrete structures with a fairly narrow distribution of aggregation numbers.

This discussion has centered on ionic micelles where repulsion between adjacent surfactant head groups plays a major role in the final shape and size taken by a micelle. With nonionic surfactants, the polar head groups are much larger in structure and, as in the case of ethylene oxide head groups, the head group may take up various structures (e.g., open coil and helical) (14). Thus, structural parameters enter into forming micelles, parameters that are not so readily accounted for as in the case of small ionic head groups. In the neutral micelles, head group repulsion is not important; closer packing of surfactant units is possible and leads to large aggregation numbers for nonionic micelles compared to ionic micelles (15).

Thermodynamics

The standard free energy change per monomer on micellization, $\Delta G°$, is given by $\Delta G° = RT \ln(CMC)$, where CMC stands for the critical micelle concentration. Because CMC values are known for many micellar systems (15) the corresponding $\Delta G°$ values are readily calculated. To break up $\Delta G°$ into $\Delta H°$ and $\Delta S°$ where $\Delta G° = \Delta H° - T\Delta S°$ is not as easy. However, the temperature dependence of some CMC values has been studied and $\Delta H°$ is then extracted via the expression (16)

$$\Delta H° = - RT^2 \frac{d \ln(CMC)}{dT}$$

Some $\Delta H°$ values have also been determined calorimetrically (17). The data show that the dominant term contributing to $\Delta G°$ is the entropy change ($\Delta S°$). This observation is explained in terms of the behavior of water molecules associated with the alkane chain of the surfactant. The water in the immediate vicinity of an alkane chain exhibits enhanced hydrogen bonding (18, 19). On micellization, the alkane chains are transferred from an aqueous environment to an alkanelike environment, and the water molecules associated with the alkane chains are released. These reactions produce a positive increase in entropy.

The general concept of the process of micellization and the nature of the thermodynamic processes associated with it appear to be quite satisfactory. However, more elaborate considerations of the variation of $\Delta H°$ and $\Delta S°$ with surfactant chain length pose problems in their quantitative explanation (6). Table I shows the variation in $\Delta H°$ for a series of surfactants as a function of alkane chain length. The data show that a change in $\Delta H°$ of about -0.6 kcal/mol accompanies an addition of a CH_2 group to the surfactant. Changes in $\Delta G°$ of -0.68 kcal/mol are known for an additional CH_2 group (16); hence, the change in $\Delta H°$ accounts for nearly all the change observed in $\Delta G°$; therefore, there is little change in $\Delta S°$. To account for these data, the freezing or structuring of water molecules by alkanes may be much more effective for small chains up to butane, and the effect may not increase progressively as the chain length increases in longer alkanes.

Sample Calculation of Thermodynamic Parameters Involved in Micellization

Thermodynamic parameters derived from other physical measurements can be used to comment in a quantitative fashion on the preceeding model of micellization (5). The process is rather empirical but nevertheless, serves to instill confidence into the preceding discussion.

Principle of Opposing Forces. Two opposing influences affect the nature of the aggregation number of a micelle and project the formation of a unique or narrow range of values for the aggregation number with

Table I.
Chain Length Dependence of the Heat of Micellization ($\Delta H_m°$) for Homologous Series of Detergents

Head Group	$\Delta H_m°$				
	6C	8C	9C	10C	12C
$(OC_2H_4)_6OH$	5.5	4.8	—	3.6	—
$N(O)(CH_3)_2$	—	4.0	4.4	2.7	2.6
$OSO_3{}^-Na^+$	—	1.5	—	1.0	-0.3
$SO(CH_2)_2OH$	2.4	1.2	—	—	—
$SO(CH_2)_3OH$	2.7	1.7	—	—	—
$SO(CH_2)_4OH$	3.4	2.0	—	—	—

NOTE: Values are given for $\Delta H_m°$ in kilocalories per mole of material micellized at 25 °C.

the absence of multimers. It has been indicated that K_n increases with n up to a critical value and then decreases. If this concept is formalized by the aid of a simple mechanism, then various properties of micelles can be calculated and compared to experimentally determined values. The attractive force contributing to micelles is the increase in free energy from the hydrophobic effect of transporting the alkane chains from water to a nonaqueous environment. The repulsive force is primarily due to the repulsion of the ionic head groups. The interplay of the two effects leads to the overall $\Delta G°$ of micellization.

A detailed calculation relating various thermodynamic parameters of the NaLS surfactant system to the micellar properties has been carried out; the details are noted here (5).

Calculation of Micellar Properties. To understand the factors involved in the formation of a micelle, the free energy of a monomer surfactant molecule must be considered as a function of the state of association. The free energy may be broken down into two main terms: a hydrophobic contribution that favors association and a term marking the repulsion between head groups of the surfactant that limits the size of the assembly. This consideration is concerned mainly with ionic micelles. With nonionic micelles, the hydrophobic term is similar to that considered in ionic micelles; however, the repulsion terms are more difficult to imagine in nonionic micelles. The shape of the micelle is important because this feature influences the packing of the head groups and, hence, the repulsion energy term. The free energy of transfer of a monomer surfactant molecule from the aqueous phase to the micelle, $\Delta G°$, is given by $\Delta G° = (-RT/m) \ln K_m$, where m is the size of the micelle, and K_m is the equilibrium constant for formation. The free energy $\Delta G°$ is broken down into two parts: $\Delta G° = \Delta G_A + \Delta G_R$, where ΔG_A and ΔG_R represent the attractive and repulsive fractions of $\Delta G°$, respectively. The values of $\Delta G°$ are estimated for all values of m, the aggregation number of the micelle. As discussed previously, the shape of the micelle is assumed to be disclike.

Attraction Component, ΔG_A. The hydrophobic interactions of ΔG_A may be considered as consisting of three distinct factors: the free energy of transfer of the hydrophobic portion of the surfactant from water to the alkane liquid state; a factor taking into account the fact that the micelle interior has more order than the pure alkane liquid (i.e., the alkane chain is more constrained near the head group region); and a term allowing for incomplete removal of the surfactant tail from the aqueous environment. The free energy of transfer is -2100 cal/mol/CH_3 group and -883 cal/mol/CH_2 group. The CH_2 group adjacent to the head group is considered to be still in contact with H_2O and is not considered in the calculation. The value for ΔG_A is expressed as

$$\Delta G_A = -(E_{CH_3} + qE^1_{CH_2} + (n - S - 1)E_{CH_2})$$

where q is the number of ordered CH_2 groups (approximately 5–8). E_{CH_3} is 2100 cal/mol, E_{CH_2} is expected to be less than 884 cal/mol, and $E^1_{CH_2}$ is $< E_{CH_2}$. The value for E_{CH_2} is expected to be 700 cal/mol, and $E_{CH_2} - E^1_{CH_2}$ is 100 cal/mol. An additional parameter (S) to allow for the residual hydrocarbon–water contact is also used; $S = 25(A - 21)$ cal/mol, where A is the surface area of contact between hydrocarbon and solvent at the surface of the micelle core per surfactant molecule. Hence,

$$\Delta G_0 = -1400 - 700n + q(E_{CH_2} - E^1_{CH_2}) + 25(A - 21)$$

Head Group Repulsion. Head group repulsion is the main factor contributing to the repulsion in ionic micelles. This interaction (ΔG_R) is given by the observed work of compression of a surfactant film on water (from infinite dilution to the area given by the head group), minus the ideal work for the same process. In actual fact ΔG_R in calories per mole is given by

$$\Delta G_R = \frac{1.432 \times 10^5}{A} - \frac{4.23 \times 10^6}{A^2} - \frac{9.65 \times 10^7}{A^3}$$

where A is in square angstroms.

By using the above data and a CMC of 1.4×10^3 M, aggregation numbers (m) of 94–98 were calculated for micelles of oblate or disc shapes for NaLS in 0.1 M NaCl. The calculated values of m agree well with the experimental number of about 90. The agreement of theory and data lends confidence to the concepts of micelle formation governed by two major driving forces, hydrophobic attraction of the $(-CH_2)_n$ chains and repulsion of the polar head group. Some reasonable assumptions regarding the trans-micelle dimension and the location of the sulfate group outside the core were made; however, the adjustment of these parameters is minimal and no freely adjustable parameters were used.

Effect of Geometry on Aggregate Formation

The geometric shape of the surfactant molecule can control the nature of the aggregate formed, that is, whether a micelle or vesicle is formed when the surfactant forms an assembly. A particularly clear treatment of this effect is available (21, 22). The fundamental postulates of the treatment are similar to those given earlier; namely, the surfactant molecules congregate because of the hydrophobic interaction between the hydrocarbon tails, with the added constraint that the hydrophilic head groups must remain in contact with the aqueous phase. Thus an optimal surface area per head group is sought, where the total interaction free energy per lipid molecule is a minimum. The considerations define the parameters that force a single chain surfactant to form micelles; double chain surfactants cannot form micelles but form bilayers or vesicles. The controlling thermodynamic parameter is entropy, which favors the formation of structures with small aggregation numbers. Structures with large aggregation numbers may be possible on energetic considerations, but they are unfavorable from entropic considerations.

   All systems considered are in thermodynamic equilibrium; hence, the chemical potential of all molecules in a system of assemblies will be identical. If the free energy of a surfactant molecule in a micelle is μ_N° then

$$\mu_N^\circ + \frac{kT}{N} \ln(X_N/N) = \text{constant}$$

where $N = 1, 2,$ or 3, and X_N is the mole fraction of molecules incorporated into micelles of aggregation number N. If M is a reference state of micelles with aggregation number M then $X_N = N(X_M/M)N/M \ e^N(\mu_M^\circ - \mu_N^\circ)/kT$. This equation shows that small assemblies are entropically favored over larger ones.

   The attractive contribution to μ_N° is given by the interfacial free energy per unit area, $\delta \simeq 50$ erg/cm^2, and the hydrophobic free energy contribution is δa where a is the molecular area measured. The repulsive contribution to γ_n° is taken in the form C/a, where C is a constant. Thus, $\mu_N^\circ = \delta a + C/a$, and, because the minimum in μ_N° is needed,

$$\frac{\delta \mu_N^\circ}{\delta a} = 0 = \delta - C/a^2$$

or

$$a = a_o = \sqrt{c/\delta}$$

or

$$\mu_N^\circ \ (\text{minimum}) = 28a_o$$

Here a_o is referred to as the optimal surface area per molecule. The free energy per molecule is now given by,

$$\mu_N{}^\circ = \delta(a + a_o{}^2/a) = 2a_o\delta + \frac{\delta}{a}(a - a_o)^2$$

The constraints that surfactant molecular geometry imposes on the forming of various structures remain to be determined; surface areas should be close to a_o and N is a minimum.

The following parameters are defined: V is the hydrocarbon chain volume, l is the hydrocarbon thickness, and R is the radius of curvature R of the structure at the interface.

For a spherical micelle $R = l$, and $4\pi R^3/3V = 4\pi R^2/a = N$, or $V/a = l/3$. For a cylindrical micelle of radius $R = l$, $V/a = l/2$. For a spherical bilayer vesicle of outer radius R_1,

$$V/a = l\left(1 - \frac{l}{R} + \frac{l^2}{3R^2}\right)$$

and applies to the outer layer.

The value for l must not exceed a maximum length l_c. Now, the constraints for the most stable structure are that N should be a minimum and $l \leq l_c$.

Twelve-Carbon Single Chain Anionic Surfactant

For a single chain anionic surfactant $a_o \simeq 70$ Å, $V = 350$ Å^3, and $V/a_o = 5$ Å. Thus, $l = 3V/a = 15$ Å. Because the 12-carbon chain can extend over this distance, these surfactants should form spherical micelles of radii ≈ 15 Å. Other structures, such as bilayers, require an increase in N compared to that required for micelles and are thus entropically unfavorable.

Cylindrical Micelles. If the ionic strength of the aqueous solvent is increased, the electrostatic head group repulsion is also decreased; the optimal area also decreases below 70 Å^2 to about 60 Å^2. The value for V/a is ≈ 5.8 Å, and l should be 17.5 Å, which is too large for a 12-carbon chain. Hence, spherical micelles are not formed. Cylindrical micelles can be formed, however, because this geometry gives $l = 2V/a = 11.6$ Å. However, the aggregation number increases under such conditions.

Vesicles. If the surfactant contains two 12-carbon chains, then $V = 700$ Å^3; however, a_o is still only 70 Å^2–60 Å^2. Thus $V/a \approx 700/60 \approx 12$ Å.

The fully extended hydrocarbon chain is 15 Å; hence, the surfactant cannot be packed into spherical or cylindrical micelles because the radii would have to be 36 or 24 Å, respectively, and much longer than the chain. Planar bilayers could be formed with a half thickness of $V/a \approx$ 12 Å, but large aggregation numbers are required. The surfactants could pack into spherical vesicles where $a = a_o + 60$ Å, $V = 70$ Å^3, $l = 15$ Å, and an outer radius of 62 Å is formed. This structure is entropically favored over the planar bilayer because the aggregation number is smaller.

This treatment also predicts the asymmetrical features of mixed lipid vesicles. Lipids that have longer head groups and larger a/R should go preferentially to the outer layer of the vesicle. This arrangement was found experimentally for mixed phosphatidylcholine–phosphatidyle-thanolamine and phosphatidylcholesterol vesicles. If one lipid is charged (e.g., phosphatidylserine) then increasing the degree of ionization, for example, by increasing the pH, increases the head group area. The charged lipid should then go progressively more to the outer layer at the expense of the noncharged lipid (23). The asymmetric distribution of charge in these vesicles results in different surface potentials on the inner and outer vesicle surfaces, a feature that could be important in membrane transport processes.

Microemulsions. A similar treatment has also been applied to microemulsion systems to investigate the effects of the geometry of the surfactant molecule on the maximum radius of the aggregate, the degree of solubilization of water in water-in-oil microemulsions, and the necessity of using cosurfactants to form certain microemulsions (24). The treatment is confined to water-in-oil microemulsions where the radius of the water pool is γ_ω and the effective length of the surfactant chain is l. The surfactant diisooctylsulfosuccinate (AOT) forms a water-in-oil microemulsion without the necessity of a cosurfactant because of the double chain structure of AOT, which in effect contains its own co-surfactant.

The number of microemulsion droplets per cubic meter is denoted by n, the number of detergent molecules by N, and the number of water molecules per droplet by N_W. The polar head groups are considered to be part of the aqueous core; hence, the volume of the nonpolar surfactant shell around the water core is given by $NV/n = 4\pi[(\gamma_\omega + l)^3 - \gamma_\omega^3]/3$, where v is the volume of a nonpolar tail. The concentration of the detergent is given by $N = 4\pi n(3l\gamma_\omega^2 + 3l^2\gamma_\omega + l^2)/3V$. It can also be shown that

$$N_\omega = n(4\pi\gamma_\omega^3/3 - N\phi_v/n)V_w$$

where v_w is the volume of a water molecule, and ϕ_v is the apparent molecular volume of the polar head group and its counterion.

The molar ratio of water to detergent R is then given by

$$R = N_w/N = \frac{1}{V_w} \left(\frac{V\gamma_\omega^3}{3\gamma_\omega^2 l + 3\gamma_\omega l^2 + l^3} - \phi_v \right)$$

If $\gamma_\omega \gg l$, then the expression becomes

$$R \approx (\gamma_\omega V/3l - \phi_v)/V_w$$

which predicts a linear relationship between R and γ_ω.

Maximum Size of Emulsion Particles

The free energy of an emulsion system consists of the interfacial free energy and the entropic effects of mixing the dispersed phase with the solvent. The latter effect is small compared to the former and is normally ignored. The free energy per detergent molecule (μ_n) in an assembly containing n detergent molecules is given by $\mu_n = \delta A + K/A + g$, where A is the area per detergent molecule at the detergent water interface and δA is the hydrophobic contribution to the free energy; δ is the hydrocarbon–water interfacial tension of 5×10^{-2} J M^{-2}. The term K/A is the electrostatic repulsion free energy and g is the bulk free energy contribution from hydrocarbon chain interactions. The above equation can be differentiated with respect to γ_ω by setting $d\mu_n/d\gamma_\omega = 0$. Therefore, the condition for the maximum radius of the aqueous core V_{max} is given where $A = \sqrt{K/\delta}$.

If the nonpolar part of each detergent molecule is looked upon as a frustrum of height l and area A at the detergent–water interface, then the volume V of the nonpolar tail is

$$V = A[(l + \gamma_\omega)^3/\gamma_\omega^2 - \gamma_\omega]/3$$

If $A = K/\delta$, then it is possible to write

$$l^3/r^2_{max} + 3l^3/r_{max} + 3l = 3V\sqrt{\delta/K}$$

and as $r_{max} \gg l$ then r_{max}, the maximum water pool size, is

$$r_{max} = l^2(V\sqrt{\delta/K} - l)$$

and R_{max}, the solubilization capacity, is calculated from

$$R = N_w/N = \frac{1}{V_\omega} \left(\frac{V\gamma_\omega^3}{3\gamma_\omega^2 l + 3\gamma_\omega l^2 + l^3} - \phi_v \right)$$

Conditions That Require Cosurfactants

A surfactant with a single long hydrocarbon chain will not normally form a microemulsion unless a cosurfactant, that is, a long chain alcohol,

is introduced into the system. The combined geometric shape of surfactant and cosurfactant at the water–oil interface then resembles that
of AOT. Essentially the cosurfactant acts as a spacer or insulator between the ionic surfactant head groups, and the single geometric parameters for AOT are now replaced by combined surfactant–cosurfactant
parameters. For example, if the subscripts S and C stand for surfactant
and cosurfactant and the ratio of cosurfactant to surfactant molecules
is M, then

$$V = mV_C + V_S$$

and

$$l = (ml_C + l_S)(l + m)$$

and

$$\phi_V = m\phi_{V,C} + \phi_{V,S}$$

The water droplet radius r_w is always greater than zero, hence, the
mean cross-sectional area of the molecule (V/l) is greater than $(K/\delta)^{1/2}$.
The term V/l is also proportional to the surface area per detergent head
group at the detergent–water interface. Reasonable values for the above
parameters can be given: $\delta = 5 \times 10^{-2}$ J M^{-2} (21, 22), and $K = 7.62$
$\times 10^{-5}$ J Å^{-2} (2).

The value for δ/K is 1.54×10^{18} M^{-2} and $V/l > 0.65$ nm^2. The result
suggests that ionic detergents, where $V/l < 0.7$ nm^2, form micelles, but
they cannot form microemulsions without incorporating a cosurfactant.
The surfactant AOT has $V/l = 0.73$ nm^2, whereas potassium oleate and
NaLS require a cosurfactant (e.g., pentanol) to increase V/l for these
systems beyond 0.70.

Variation of Droplet Radius with R

Several measurements of microemulsion size have been carried out by
a variety of techniques. The radii for various R calculated by the considerations discussed previously agree quite well provided R is not
greater than 22 (24). Typical examples are shown in Figure 1 and Table
II. Droplet radii vary with temperature for the AOT system. The model
given does not predict any significant temperature dependence. However, to be able to calculate assembly radii that agree well with experimental values for such diverse microemulsions as AOT systems, po-

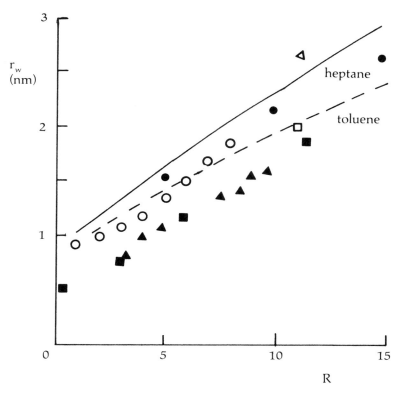

Figure 1. Radii of microemulsion droplets of AOT–water in heptane or toluene at different water–detergent ratios (R). Conditions under which the curves were calculated: ●, reaction diffraction in heptane (42); ○, viscosity in toluene (43); ▲, dynamic light scatter in toluene (43); △, dynamic light scatter in toluene (44); ■, ultracentrifuge in heptane (43); and □, ultracentrifuge in isooctane (44) (24).

Table II.
Values for Hydrodynamic Radii of Microemulsion Droplets in the Water–OT/Isooctane System

	Radius (nm)				
Method	11	22	33	44	56
Estimated by dynamic light scattering	3.6	5.0	8.6	12.0	14.5
Estimated by ultracentrifugation	2.9	4.3	—	9.6	13.6
Calculated	3.3	4.7	6.1	7.5	9.0

NOTE: All measurements were taken at 298 K.
SOURCE: Ref. 24.

tassium oleate–hexanol–benzene, and NaLS–pentanol–heptane systems is a triumph.

The relationship, $r_{max} = l^2(V\sqrt{\delta/K} - l)$, can be used to predict the maximum amount of water solubilized by the system in both unadulterated surfactant systems and in the presence of additives such as benzene and cyclohexane. The additives congregate in the interface region and act as cosurfactants in the AOT systems. Typical data are shown in Figure 2 and Table III. In the AOT–isooctane systems, R_{max} increases with addition of cyclohexane up to a mole fraction of cyclohexane of 0.1; thereafter it shows a continuous decrease on addition of further amounts of additive. The calculated and experimental R_{max} values are in excellent agreement.

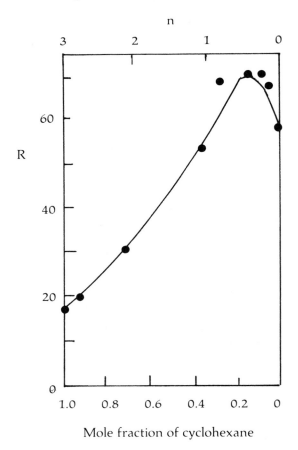

Figure 2. Solubilization capacity (R) of AOT in isooctane at different concentrations of C_6H_{12}; the curve is calculated (24).

Table III.
Calculated and Observed Values of the Solubilization Capacity (R_{max}) of Some Microemulsion Systems

Detergent	Cosurfactant	R_{max} Observed	R_{max} Calculated
AOT	—	56	56
	hexanol	19	22
	toluene	10	10
	chlorobenzene	8	9
Aerosol-TR	hexanol	11	11
Potassium oleate	hexanol	30	21
Sodium dodecysulfate	pentanol	>70	84

SOURCE: Ref. 24.

Solubilization Aspects

Aggregates formed from organic amphiphiles are, for the most part, dynamic in nature. Hence the solubilization of a molecule by a micelle, microemulsion, or vesicle is a dynamic event; the molecule moves from the assembly into the aqueous phase and vice versa. This section discusses systems that consist mainly of water. Reversed micelles and water-in-oil microemulsions will be discussed later; the only data available on these systems are directly connected with kinetic studies and are best discussed with the studies from which they are derived.

Solubilization Site. The site of solubilization is worth reemphasizing. Ionic species will locate close to the oil–water interface in the surfactant head group region; although nonpolar molecules such as benzene and pyrene will also locate on the surface region of micelles, they will penetrate to the core of oil-in-water microemulsions. In some cationic micelles (e.g., CTAB) weak complexes of arenes and the quaternary ammonium group occur, with $K \sim 1$ (25); this complex promotes surface solubilization of arenes.

Dynamic Aspects. The kinetic events controlling solubilization of a molecule P by a micelle M are depicted by

$$ P + M \underset{k_-}{\overset{k_+}{\rightleftharpoons}} MP $$

bulk phase   micellar
location of P   location of P

and an equilibrium constant K_{eq}, which is k_+/k_-, can be written as a statement of the distribution of P between the micelle and the aqueous bulk. The entry of P into the micelle with rate constant k_+ is diffusion controlled and is measured to be $\sim 10^{10}$ $M^{-1}s^{-1}$ for many molecules. Diffusion theory also predicts values for k_+ of this order. Unlike k_+, exit rate varies quite markedly with each solute (see Table IV) and is controlled entirely by k_-. Generally, the larger the molecule is, the lower the solubility in water, and the slower the exit rate k_-. An interesting correlation of K is given in Figure 3 and Table V where ln $K \propto T_B$ (T_B is the boiling point of the solute) (25, 26). The best solvent for aromatic solutes is CTAB, presumably because of the weak complexing of the quaternary ammonium group with the solute (25).

The solubilities of aromatic solutes in water follow the same behavior as that noted for micelles, that is, a linear increase in log(solubility) with the boiling point of the solute. The behavior can be rationalized by considering the surface area of the solute. For example, for many solutes the log(solubilization in water) is a linear function of the solute surface area (27–29).

The free energy change from the gas to the liquid phase, $\mu_L°(T)$ − $\mu_g°(T)$ is given by $RT \ln (P/P_0) = \mu_L°(T) - \mu_g°(T)$, where P and P_0 are the vapor pressures over the pure liquid. The free energy change is expected to be a linear function of the molecular surface area. Assuming Trouton's rule to be correct, we can write $RT \ln (P/P_0) = -\Delta S°(T_b - T)$, where $\Delta S°$ is 21.2 cal $\deg^{-1}$ mol^{-1}. Hence, at constant T, $\ln (P/P_0)$ varies linearly with T_b (the boiling point of the solute). From the earlier relationship it is now possible to say that $\mu_L°(T) - \mu_g°(T)$ is

Table IV.
Exit Rates of Aromatic Hydrocarbons from NaLS and CTAB Micelles

	Equilibrium k_- (s^{-1})		Direct k_- (s^{-1})	
Arene	NaLS	CTAB	NaLS	CTAB
Perylene	4.1×10^2	2.6×10^2		
Anthracene	1.7×10^4	3.2×10^3		$\leq 10^3$
Pyrene	4.1×10^3	1.7×10^3		
1-Bromonaphthalene	3.3×10^4	4.1×10^3	2.5×10^4	
Biphenyl	9.6×10^4	1.6×10^4	1.2×10^5	
			$>5 \times 10^4$	
1-Methylnaphthalene	1×10^5		$>5 \times 10^4$	
Naphthalene	2.5×10^5		$>5 \times 10^4$	
p-Xylene	4.4×10^5			
Toluene	1.3×10^6			
Benzene	4.4×10^6	7.5×10^5	$\geq 10^4$	

SOURCE: Ref. 25, 26.

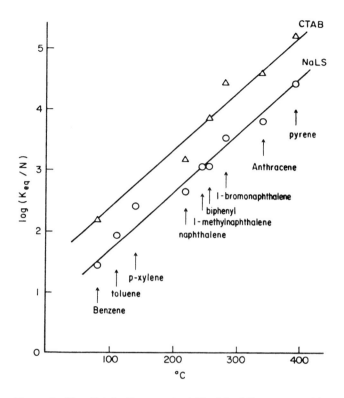

Figure 3. The distribution constant K_{eq}/N of the arenes with NaLS and CTAB plotted as a function of the boiling point of the arenes. (Reproduced from Ref. 25. Copyright American Chemical Society, Washington, D. C. 1979.)

also a linear function of T_b, together with the solute surface area. The initial statement (27–29) that the surface area varies as log(solubility) can then be used to predict that log(solubility) should vary as T_b; this statement is supported experimentally.

Solute Distribution Among Micelles. Any theoretical or kinetic treatment of the solubilization of a solute among micelles has to consider the unique features of the system as compared to the homogeneous phase. Homogeneous solution is the situation most familiar to all chemists; however, in the microcosmic world of micelles, each micelle may act out a complete kinetic event without communication with other micelles. For example, if 10^{-3} mol/L of solute are distributed among 10^{-3} mol/L of micelles, then a uniform distribution of one solute per micelle does *not* exist; but, a specialized distribution occurs.

Table V.
Solubility Data for Arenes in Micellar Solution

Solvent	Solubility[a] (M)	$\overline{n}^b$	Minimum Fraction at Surface[c]	K^b (M^{-1})
	Pyrene			
H_2O	6×10^{-7}			
Hexane	7.1×10^{-2}			
Dodecane	7.4×10^{-2}			
$NaC_{10}S$ (0.08 M)	4.0×10^{-2}	0.4		6.7×10^5
$C_{10}TAB$ (0.09 M)	8.2×10^{-2}	0.8	0.15	13×10^5
NaLS (0.06–0.04 M)	7.0×10^{-2}	1.0		17×10^5
$C_{12}TAB$ (0.04 M)	22×10^{-2}	2.4	0.67	40×10^5
$C_{12}AC$ (0.04 M)	17×10^{-2}	3.8	0.57	63×10^5
$NaC_{14}S$ (0.04 M)	10.5×10^{-2}	2.2	0.30	37×10^5
$NaC_{16}S$ (0.04 M)	10.9×10^{-2}	3.2	0.33	53×10^5
CTAB (0.04 M)	41×10^{-2}	7.3	0.82	102×10^5
	4-Bromo-p-terphenyl			
H_2O	$<10^{-9}$			
NaLS (0.05 M)	4.2×10^{-4}	0.006		$>6 \times 10^6$
	Perylene			
H_2O	$\sim1.6 \times 10^{-9}$			
NaLS (0.01 M)	1.9×10^{-3}	0.027		1.7×10^7
CTAB (0.01 M)	2.4×10^{-3}	0.043		2.7×10^7
	Anthracene			
H_2O	2.2×10^{-7}			
Hexane	1.4×10^{-2}			
NaLS (0.02 M)	0.63×10^{-2}	0.09		4×10^5
CTAB (0.02 M)	3.3×10^{-2}	0.58	0.6	26×10^5
	Naphthalene			
H_2O	2.2×10^{-4}			
Hexane	0.91			
NaLS (0.04 M)	0.38	5.4		2.5×10^4
NaLS (0.04 M, 0.1 M NaCl)	0.36	7.1		3.6×10^4
CTAB (0.02 M)	1.11	20.0	0.2	9.1×10^4
	1-Methylnaphthalene			
H_2O	2.1×10^{-4}			
NaLS (0.025 M)	1.06	15		7.1×10^4

Table V.
Solubility Data for Arenes in Micellar Solution—Continued

Solvent	Solubility[a] (M)	$\bar{n}^b$	Minimum Fraction at Surface[c]	K^b (M^{-1})
	1-Bromonaphthalene			
H_2O	4.5×10^{-5}			
NaLS (0.05 M)	0.66	9.4		2.1×10^5
CTAB (0.05 M)	4.3	76		1.7×10^6
$NaC_{14}S$ (0.025 M)	0.94	19.5		4.3×10^5
	Biphenyl			
H_2O	4.1×10^{-5}			
Dodecane	0.72			
NaLS (0.05 M)	0.21	3		7.3×10^4
CTAB (0.05 M)	1.00	17.7	0.28	4.3×10^5
	Benzene			
H_2O	2.3×10^{-2}			
NaLS (0.05 M)	2.5	35		1.6×10^3
CTAB (0.05 M)	12.5	215		9.3×10^3
	Toluene			
H_2O	6.8×10^{-3}			
NaLS (0.05 M)	2.5	35		5.3×10^3
	p-Xylene			
H_2O	1.84×10^{-3}			
NaLS (0.05 M)	2.1	30		1.6×10^4

NOTE: Values were determined in solutions at $21 \pm 1\ °C$.
SOURCE: Ref. 25, 26.
[a] The solubilities in the micelles have been calculated from $([P]_{tot} - [P]_{aq,sat})/\{([surf] - cmc)V_{hc}\}$, where V_{hc} is the molar volume of the hydrocarbon from which the tail of the surfactant is derived.
[b] The aggregation numbers used were 50 ($NaC_{10}S$), 48 ($C_{10}TAB$), 62 (NaLS), 100 ($C_{12}AC$), 50 ($C_{12}TAB$), 80 ($NaC_{14}S$), 100 ($NaC_{16}S$), and 60 ($C_{16}TAB$).
[c] These values assume that the solubility in the core is equal to that in a hydrocarbon solvent.

The nature of the statistical distribution will depend on whether solubilization of the solute by the micelle disturbs the micelle, whether a solute already in the micelle affects solubilization of another solute, or whether the interaction is completely random. In the latter case, which has been shown to be correct for a low degree of solubilization in micelles (30, 31) and microemulsions (31), the distribution of solute among micelles follows Poisson statistics. Other distributions have been suggested (32, 33) and challenged (34). For this discussion Poisson statistics will be used.

A simple statement of the fraction P_i of micelles with i solubilized molecules is given by

$$P_i = \frac{(MP_i)}{(M)_{tot}} = \frac{\bar{n}^i e^{-(\bar{n})}}{i!}$$

where (MP_i) and $(M)_{tot}$ refer to the concentration of micelles with i solutes per micelle and the total micelle concentration, respectively. The symbol $\bar{n}$ is defined by $\bar{n}$ = [solute]/[micelle]. Thus, in the original example of 10^{-3} mol of solute per 10^{-3} mol of micelle, the following P_i values can be calculated: For solute/micelle = 0, 1, 2, 3, and 4, P_i = 0.37, 0.37, 0.18, 0.06, and 0.015, respectively. Thus, in this solution, a significant number of micelles contain more than one solute per micelle while a large fraction contain no solute whatsoever. This situation controls any kinetic considerations in assembly systems.

Solubilization Model. A simple model for micellar solubilization involves the following concepts:

1. The micelles are considered as oil droplets in water.

2. The oil droplets provide a spherical free-energy well for solubilizing the molecules.

3. The micelles are considered as static entities. This assumption is not strictly correct, but the kinetic notions of solute entry and exit are in the time range $10^{-8} - 10^{-3}$, and monomer units are in the time range $10^{-8} - 10^{-5}$ s. Thus the solute does experience an overall static assembly of surfactant molecules.

Figure 4 shows a diagram of the spherical free-energy potential experienced by a molecule as it approaches and enters a micelle. The probability $p(r)dr$ of finding the probe between r and $r + dr$ in a spherical well is shown as $p(r)$ versus r, where $\bar{r}$ represents the mean distance from the center of the micelle or sphere given by $\bar{r} = 3R_c/4$, and the median r_m is $0.797R_c$.

Surface solubilization is also very important, as well as the interior hydrocarbon solubilization outlined previously. Surface solubilization can be incorporated into the model if a free-energy barrier, much larger than RT, is imposed on the solute as it tries to pass from the surface to the micelle interior (Figure 4). The depth of the surface well is 1 × RT, the hydrocarbon core radius is 16.5 Å (typically C_{12}), and the width of the surface well is 2 Å. The surface/core solubilization ratio in this

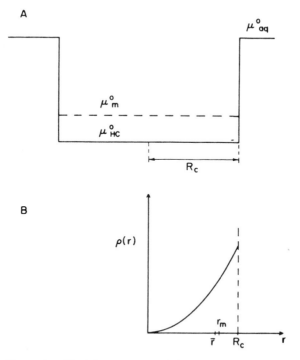

Figure 4a. (Top) Free-energy well for a solubilized molecule in a micelle regarded as a spherical oil droplet. The standard free energy μ^0_{hc} of the molecule in the hydrocarbon solvent is lower than that in the spherical micelle hydrocarbon because of the Laplacian pressure in the micelle and the entropy factors resulting from restrictions on the freedom of motion of the probes.

(Bottom) The probability p(r)dr of finding the probe between r and r + dr in a spherical well (see top figure) depends parabolically on r. The mean distance from the center of the sphere is $\bar{r} = 3R_c/4$, and the median is $r_m = 0.797 R_c$.

(Reproduced from Ref. 25. Copyright American Chemical Society, Washington, D. C. 1979.)

"two site" model is 1.1, and a molecule in the core should spend 50% of its time at 3.5 Å or less from the surface (i.e., 50% of its time at the surface).

Other factors will affect the relative site of solubilization, for example, specific interaction with the head group region, as with CTAB, and the Laplace pressure of the micelle core. Both effects increase the solubilization at the surface. The Laplace pressure P_i in the core is given by $P = 2\delta/R$, where δ is the interfacial tension and R is the micelle radius; pressures of a few hundred atmospheres have been calculated

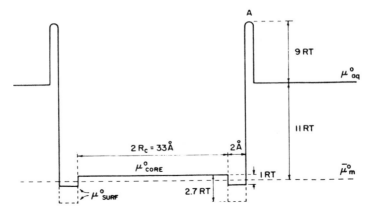

Figure 4b. *A spherical free-energy well with surface solubilization drawn to represent pyrene in NaLS micelles assuming that surface and core solubilization are equally important.*

The mean standard chemical potential $\bar{\mu}_m{}^\circ$ of pyrene in the micelle phase is about equal to that in a hydrocarbon solvent according to solubilization measurements. The surface well is assumed to be 2 Å wide. A depth of about RT of the surface well compared to μ°_{core} is then sufficient to place half of the solubilized molecules at the surface. The potential barrier A is a kinetic barrier as estimated in Appendix II. The deeper surface well, indicated by the broken line, represents the depth required to obtain a solubilization as in C^{12} AC. (Reproduced from Ref. 25. Copyright American Chemical Society, Washington, D. C. 1979.)

for NaLS micelles. This effect decreases the core solubility of the probe by a factor of exp $(P\bar{V}/RT)$, where $\bar{V}$ is the molar volume of the probe. Hence at 200 atm and with $\bar{V} = 0.1$ L/mol this factor is 2.5.

In Figure 4, the rate of hopping over the energy barrier from a position outside to one just inside, a distance λ away, is given by

$$J_+ = 4\,\pi R_o{}^2 \lambda k_o C(R_+)$$

where $C(R_+)$ is the concentration just outside the barrier,

$$k_0 = \frac{kt}{h} \exp\left(\frac{-\Delta G^*}{RT}\right)$$

and R_o is the micelle radius. The inward flow $J_+{}^1$ is given by $J_+{}^1 = 4\pi R_o D\{[P]_{aq} - C(R_+)\}$, where $C_\infty = [P]_{aq}$.

At steady state with no solute within the micelle $J_+ = J_+{}^1$. Similar considerations show that the rate of outward jumping J_- is given by

$$J_- = \frac{3\bar{n}}{R_o} \lambda k_o \exp\left(\frac{-\Delta G_M{}^0}{RT}\right) - 4\pi R_o{}^2 \lambda k_o C(R_+)$$

where $\bar{n}$ is the mean number of probes per micelle. At equilibrium J_- is equal to J_+.

The kinetic experimental description is given by

$$P + \left(\frac{\text{micelle}}{P_{n-1}}\right) \underset{nk-}{\overset{k_+}{\rightleftharpoons}} \left(\frac{\text{micelle}}{P_n}\right)$$

where n is the number of solute molecules in a particular micelle.

Hence,

$$k_+ = 4\pi R_o L \left(\frac{1}{D} + \frac{1}{\lambda R_o k_o}\right)^{-1}$$

and

$$k_- = \frac{3}{R_o} = \exp\left(\frac{\Delta G_M^\circ}{RT}\right)\left(\frac{1}{D} + \frac{1}{\lambda R_o k_o}\right)^{-1}$$

For a situation where the rates of the processes are controlled by diffusion to and from the micelle, k_+ is $4\pi R_o DL$, and k_- is $(3D/R_o^2)$ exp $(\Delta G_M^\circ/RT)$. If $D = 8 \times 10^{-5}$ cm²/s, and $R = 20$ Å, the k_+ is calculated as 1.2×10^{10} M⁻¹s⁻¹. With $D = 8 \times 10^{-5}$ cm²/s and $\lambda = 1$ Å, ΔG^* is calculated as $9\ RT$.

Long Chain Alcohols. The solubilities of long chain alcohols in micelles are of interest because these two components form the basis of microemulsion systems. In the solubilization process, a dynamic equilibrium between aqueous phase and micelle M is written as

$$\text{ROH} + \text{M} \underset{k_-}{\overset{k_+}{\rightleftharpoons}} \text{MROH}$$

and

$$K = \frac{k_+}{k_-} = \frac{[\text{MROH}]}{[\text{ROH}][\text{M}]}$$

One problem with this mechanism is that the aggregation number N varies with alcohol concentration (36). For example, for the system pentanol–NaLS, K is 675 L mol⁻¹ at 0.1 M alcohol and 1140 L mol⁻¹ at 0.31 M alcohol. The corresponding k_+ and k_- values are 1.4×10^{10}

$M^{-1}s^{-1}$ and 2.1×10^7 s^{-1}, respectively, and 2.0×10^{10} $M^{-1}s^{-1}$ and 1.8×10^7 s^{-1}, respectively, at these two alcohol concentrations. The entrance rate constant k_+ is larger in the more swollen micelle at higher alcohol concentration; the exit rate k_- does not change (35).

Counterion Kinetics

Counterions, such as alkali metal ions on anionic micelles and halide ions on cationic micelles, exchange rapidly with the aqueous medium (37). The situation is described by the conventional equilibrium,

$$\text{ion} + \text{micelle} \underset{k_-}{\overset{k_+}{\rightleftharpoons}} \text{ion (micelle)}$$

where $K = k_+/k_-$.

A simple rationale (38) states that the more highly charged metal cations bind more strongly to anionic micelles, mainly because the binding force is electrostatic. The following progression can be written where Eu^{3+} binds stronger than all ions to NaLS: $Eu^{3+} > Cu^{2+} > Ni^{2+} > Mn^{2+} > Fe^{2+} > Tl^+ > Ag^+$. Larger ions bind stronger than small ions so that for CTAB the order of binding is $I^- > Br^- > Cl^-$. The binding constants K for Ag^+ and Tl^+ and Cu^{2+} to NaLS micelles are 1.3×10^3 M^{-1}, 1.1×10^3 M^{-1} and 6×10^4 M^{-1}, respectively; k_- values for Ag^+ and Cu^{2+} are 4.5×10^7 s^{-1} and 4.8×10^5 s^{-1}, respectively. The residence time of the complex ion Br_2^- is 1.5×10^{-5} s on CTAB micelles (39).

A second mechanism for counterion exit can be operative whereby two micelles, M_1 and M_2, approach so closely that they exchange counterions.

$$M_1(\text{ion}) + M_2 \xrightarrow{k_E} M_1 + M_2(\text{ion})$$

The bimolecular rate constant for two micelles approaching each other is 1.3×10^{10} $M^{-1}S^{-1}$ when calculated from simple diffusion theory; this value is the maximum value of k_E. Values approaching the theoretical limit have been observed; for example, for NaLS micelles $k_E(Ag^{2+}) = 2.8 \times 10^8$ $M^{-1}s^{-1}$ (40), $k_E(Eu^{3+}) = 2.6 \times 10^9$ $M^{-1}s^{-1}$ (41), and $k_E(Cu^{2+}) = 3.3 \times 10^9$ $M^{-1}s^{-1}$ (38). The exit rate constant k_E determines charge separation following photoinduced electron transfer and is discussed later.

Literature Cited

1. Cordes, E. H. "Reaction Kinetics in Micelles"; Plenum Press: New York, 1973.
2. Tanford, C. "The Hydrophobic Effect"; Wiley: New York, 1973.
3. Fendler, J. H.; Fendler, E. J. "Catalysis in Micellar and Macromolecular Systems"; Acad. Press: New York, 1975.
4. Mukerjee, P. *J. Phys. Chem.* **1978**, *82*, 931.
5. Tanford, C. In "Micellization, Solubilization, and Microemulsions"; Mittal, K. L., Ed.; Plenum Press: New York, 1977; p. 119.
6. Muller, N. p. 1 in Ref. 1.
7. Hartley, G. S. "Aqueous Solutions of Paraffin Chain Salts"; Herman et Cie: Paris, 1980.
8. Hall, D. G.; Pethica, B. A. "Nonionic Surfactants"; Schick, M. J., Ed.; M. Dekker: New York, 1967.
9. Anacker, E. W. "Cationic Surfactants"; Jungermann, E., Ed.; Dekker: New York, 1970.
10. Menger, F. *Acc. Chem. Res.* **1979**, *12*, 111.
11. Mukerjee, P.; Cardinal, J. R.; Desai, N. R. p. 131 in Ref. 5.
12. Tanford, C. p. 119 in Ref. 5.
13. Stigter, D. *J. Phys. Chem.* **1966**, *70*, 1323.
14. Kalyanasundaram, K.; Thomas, J. K. *J. Phys. Chem.* **1976**, *80*, 1462.
15. Mukerjee, P.; Mysels, K. J. Critical Micelle Concentrations. N.S.R.D.S.-N.B.S. 36, 1971.
16. Mukerjee, P. *Adv. Colloid Interface Sci.* **1967**, *12*, 241.
17. Coskill, J. M.; Goodman, J. F.; Tali, J. R. "Hydrogen Bonded Solvent Systems"; Covington, A. K.; Jones, P., Eds.; Taylor and Francis: London, 1968.
18. Kauzmann, W. *Adv. Protein Chem.* **1959**, *14*, 1.
19. Nemethy, G.; Scheraga, H. A. *J. Chem. Phys.* **1962**, *36*, 3401.
20. MacAuliffe, C. *J. Phys. Chem.* **1966**, *70*, 1267.
21. Israelachvili, J. N.; Mitchell, D. J.; Ninham, B. W. *Biochim. Biophys. Acta* **1977**, *470*, 185.
22. Israelachvili, J. N.; Mitchell, D. J.; Ninham, B. W. *J. Chem. Soc., Faraday Trans. 2* **1976**, *72*, 1525.
23. Berden, J. A.; Barker, R. W.; Radda, G. K. *Biochim. Biophys. Acta* **1975**, *375*, 186.
24. Oakenfull, D. *J. Chem. Soc., Faraday Trans. 1* **1980**, *76*, 1875.
25. Almgren, M.; Grieser, F.; Thomas, J. K. *J. Am. Chem. Soc.* **1979**, *101*, 279.
26. Almgren, M.; Grieser, F.; Powell, J. R.; Thomas, J. K. *J. Chem. Eng. Data* **1979**, *24*, 285.
27. Hermann, R. B. *J. Phys. Chem.* **1972**, *76*, 2754.
28. Harris, M. J.; Higuchi, T.; Rytting, J. H. *J. Phys. Chem.* **1973**, *77*, 2694.
29. Reynolds, J. A.; Gilbert, D. B.; Tanford, C. *Proc. Nat. Acad. Sci.* **1974**, *71*, 2925.
30. Kalyanasundaram, K.; Grätzel, M.; Thomas, J. K. *J. Am. Chem. Soc.* **1974**, *96*, 7869; *J. Am. Chem. Soc.* **1975**, *97*, 3915.
31. Atik, S.; Thomas, J. K. *J. Am. Chem. Soc.* **1981**, *103*, 3543.
32. Dorrance, R. C.; Hunter, T. F. *J. Chem. Soc., Faraday Trans. 1* **1972**, *11*, 1312.
33. *Ibid.* **1974**, *70*, 1572.
34. Almgren, M.; Aniansson, E. A. G. Unpublished data.
35. Almgren, M.; Grieser, F.; Thomas, J. K. *J. Chem. Soc., Faraday Trans. 1* **1979**, *75*, 1674.
36. Hayase, K.; Hayano, S. *J. Colloid. Interface Sci.* **1978**, *63*, 446.
37. Aniansson, E. A. G.; Wall, S. *J. Phys. Chem.* **1974**, *78*, 1024; **1975**, *79*, 857.
38. Grieser, F.; Tausch-Treml, R. *J. Am. Chem. Soc.* **1980**, *102*, 7258.

39. Proske, R.; Henglein, A. *Ber. Bunsenges Phys. Chem.* **1978**, *82*, 711.
40. Henglein, A.; Proske, R. *Ber. Bunsenges Phys. Chem.* **1978**, *82*, 471.
41. Moroi, Y.; Infelta, P. P.; Grätzel, M. *J. Am. Chem. Soc.* **1979**, *101*, 573.
42. Fletcher, P. D. I.; Robinson, B. H.; Bermejo, F.; Dore, J. C.; Steytler, D. C. *J. Colloid Interface Sci.*, in press.
43. Day, R. A.; Robinson, B. A.; Clarke, J. H. R.; Doherty, J. V. *J. Chem. Soc., Trans Faraday 1* **1979**, *75*, 132.
44. Zulauf, M; Eicke, H. F. *J. Phys. Chem.* **1979**, *83*, 480.

Reactions in Micellar Systems

MICELLAR SYSTEMS HAVE RECEIVED THE MOST ATTENTION of the many assembly systems available for study. This statement is particularly true for radiation-induced studies. The simple nature of the micelle and its relative ease of preparation in large quantities promote its use as an assembly for influencing many types of reactions. A description and review of some of the early thermal reaction kinetics work and extrapolations to more recent photoinduced reactions are well documented (1–7). No particular distinction will be made between reactions initiated photochemically and those initiated by high-energy radiation. However, the work is divided into four broad classes: reactions of free radicals, reactions of excited states (quenching), energy transfer reactions, and electron transfer reactions (photoionization).

Reactions of Free Radicals

Emulsion polymerization is the earliest example of a free-radical reaction that is promoted by surfactants. The hydrophobic monomer to be polymerized is solubilized as an emulsion in an aqueous environment by means of a suitable surfactant. The polymerization is then started by a free-radical initiator. The polymerization is confined to small emulsion particles, and the reaction proceeds rapidly because of the locally high solute concentration. High molecular weights are also achieved because the polymerizing centers proceed independently of each other in separate particles.

Hydrogen Abstraction Reaction and Radiolysis. Figure 1 (1) dramatically illustrates the effects of micelles on free-radical reactions. This figure shows the yield of nitroxide radicals produced, as a function of surfactant concentration, in the reaction of a surfactant radical (R') with a nitroso spin trap (BN = 0) (8).

$$R^{\cdot} + BN = 0 \rightarrow R-\overset{\displaystyle B}{\overset{\displaystyle |}{N}}-\dot{O}$$

0065-7719/84/0181-0149$11.50/1

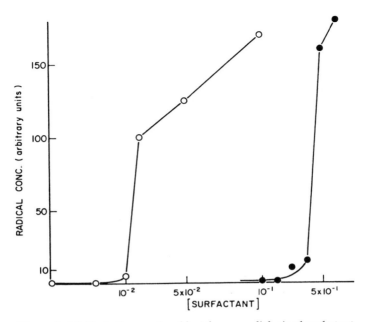

Figure 1. Yield of nitroso spin adduct from γ radiolysis of surfactant in the presence of 10^{-2} M nitroso-tert-butane. CMCs are 8×10^{-3} and 1.4×10^{-1} M for NaLS (○) and NaOS (●), respectively. The experimental variation in the data is ±5% (8).

The surfactant radical is produced via radiolysis of the N_2O-saturated aqueous phase. The radiolysis produces e_{aq}^- and OH radicals, which carry out the following reactions,

$$H_2O \rightsquigarrow e_{aq}^- + OH + H^+$$

$$e_{aq}^- + N_2O \xrightarrow{H^+} OH + N_2$$

$$OH + surfactant \longrightarrow R' + H_2O$$

The unstable R' is stabilized by reaction with the spin trap and produces the radical R–N(–B)–O'. This reaction does not proceed until a surfactant concentration of less than 10^{-2} M is achieved with sodium lauryl sulfate (NaLS) or above 0.2 M for sodium octyl sulfate. Beyond these concentrations, which are the critical micelle concentrations (CMC) of the surfactants, efficient reaction is observed. The data show conclusively that the monomeric form of the surfactant does not react with the spin trap. However, the confinement of R' and the spin trap to a micelle greatly promotes the reaction of the two species.

Useful information on surfactant monomer exchange with micelles,

$$S + (S)_{n-1} \rightleftharpoons (S)_n$$

can be obtained from pulse radiolysis studies of micellar systems. For example, pulse radiolysis of H_2O-saturated solutions of sodium 4-(6-dodecyl)benzenesulfonate produces OH radicals. These radicals subsequently attack and add to the benzene ring of the surfactant and form a hydroxycyclohexadienyl radical (9). Pulse radiolysis of homogeneous solutions of benzene in water (10) has shown that the OH–benzene adduct has a strong absorption at λ = 3200 Å. Derivatives of benzene behave in a similar fashion. In direct observations of the surfactant radical, the radicals dimerized below the CMC with a rate constant of $2k = 4.2 \times 10^8$ $M^{-1}s^{-1}$. The rate decreases above the CMC as the radicals became associated with individual micelles. Analysis of the data gives a rate constant $k = 4 \times 10^8$ $M^{-1}s^{-1}$ for the association of the free-radical surfactant with a micelle and $k = 8 \times 10^5$ $M^{-1}s^{-1}$ for the reverse reaction, that is, exit of the free radical from the micelle. The data agree with the corresponding rates for association and dissociation of a monomer surfactant and micelles. Similar studies were carried out with the surfactants, $C_{16}H_{33}(OCH_2CH_2)_{21}OH$ and $C_{14}H_{29}(OCH_2CH_2)_{23}SO_3Na$. Radiolysis of these surfactants leads to cross-linking of the radicals in the micelles and eventually leads to polymerized material (11, 12). Enhanced radical dimerization in a single micelle was also observed and produced radical lifetimes of 10^{-6} s or less, as compared to 10^{-3} s in a homogeneous solution.

The reactions of the surfactant radicals produced in the aqueous bulk of a system with added free-radical scavengers such as $Fe(CN)_6^{3-}$ or benzoquinone are also modified by micelles. This modification is due to competition of micelles and the scavengers for the radicals.

$$R^{\cdot} + \text{micelle} \underset{}{\overset{k_1}{\rightleftharpoons}} (R^{\cdot} \text{ micelle})$$

$$R^{\cdot} + Fe(CN)_6^{3-} \xrightarrow{k_{-1}} R^+ + Fe(CN)_6^{4-}$$

$$R^{\cdot} + \text{quinone} \xrightarrow{k_2} R^+ + (\text{quinone})^-$$

Radicals captured by the micelle react much more slowly with solutes that dissolve primarily in the aqueous phase. Typical data for the reaction of NaLS molecules with benzoquinone are shown in Figure 2. In an aqueous solution and in the absence of micelles, the growth of the reaction product, the quinone anion BQ^-, is a smooth curve. In the presence of micelles, an initial rapid production of BQ^- is followed

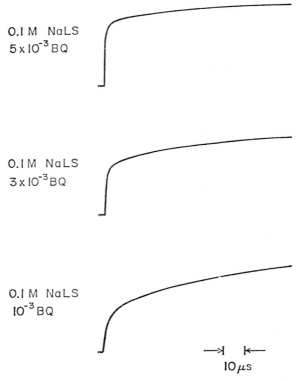

Figure 2. Oscilloscope traces of the reaction of NaLS radicals with benzoquinone in NaLS micelles. The reaction
product, the benzoquinone anion, was observed at λ =
430 nm.

by a slow growth rate. These rates are described by the following
equations:

$$\text{rate} = \frac{d(BQ^-)}{dt} = k_2(R^·)[BQ] \text{ for a homogeneous system}$$

$$\text{rate} = \frac{d(BQ^-)}{dt} = k_2(R^·)[BQ] + k_1(R^·)(\text{micelle}) \text{ for a micellar system}$$

The subsequent slow reaction rate is due to the reaction of micellar R$^·$
and BQ. The extent of the two processes depends on the association
rate constants k_1 and k_2 and the micelle and BQ concentrations. This
dependence is a general phenomenon of micellar systems; by using the
scavenger $Fe(CN)_6^{3-}$ we can measure the association rate of R$^·$ with
the NaLS micelle. In this case, k_1 is 1.5×10^9 M^{-1}s^{-1}, and the dissociation rate constant (k_{-1}) is 1.8×10^5 s^{-1}, in agreement with other

data (14). This system has also been used to measure the distribution of pentanol between NaLS micelles and water, and a K of $25N$ is obtained, where N is the aggregation of the micelle ($N = 60$).

In some instances, the counterion of the surfactant is reactive with radicals. A typical example is the pulse radiolysis of H_2O-saturated CTAB solutions. The OH radicals thus produced react with Br^- and form Br atoms,

$$OH + Br^- \rightarrow Br + OH^-$$

and the Br atoms form an equilibrium with Br^- ions to produce Br_2^-.

$$Br + Br^- \rightleftharpoons Br_2^-$$

The Br_2^- radical ion is readily identified by its strong absorption at 3650 Å (15); other halides behave similarly (16, 17). Below the CMC of CTAB the Br_2^- is formed uniformly and decays via second-order kinetics,

$$Br_2^- + Br_2^- \rightarrow 2Br^- + Br_2$$

Above the CMC, Br_2^- is formed rapidly; and, depending on the micelle and Br_2^- concentrations, an initial rapid decay of Br_2^- is followed by a slower decay (18). The initial rapid decay is due to the reaction of two or more Br_2^- radicals at one micelle. These anions are held at the positive surface of the micelle, where their local concentration is very high. The slower decay is due to the reaction of Br_2^- in the bulk of the system. The residence time of Br_2^- on the micellar surface is 1.5 $\times 10^{-5}$ s (19).

Hydrogen Abstraction Reactions; Photochemical Systems. Surfactant radicals are also produced when benzophenone derivatives are photolyzed in micellar media (20, 21). Photolysis produces the triplet excited state of the benzophenone derivative, which subsequently abstracts an H atom from the surfactant methylene chain.

Photolysis in CTAB micelles leads to attack on all methylene groups except the last three C atoms of the chain. A more pronounced attack is seen at the methylene group β to the head group. However, this effect is more pronounced below the CMC. This finding indicates that the COO⁻ group locates the benzophenone derivative in the Stern layer and leads to pronounced attack at the β cation. However, a complex of the ketone and the monomeric surfactant must also occur by electrostatic attraction, which leads to an even more precise location of the ketone close to the surfactant head group. Photolysis of CTAB and derivatives of benzophenone with longer methylene chains connecting the ketone to the COO⁻ group, leads to less specificity of reaction,

probably a result of chain folding of the probe molecule. Photolysis of the benzophenone derivative in sodium tetradecanoate micelles leads to strong attack on the fifth, sixth, seventh, and eighth methylene groups of the C_{14} chain, but attack at the head group region and the tail is very small. This finding indicates an extension of the C=O chromophore into the micelle in a specific fashion with little random interaction.

Intramolecular Hydrogen Abstraction. Excitation of ketones containing one lone methylene chain leads to Norrish type II reactions, where the excited triplet carbonyl abstracts an H atom from the methylene chain (22),

The products depend markedly on solvent. Photolysis of phenyl heptyl ketone in CTAB micelles leads to a product quantum yield of $\phi = 0.7$, and the ratio of the cyclobutanols I/II is 1.2 (23). The data compare well to a product quantum yield of $\phi = 1.0$ and I/II = 1.5/1 found for photolysis of this ketone in *tert*-butyl alcohol, and are quite unlike the data in benzene, where $\phi = 0.33$, and I/II = 4.7/1. The data suggest a polar environment for the ketone, probably in the surface area of the micelle.

Very long chain derivatives are often coiled up in a micelle. For example, photolysis of the molecule 16-oxo-16-(*p*-tolyl)hexadecanoic acid in sodium lauryl sulfate leads to a type II reaction with $\phi = 0.8$ (compared to 0.2 in benzene); the product was 4-methylacetophenone (24). Both the COO^- group and the tolyl ketone group are probably located at the micelle surface, and the methylene chain considerably bends in the micelle. Photolysis of this compound in the stretched configuration, as in a bilayer system, leads to $\phi = 10^{-4}$.

Radical Reactions in Micelles and Isotope Enrichment. Photolysis of aromatic ketones in micellar solutions has been used to obtain $^{13}C/^{12}C$ isotope enrichment by use of the well-known chemically induced dynamic nuclear polarization (CIDNP) effect (25). Photolysis of dibenzyl ketone (DBK) in benzene solution leads to a small $^{13}C/^{12}C$ isotope enrichment. The reactions cited are as follows:

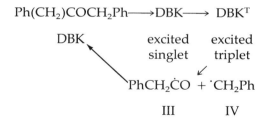

Radicals III and IV subsequently recombine and reform DBK. However, other products are possibly formed by a scrambling of the radicals III and IV. If the radicals are confined to a cage-type environment, a micelle, then radical recombination is much more rapid than in a homogeneous solution. The radicals containing ^{13}C relax their spin configurations more rapidly than those containing ^{12}C, and a DBK product enriched in ^{13}C is formed. The ^{12}C radicals, which recombine more slowly, do so in the solvent bulk, and give rise to products enriched in ^{12}C. The recombination of radicals III and IV in the micelle must involve an intersystem crossing from triplet character to singlet character in order for a stable bond to be formed in the final products. The nuclear hyperfine coupling (HFC) in the cage for III and IV is due to the coupling of the odd electron of the radical pair to a ^{13}C nucleus. This coupling enhances intersystem crossing and promotes a singlet product. In the absence of HFC, the radicals are scavenged, that is, they decarboxylate at rates faster than they are converted to products.

A substantial micelle decarboxylation that produces a triplet gem-inate radical pair (PhĊH$_2$—ĊH$_2$Ph) occurs. After intersystem crossing, these radicals form diphenylethane. Some radicals exit the micelle and can be scavenged by Cu^{2+} radicals in the aqueous bulk. In the presence of a small magnetic field (~200 gauss) the geminate-cage reaction drops from a yield of 0.06 to 0.02. The magnetic field splits the triplet levels of the radical pair and decreases intersystem crossing; hence, it de-creases the probability of cage recombination. Micelles can thus provide a useful vehicle for the promotion of enrichment of important isotopes.

Other Radical Reactions. Many intermolecular cycloaddition reactions are facilitated in micelles because of the local high concentration of reactants. For example, excitation of the 2-methylpropene–cyclohex-enone micelle system leads to addition products V and VI, whereas excitation in homogeneous solution leads to reaction of the excited enone with the solvent (26).

Other examples (27–30) of enhanced photoinduced cyclization reac-tions in micellar systems exist. The micelle exhibits proximity effects for the reactants and promotes cyclization reactions at the expense of bulk reactions. The proximity effect is also evident in the photoinduced substitution of CN$^-$ and 4-methoxy-1-nitronaphthalene.

The quantum yield of this process is almost zero in water and NaLS micelles. A quantum yield of 0.2 is achieved in CTAC micelles where the naphthalene compound is solubilized in the micelle surface and the CN^- is held close to the surface by electrostatic attraction to the quaternary ammonium head group.

Reactions of Singlet Oxygen

Singlet oxygen (ΔO_2^1) may be used as a probe molecule for micellar systems, and the information gained may be carried over to oxygen itself. This fact is an asset in many instances because the well-established reactions of ΔO_2^1 enable an easy observation of its kinetic progress. Photolysis of methylene blue in oxygenated water leads to the triplet excited state and a transfer of energy to O_2. This reaction produces the singlet O_2 (ΔO_2^1).

Triplet Excited State

$$(CH_3)N_2 \quad \text{—} \quad ^+N(CH_3)_2 \, Cl^- + O_2$$

$$\Delta O_2' + \quad \text{Ground state of methylene blue}$$

The ΔO_2^1 is observed by its specific reaction with 2,5-diphenylisobenzofuran, which leads to reduction of this compound. Photolysis of the system in micellar solution establishes a system where the benzofuran is in the micellar phase and the O_2 and methylene blue are in the aqueous phase. Reduction of benzofuran by ΔO_2^1 produced in the aqueous phase shows that ΔO_2^1 and, presumably, ΔO_2^1 also rapidly enter and exit the micellar system (31,32). The data are in agreement with other O_2-quenching studies.

Energy Transfer

The close proximity of two reactants in a micelle leads to efficient energy transfer from an excited molecule to an acceptor. A good example of such a process is provided by energy transfer from the excited phenyl group of sodium phenyl undecanoate micelles to a guest molecule such as naphthalene. In Figure 3, the naphthalene fluorescence intensity is

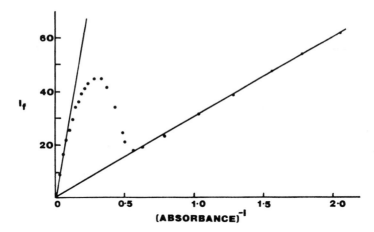

Figure 3. Energy transfer from the phenyl group of phenylundecenoic acid to naphthalene (108).

plotted versus the inverse of the total absorbance of the solution at an excitation wavelength of 2600 Å, where both chromophores absorb. The naphthalene concentration is kept constant, and the surfactant or phenyl group concentration is varied. Increasing the phenyl concentration at low surfactant concentration (to the right of the figure) decreases the naphthalene fluorescence because the excitation light is increasingly absorbed by the phenyl group. At the CMC, the fluorescence suddenly increases when naphthalene is incorporated into the micelles in close proximity to the excited phenyl group; this situation produces an efficient energy transfer. At higher surfactant concentrations (as the micelle concentration increases) the mean occupancy of the micelle by naphthalene molecules drops below unity, and some micelles do not contain naphthalene, a situation that produces inefficient energy transfer. Only micelles that contain naphthalene can produce energy transfer. The situation is a system in which an antenna molecule, such as the phenyl group, increases the amount of energy centered in the guest molecule, naphthalene. It is analogous to the photomechanism of the photosynthetic unit.

Similar studies have been carried out with Igepal EO-630 and Triton X-100 micelles, which incorporate pyrene as a guest molecule (33). A strong overlap of the micellar phenoxyl fluorescence exists within the pyrene absorption spectrum, a situation that leads to efficient energy transfer via a Förster mechanism. The average separation of the phenoxyl group and pyrene is approximately 15 Å. This separation locates the pyrene within the micelle and may suggest that the guest molecule is contained in the mantle region of the particle, because the maximum radius of the hydrocarbon core is only 15 Å. The pyrene III/I fluorescence ratio also indicates a polar environment for the guest molecule, more polar than that observed in NaLS micelles. Because the III/I ratio is similar to the ratio observed in polyethylene oxide solvents, the data confirm a location for guest molecules primarily in the mantle of the micelle.

Photosynthetic pigments, such as chlorophyll a or b and lutein, may be solubilized in micelles up to concentrations approaching those in photosynthetic membranes. Energy transfer processes similar to those postulated in chloroplasts have been observed in these systems (34, 35). Excitation energy transfer for methylene blue to thionine, both of which are solubilized by anionic micelles, is quite efficient, and each acceptor may quench five to seven donor molecules (36, 37).

Micelles enhance the efficiency of energy transfer from excited triplet hydrocarbons to inorganic ions such as Tb^{3+} and Eu^{3+} (38, 39). The following rate constants for triplet energy were measured in NaLS micelles:

$$k \text{ (1-bromonaphthalene} \rightarrow Eu^{3+}) = 5 \times 10^5 \text{ M}^{-1}\text{s}^1$$
$$k \text{ (1-bromonaphthalene} \rightarrow Tb^{3+}) = 1.8 \times 10^5 \text{ M}^{-1}\text{s}^1$$
$$k \text{ (biphenyl} \rightarrow Tb^{3+}) = 4 \times 10^5 \text{ M}^{-1}\text{s}^1$$

The rate constants were calculated from the observed rates of decay of the excited aromatic triplet states, or the observed rate of growth of Tb^{3+} emission, together with the effective concentration of metal ion at the micelle surface. The latter perimeter is calculated by assuming that the rare earth ions are contained in a spherical shell of width 2 Å centered around a sphere of 19.5 Å. The number of ions in this vicinity is calculated from the known binding constant of the ion to the anionic micelle. The distribution of ions at the micelle surface (shown in Figure 4) is calculated from a micelle surface potential suggested by Aniansson (40). This figure also shows the limits of contact between aromatic and rare earth ions; 50% of the aromatic molecules are assumed to be at the micelle surface. The solid line indicates the distribution of aromatic molecules in the micelle, from the surface to the core center. A steric factor of one-third indicates that the reactants are somewhat shielded from one another by the surfactant. The rate constants are very low and in agreement with similar reactions reported in homogeneous media (41). The low rate of energy transfer could be due to the weak interaction between the unpaired electrons of the aromatic triplet and the rare earth ion, which leads to a weak resonance exchange (42). Transfer of energy from aromatic ketones and aldehydes to rare earth ions involves an exchange of a water molecule in the hydration sphere of the ion (43–45a). Such a mechanism could be quite slow in the case

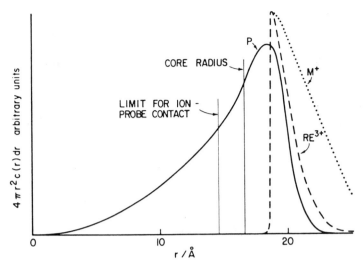

Figure 4. A schematic representation of the possible distributions in and at a NaLS micelle of a probe with slight preference for the surface (—), and ions with a charge of +3 (---), and +1 (···). The probe radius is assumed to be 2 Å and the ion radii are 2.3 Å (39).

of excited arene–rare earth ion energy transfer. Convenient observation of these slow energy-transfer processes in homogeneous solution is not possible because of quenching of the aromatic triplet in the solution bulk by impurities. The micelles protect the aromatic triplet and selectively promote the desired energy-transfer process.

Studies of metal-ion quenching of aromatic excited states have been reported (45a); however, only efficient quenching reactions have been studied, the quenching of excited naphthalenes by Cu^{2+}, Ag^+, etc. In anionic micelles, the quenching reaction occurs at the micelle surface, the site of binding of the inorganic cation. The binding process is electrostatic in nature (45a); the more tightly charged cations are more strongly bound to the anionic surface. The following progression is suggested for the efficiency of binding: $Eu^{3+} > Cu^{2+} > Ni^{2+} > Mn^{2+}$, $> Fe^{2+} > Tl^+$, $> Ag^+$. The following equilibrium constants are reported for the binding of Ag^+, Tl^+, and Cu^{2+} to NaLS micelles: $K_{Ag^+} = 1.3 \times 10^3 \ M^{-1}$, $K_{Tl^+} = 1.1 \times 10^3 \ M^{-1}$, and $K_{Cu^{2+}} = 6 \times 10^4 \ M^{-1}$. The quenching process is somewhat complicated by migration of the inorganic cation from one micelle to another upon micellar collision. An exchange of Ag^{2+} from one micelle to another occurs with a k_e of 2.8 $\times 10^8 \ M^{-1}s^{-1}$, as defined by the following equilibrium (45d):

$$(micelle)Ag^{2+} + (micelle) \xrightarrow{k_e} (micelle) + Ag^{2+}(micelle)$$

The exchange rate constant (k_e) is somewhat faster for Cu^{2+} and Eu^{3+} (45c, 45d), where $k_e = 3.3 \times 10^9$ $M^{-1}s^{-1}$, and $k_e = 2.6 \times 10^9$ $M^{-1}s^{-1}$, respectively. These values are lower than the calculated rate constant for collision of micelles, which is 1.3×10^{10} $M^{-1}s^{-1}$. However, several assumptions, such as the rate of micelle diffusion, effective collision radius, etc., have been used in this calculation; hence, the final value is not precise.

Triplet energy transfer from triplet N-methylphenothiazine $(MPTH^T)$ to trans-stilbene and naphthalene has been observed in CTAB micelles (45b). In the $MPTH^T$–trans-stilbene pair the transfer is irreversible, and the transfer rate constant is $k_q = 1.5 \times 10^7$ $M^{-1}s^{-1}$ for a micelle containing one of each reactant. This rate compares well to $k_q = 1.4 \times 10^7$ $M^{-1}s^{-1}$ for the annihilation of two 1-bromonaphthalene triplets and indicates a value for an efficient diffusion-controlled triplet reaction in CTAB micelles. The $MPTH^T$–naphthalene system is reversible; the forward rate-of-energy transfer, k_q, is 2.8×10^6 $M^{-1}s^{-1}$, and the reversed reaction is $k_{-q} = 3.3 \times 10^6$ $M^{-1}s^{-1}$. Compare these rate constants to the following data in homogeneous ethanol–water solution: $k_q = 1.8 \times 10^9$ $M^{-1}s^{-1}$ and $k_{-q} = 1 \times 10^9$ $M^{-1}s^{-1}$. The ratio k_q/k_{-q} in the micelle system is about twice the ratio in ethanol–water mixtures. The energy of $MPTH^T$ may be lower by 1–9 kJ/mol in a micelle compared to a water–ethanol mixture; such an energy difference would account for the difference in the rate-constant data.

Quenching of Excited States

The fluorescence of an aromatic probe in a micelle is often quenched without energy transfer. However, in some instances electron transfer does occur and a special product of the quenching reaction is observed, as in the case of excimer formation.

Excimer Formation. The interaction of the excited singlet state of a molecule S* with the ground state S can give rise to an excited dimer or excimer S_2* (46).

$$S^* + S \rightleftharpoons S_2^*$$

These entities are characterized by broad, structureless emissions that are red shifted from that of the monomer fluorescence. Figure 5 shows data for the pyrene system, which is one of the best examples of this type of effect. The highly structured monomer emission is centered around 3900 Å, and the broad excimer emission is at 4800 Å. The formation of an excimer requires the mutual approach of the excited state and ground state to form an encounter complex. Excimer formation

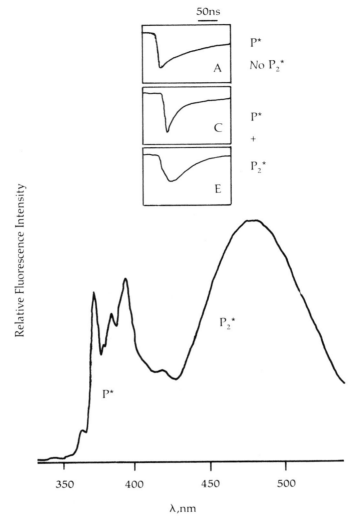

Figure 5. Fluorescence spectra of pyrene monomer (~ 400 nm) and pyrene excimer (~ 450 nm) in CTAB micelles. Insert: Rate of decay of monomer at 400 nm and growth and decay of excimer at 480 nm.

can be very efficient and may occur at every collision, as in the case of pyrene. Excimer formation often competes successfully with other dynamic processes of the excited singlet state such as fluorescence and intersystem crossing. Hence, the yield of excimer can be used to reflect the restrictions imposed on solute movement by the solvent, or the viscosity in which S^* and S are diffusing.

Excimer emission was one of the earliest photochemical observations in micellar systems (47); excimer formation is often enhanced in micellar systems where S* and S are in close proximity. Initial interpretations of excimer yields for micellar microviscosities (η) gave unusually large η values (48), especially when compared to fluorescence polarization work (49). These techniques actually measure different motions of the probe molecules; the excimer method monitors lateral diffusion, and the polarization method measures mainly rotational diffusion. A second feature that clouds the results is the failure to take into account a random distribution (e.g., Poisson) of probe molecules among the micelles. This latter effect has been studied with success recently and will be discussed.

Probe molecules, such as pyrene, will be distributed among micelles according to a Poisson distribution; possibly some micelles will contain many pyrenes, some will contain a single pyrene, and some will be empty. Excimer formation is extremely rapid in multiply occupied micelles, but quite inefficient in other micelles, because transport of pyrene from one micelle to another is very slow compared to the excited-state lifetime. Typical data illustrating this point are shown in Figure 5 (insert) for pyrene, where the initial decay of P*, giving P_2* via

$$P^* + P \rightleftharpoons P_2^*$$

occurs in multiply occupied micelles and leads to an initial rapid decay of P* followed by a slower decay for the situation of one P* per micelle. The system may be analyzed exactly to give several parameters of interest. However, a more detailed discussion will follow shortly.

General Quenching. Table I shows the rate constants for quenching of excited states in various micelles and with several different types of quencher molecules. Several features are immediately ascertained from the data. The quenching-rate constant of a probe molecule, located in the micelle, by a noncharged molecule, located primarily in the aqueous phase, is less than that observed in a homogeneous solvent such as water or methanol. These effects are not particularly large, certainly less than an order of magnitude, and could be readily explained in terms of a screening of the reactants by the surfactant molecules (Table I).

Reactions involving charged quenching molecules show much larger effects. If the charge of the quencher is the same as that of the micelle, then the micellar quenching rate constant is many orders of magnitudes smaller than that observed in homogeneous solution. Addition of an inert salt (e.g., NaCl) to NaLS micelles markedly reduces this retardation effect of the micelle. On the other hand, if the charge of the quenching

Table I.
Rate Constants (10^9 M^{-1} s^{-1}) for Reaction in Micelles

Reaction	Water or Alcohol	NaLS	NaTC	NaTC/3% Cholesterol	Sodium Oleate	CetMe$_3$NBr	Brij 35	Igepal Co 630	Triton X-100	Bilayer Distearyl-lecithin	Lyso-lecithin Micelle
Py* + O$_2$	20.9	9.2	4.6	7.3	7.0	3.5 (4.3)[a]	2.4	5.5	5.7	2.7	7.2
Py* + CH$_3$NO$_2$	8.1	3.0	0.12			1.8	1.7	0.53	0.51	0.055	0.57
Py* + Cu^{2+}	20.00	$<5 \times 10^8$ s^{-1}				$<10^{-3}$	2.7				
Py* + I$^-$	3.0 (1.9 × 10^{-2})	2.6 × 10^{-3}	3.6 × 10^{-2} (0.38)[b] (2.2 × 10^{-2})[c] (7.3 × 10^{-2})[d]	5.4 × 10^{-2}	8.2 × 10^{-2} [NaCl = 0] 0.23 (NaCl = 0.15) 0.13	66	0.54	0.40	0.69	$<10^{-2}$	0.51
Py* + triethylamine	0.3	0.25 0.04[c] 0.40[b]	1.4 × 10^{-2} (0.16)[b]	4.2 × 10^{-2}							
Py* + Tl$^+$	5.0		$r < 6 \times 10^{-8}$								
Py* + methyl viologen	7.2					3.1 × 10^{-2}					
Py* + cetylpyridinium chloride	7.4 (9 × 10^{-3} s^{-1})										
Py* + Py	1.7 × 10^{-2} s^{-1}					1.2 × 10^{-2} s^{-1}					
Py* + Py						4.5 × 10^{-3} s^{-1}					
PBA* + PBA						1.0 × 10^{-2} s^{-1}					
PBA* + I$^-$	0.5										
Py* + RNO	1.5 × 10^{-2} s^{-1}										

Bromonaphthalene* + Eu^{3+}	5×10^{-4}	
Tb^{3+}	1.8×10^{-4}	1.7
Bromobiphenyl* + Tb^{3+}	4.5×10^{5}	
Duroquinone† + DPA	6×10^{6} s^{-1} (monomer)	
OH	7.6 (monomer)	
	0.5 (micelle)	
OH + Py	~2	~3
OH + benzene	1.2×10^{-2}	1.6×10^{-2}
	(monomer and micelle)	(monomer and micelle)
H	6.0	6.0
H + biphenyl	6.0	6.0
O$_2$* + α-tocopherol (singlet O$_2$)	0.64	
e^{-} Transfer		
Py* + DMA	>1.0 s^{-1} >1.0 s^{-1}	>1.0 s^{-1}
Cu^{+} + (methylphenothiazine)$^{+}$	9.0 s^{-1}	
Ni^{2+} + (methylphenothiazine)$^{+}$	2.1×10^{-2} s^{-1}	
Co^{2+} + (methylphenothiazine)$^{+}$	4.2×10^{-2} s^{-1}	
Eu^{2+} + (methylphenothiazine)$^{+}$	~10^{-2} s^{-1}	
(RuP^{2+})* + methylphenothiazine	3×10^{-4} s^{-1}	

NOTE: Some rate constants are given in units of s^{-1} when the reactants are both in the micelle and the concentration is uncertain.

[a] Rod, micelle.
[b] 0.2 M benzyl alcohol.
[c] 4×10^{-2} M MgCl$_2$.
[d] 0.1 M NaCl.

molecule is the opposite of that in the micelle, then the quenching-rate constant is larger than that observed in homogeneous solution. This result is seen because the rate constants are calculated from the experimentally observed rate of decay of the excited state and the bulk, or overall, quencher concentration. For micelles and quenchers of opposite charge or an ion-charged quencher that is solubilized by the micelle, locally high quencher concentrations are obtained close to the micelle (i.e., the site of residence of the probe molecule).

The decreased rate of reaction for a micelle and quencher of similar charge can be dealt with by a simple electrostatic repulsion calculation. For example, the rate of the reaction should be decreased by a factor

$$\frac{(Z_e)^2}{rEkT} \bigg/ \left(\exp \left[\frac{(Z_e)^2}{rEkT} \right] - 1 \right)$$

where Z_e is the effective charge of the micelle and r is the interaction distance. The choice of Z_e and r represents some problems, and introduces uncertainty into any a priori consideration of the system. For example, the exact location of the probe molecule in the micelle, whether on the surface or further away from the surface, can change r dramatically. Such effects are reflected by large changes in the calculation. The effect is readily seen experimentally for the reaction of e_{aq}^- with pyrene and aminopyrene. In the reaction the ratio of the rate constant in a homogeneous solution to that in a micelle [k (homo)/k (micelle)] is > 1000 for pyrene and about 20 for aminopyrene. The aminopyrene is located closer to the water bulk than pyrene; hence, e_{aq}^- does not have to penetrate as close to the micelle for reaction to occur and experiences less electrostatic repulsion.

A more conventional interpretation may be given in terms of a double-layer theory. The electrical charge ψ in a solution surrounding a spherical charged micelle is given by

$$(r^{-2}d/dr)\,(r^2 d\psi/dr) = (8\pi e n_{\pm} Z_{\pm}/\varepsilon)\,\sinh\,(Z_{\pm}e\psi/kt)$$

where r is the distance from the micelle center; e is the elementary charge; $n_{\pm}$ is the number of ions per milliliter; Z_{+} and Z_{-} are the charges on the ions; and ε is the dielectric constant. Solutions to this equation lead to the construction of the curves in Figure 6. This figure shows the potential $\psi e/_{kT}$ as a function of distance in ångströms from the micelle surface ($r - r_o$) for several different solution ionic strengths; r_o is the micelle radius and r is the distance from the micelle center. The parameter C_r/C_b is also shown where C_r is the ion concentration

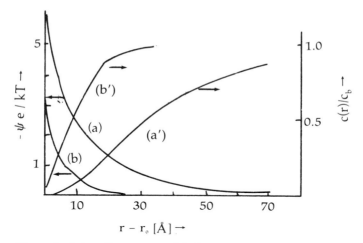

Figure 6. Left ordinate, potential distance function in the diffuse double layer of NaLS micelles: (a) μ = 0.008 M; (b) μ = 0.1 M. Right ordinate, distribution function for monovalent negative ions: (a') μ = 0.008 M; (b') μ = 0.1 M (66).

at distance r, and C_b is the bulk ion concentration for an ion with similar sign charge to the micelle surface.

Increasing the Ion Strength. To calculate C_r/C_b we used the following equation:

$$C_r = C_b \exp\left[(\mp) Z_{\pm} e\psi/kT\right]$$

The ratio C_r/C_b represents the probability of finding an ion at $r - r_o$ from the micelle surface compared to that in the bulk phase. Increasing ionic strength makes the potential much stronger at the micelle surface and markedly increases C_r/C_b at small values of $r - r_o$. Hence, the approach of an ion of similar charge to the micelle to the micelle surface is greatly enhanced by an increase in ionic strength.

The previous discussion was simplified by the assumption that a quencher molecule was located primarily in the aqueous phase and approached the micelle for reaction to occur. This situation is often approximated by the use of very polar molecules such as nitromethane or ions. However, in many instances the approximation is not valid, and a model is necessary in which the quencher molecule may reside in the micelle and in the aqueous phase. In some instances, the quencher molecule may reside primarily in the micellar phase, and reaction only occurs by movement of the probe and quencher in the micellar phase.

A General Picture of Micellar Reaction

A solubilized molecule is distributed among the micelles according to a Poisson distribution, where P_n gives the probability of finding n molecules per micelle and

$$P_n = \frac{\langle Q \rangle^n}{n!} e^{-\langle Q \rangle}$$

$\langle Q \rangle$ is the ratio [probe molecule]/[micelle]. For example, if $\langle Q \rangle$ is 1 (i.e., one probe molecule per micelle), a situation often encountered in this work, then the probabilities of finding 0 (i.e., an empty micelle), 1, 2, and >2 molecules/micelle are 0.37, 0.37, 0.18, and 0.08, respectively. This complex situation has to be built into any interpretation of the kinetics either by the above Poisson equation, or by a kinetic sequence of reactions that describe such a distribution.

In the situation where both the fluorescent probe molecule P and the quencher molecule $\langle Q \rangle$ reside in the micelle, the efficiency of the quenching kinetics depends on the ratio of the concentration of quencher to micelle and on the efficiency of reaction of the two reactants in the micelle. This latter effect depends on the efficiency of reaction on encounter and on the rate of diffusion of the reactants around the micelle. The chance of encounters of probe and quencher is increased by an increase in the number of quencher molecules per micelle. At high values of $\langle Q \rangle$, the rate of reaction should vary as [Q] varies (51, 52). These important observations indicate that the micelle properties do not significantly change over a range of [Q] and $\langle Q \rangle$ used, and that the micelle reaction system can be looked upon as a true microscopic representation of the whole system. Furthermore, reaction parameters determined in the bulk phase may be adapted for a description of the micromicelle assembly.

At low $\langle Q \rangle$, a large fraction of micelles are empty. For $\langle Q \rangle = 1$, 37% of the micelles contain no Q, and 37% contain one Q molecule. Excited solutes present in micelles where [Q] = 0 or $\langle Q \rangle = 0$ will be unperturbed, and varying kinetic conditions will prevail in micelles where $\langle Q \rangle \geq 1$. The decay of the probe molecule will be a series of exponentials and not a single exponential. This result is compared to the condition where [Q] = 0, or a homogeneous solution where [Q] >> [P] and where pseudo first-order conditions prevail. Typical data for a homogeneous solution and for a micellar system for the following reaction are illustrated in Figure 7.

$$P^* + Q \rightarrow P + Q$$

where [Q] >> [P*]. The figure also shows the effect such conditions

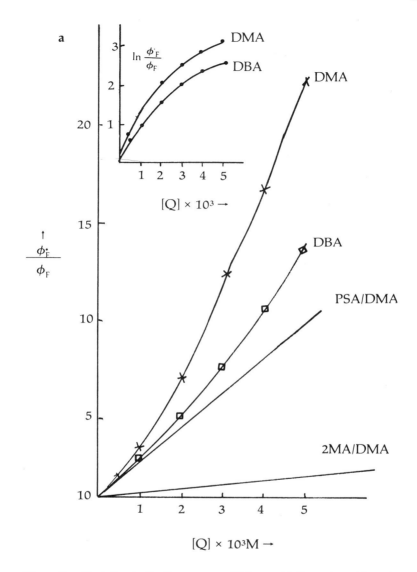

Figure 7a. Variation in the fluorescence efficiency, ϕ^0_F/ϕ_F, *where* ϕ^0_F *and* ϕ_F *are the fluorescence efficiencies in the absence and presence of scavenger, respectively, vs. the scavenger concentration. The luminescence probes used are pyrene, pyrenesulfonic acid, and 2-methylanthracene. The quenchers are dimethylaniline (DMA) and dibutylaniline (DBA). Insert: Plot of* ln ϕ^0_F/ϕ_F *vs.* [DMA] *and* [DBA] *(80).*

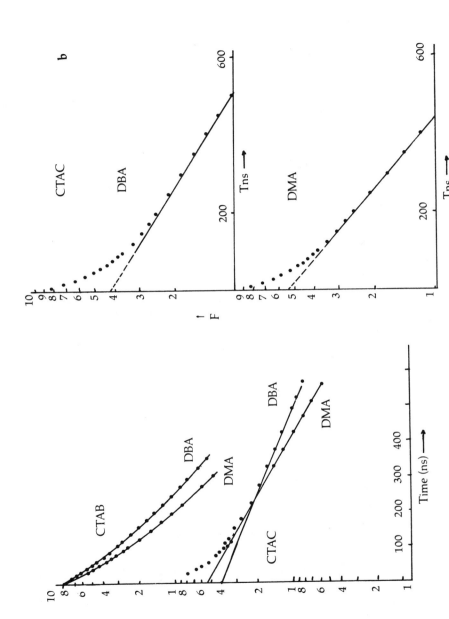

Figure 7b. Rate of decay of pyrene fluorescence plotted as log F vs. time for CTAB and CTAC micelles and for the systems pyrene–DBA and pyrene–DMA (80).

have on the typical Stern–Volmer plot for steady state observation

$$I_o/I = 1 + \frac{\beta\,[Q]}{\alpha}$$

where I_o and I are the intensities of probe fluorescence in the absence and presence of Q and β and α are the rate constants for reaction of Q with P* and the natural decay of P*, respectively.

The probe intensities may also be expressed as a function of time, t, as follows (53).

$$I = I_o \exp\left[-k_1 t - \bar{n}(1 - e^{-k_q t})\right]$$

where

$$k_1 = \alpha + \beta\,[Q]_{bulk}$$

and $[Q]_{bulk}$ is the concentration of Q in the bulk phase. This expression takes into account homogeneous quenching of P* by the bulk phase. The other parameters are $\bar{n} = [Q]_m/[micelle]$, where $[Q]_m$ is the [quencher] in the micelle, and k_q is the first-order intramicellar quenching rate constant for a micelle containing one quencher and one excited probe. This expression faithfully portrays many quenching reactions in micellar systems. The first-order parameter k_q is particularly useful because it may be converted to a second-order parameter by the assignment of a value $[Q]_M$. Direct comparison of the converted rate, $k_q/[Q]_m$ with β, the homogeneous quenching-rate constant, demonstrates the obstacles to diffusion that the micelle presents to P* and Q. The data can also be used to measure $\bar{n}$ and, hence, the aggregation number of the micelle.

A special condition can sometimes be achieved to aid in the measurement of the micellar aggregation number. If the intramicellar quenching rate is faster than the natural decay of the fluorescent probe, a condition fulfilled in the tris(bipyridyl)ruthenium and methylanthracene (MA) system (55),

$$(RuII)^* + MA \rightarrow RuII + MA^*$$

then simple steady state measurements gave a reliable estimate of the micellar aggregation number N. The Poisson distribution gives the probability that a micelle will contain zero quencher molecules,

$$P_o = \exp -\langle Q \rangle$$

where $\langle Q \rangle$, as before, is the

$$\frac{[Quencher]}{[Micelle]} = [Quencher] \Big/ \frac{[Surfactant] - CMC}{N}$$

This probability P_o is also I/I_o, where I and I_o are the (RuII)* luminescence intensities in the presence and absence of Q, respectively. Hence,

$$I/I_o = exp-\langle Q \rangle$$

$$\ln(I/I_o) = \frac{[Q]N}{[\text{Surfactant}] - \text{CMC}}$$

The condition that the rate of quenching is much faster than the natural decay rate of the probe is a stringent one. In some cases the conditions are not met, but N can still be obtained by a more rigorous examination of the fluorescence data (54, 56, 67).

Kinetic Approach. Micellar dynamics may also be approached with a completely kinetic explanation (58, 59). The quencher Q is considered to be in dynamic equilibrium with the micelle M and water phases,

$$Q + M \underset{}{\overset{\alpha}{\rightleftharpoons}} QM$$

and

$$P^* + Q \xrightarrow{k_1} P + Q$$

the rate constants for the processes are k_1, α, and β, where the distribution constant K for pyrene between the water and micelle is α/β.

The excited state P* decays via fluorescence emission $F_t \, dt$, and quenching $Q_t dt$

$$-\frac{d\,P^*}{dt} = F_t \, dt + Q_t \, dt$$

where

$$F_t = k_1 \, P^* \, dt$$

and $Q_t \, dt$ is the number of moles of P* that disappear during the time window t to $t + dt$ as a result of the quenching reaction. The number of molecules of Q that enter the micelles in the time interval t to $t_1 + dt_1$ is $\alpha \, [Q]_w \, [M] \, dt$, where $[Q]_w$ is the quencher concentration in the water phase. The number of moles of P* per mode of micelles is $P^*_t/[M]$. The probability that Q, which entered the micelle at time t_1, is still in the micelle at t is given by exp $[-\beta (t - t_1)]$. The probability that P^* is quenched by Q, which entered at time t, between t and $t + dt$ is

$$k_q \exp [-k_q(t-t_1)] \, dt \text{ if } t_1 > 0$$

and

$$k_q \exp [-k_q t] \, dt \text{ if } t_1 < 0$$

The rate constant k_q describes the quenching in the micelle as a first-order event. Hence,

$$Q_t \, dt = \frac{P^*_t}{[M]} \int_\infty^0 \exp [-\beta (t - t_1)] k_g \exp [-k_q t] \, \alpha \, [Q]_w \, [M] \, dt_1 \, dt$$

$$+ \frac{P^*_t}{[M]} \int_0^t \exp [-\beta (t - t_1)] k_q \exp [-k_q (t - t_1)] \, \alpha \, [Q]_w \, [M] \, dt_1 \, dt$$

which gives

$$Q_t \, dt = \frac{k_q \, \alpha \, [Q]_w \, P^*_t}{\beta + k_q} (k_q \exp \frac{[(-\beta + k_q)t]}{k}) \, dt$$

At $t = 0$, $P^*_t = P_o$, so,

$$P^*_t = P_o \exp \left[-(k_1 + \frac{k_q \, \alpha \, [Q]_w}{k + k_q} t - \frac{K k_q^2 \, [Q]_w}{(\beta + k_q)^2} \{1 - \exp [-(\beta + k_q)t]\} \right]$$

The following simplifications help in our discussion of this equation.

$$C = k_q \, \alpha/(\beta + k_q)$$
$$B = K \, k_q^2/(\beta + k_q)^2$$

If the value of B is low, then the first term of the quenching expression is dominant and pseudo first-order kinetics are expected. The rate constant for the quenching reaction is represented by C and is proportional to the entry rate of the quencher into the micelle.

With large values of B, complex kinetics are initially observed; but at longer times when exp $[-(\beta + k_q)t] << 1$, first-order kinetics are again observed. The following parameters can be obtained:

$$\beta = C/\sqrt{KB}$$
$$\alpha = C\sqrt{K/B}$$
$$k_q = C/(K - \sqrt{KB})$$

For the quenching of pyrene fluorescence by CH_2I_2 in NaLS micelles, $\alpha = 2.5 \times 10^{10}$ M^{-1} s^{-1}, $\beta = 9.5 \times 10^6$ s^{-1}, and $k_q = 7.5 \times 10^7$ s^{-1}.

Influence of Additives on Fluorescence Quenching in Micelles. Quite often additives, which do not participate directly in the fluorescence quenching reaction, nevertheless dramaticaliy affect the kinetic data in micellar systems. These effects are consistent with the established properties of micelles and the effects of these additives on the micellar properties. Figure 8 shows the effects of various additives on the rate

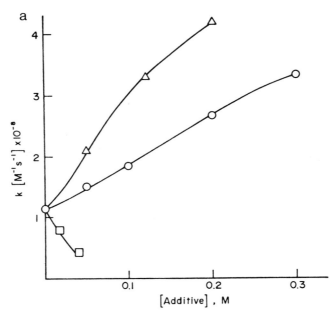

Figure 8a. Influence of additives upon the rate constant for quenching of pyrene fluorescence by triethylamine (7 × 10⁻³ M) solubilized in 10⁻¹ M NaLS. Key (additives): △, benzyl alcohol; ○, Na₂SO₄; and □, MgCl₂ (60).

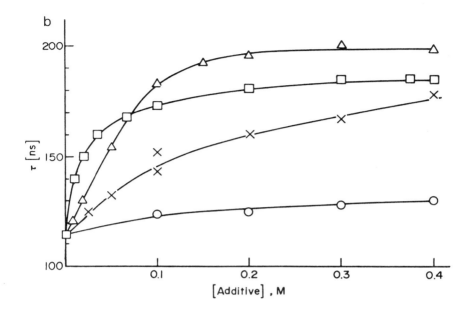

Figure 8b. Effect of additives upon the fluorescence decay time of pyrene (2.5 × 10⁻⁵ M) solubilized in 0.01 M CTAB. Key (additives): △, NaOH; □, NaCl; ×, Na₂SO₄; and ○, benzyl alcohol. The fluorescence wavelength was 400 nm (60).

constant for quenching of pyrene fluorescence by triethylamine (Et_3N) in NaLS micelles (50).

$$P^* + Et_3N \rightarrow P + Et_3N$$

The rate constant is calculated from the quotient of the observed rate of decay of P^* and the bulk concentration of Et_3N. Both benzyl alcohol and sodium sulfate increase the rate constant significantly. These solutes interact with the head group of the micelle; the benzyl alcohol acts as a cosurfactant, and the Na_2SO_4, by introducing excess Na^+ counterions at the micelle surface, reduces the surface charge. Fluorescence polarization studies show that both effects reduce the rigidity of the micelle surface and enhance the rate of diffusion of P^* and Et_3N. Magnesium sulfate reduces the rate of reaction and increases the rigidity of the head group region, presumably by binding strongly to the sulfate head groups and thus restricting their motion. The observed kinetic effects are readily explained by changes in the rigidity of the micellar structure.

Figure 8 shows similar data for pyrene in cetyltrimethylammonium bromide (CTAB) micelles. Here, however, the quenching occurs by interaction of the Br^- counterion with the P^*. Addition of other non-quenching counterions, that is, Cl^-, SO_4^{2-}, and OH^-, decreases the Br^- quenching of P^* by replacing it at the micelle surface. The quenching effectiveness of the ion varies directly as its efficiency of binding to the cationic micelle. Again, benzyl alcohol has a marked effect on the quenching process. Addition of the alcohol does increase the degree of ionization of the micelle, but this increase effect is not sufficient to account for the observed data. The effect is quite opposite to the one observed for P^* quenching by Et_3N in NaLS micelles. Other alcohols show similar effects and will be discussed subsequently. Benzyl alcohol may push the pyrene farther into the micelle structure and away from the surface where Br^- quenching takes place. This situation leads to a dramatic decrease in reaction rate. A similar effect takes place in the $P/Et_3N/NaLS$ system, but in this case Et_3N is also located farther into the micelle as well as the P. This finding is quite unlike the CTAB system where Br^- cannot readily leave the positively charged surface of the micelle to enter the hydrophobic interior.

Similar effects are noted in sodium taurocholate micelles (61) and oleate micelles (62). The taurocholate micelles have a rather small aggregation number (4) (63), but exhibit a much larger rigidity than NaLS or CTAB micelles. This result is possibly due to compact packing of the cholesterol-like structure of the micelle. The fluidity of the micelle is enhanced by the addition of NaLS, which forms mixed micelles with the bile acid (61).

Charge Transfer Reactions

Of the many examples of photoinduced charge transfer reactions in micelles, the simplest example is the reaction of hydrated electrons e_{aq}^- with solutes in micelles.

Reactions of e_{aq}^-. In micellar systems e_{aq}^- may be produced by photoionization of solutes contained in the water or micellar region, or by radiolysis of the bulk phase. Radiolysis is usually more useful; however, this technique is not specific for e_{aq}^- because other reactive species are also produced. Spectroscopically e_{aq}^- is quite clearly observed via its absorption in the red portion of the spectrum because the other radiolytic fragments (e.g., OH) absorb in the UV. Photoionization of a solute leads to e_{aq}^- and a solute cation, but the latter species often has a strong absorption spectrum that overlaps the e_{aq}^- spectrum. Thus, experimental observations are somewhat complicated in photoionization. Many e_{aq}^- + solute reactions reported have been carried out by pulse radiolysis of the aqueous phase while the solute is located in the micelle region. A list of rate constants is shown in Table II.

Table II.
Rate Constants for Reaction of e_{aq}^- (10^9 M^{-1} s^{-1})

Solute	Homogeneous Soln	NaLS	NaTC	CTAB	Igepal	Triton	Lysolecithin	Distearyl-lecithin
Micelle		$<10^{-4}$ (5,116)		$<10^{-3}$ (5,116)	1.7×10^{-2} (5,116)	3.8×10^{-2} (5,116)	3.9 (86)	$<10^{-3}$
Pyrene	10 (116)	$<10^{-2}$ (116) 4.8×10^{-2} (0.2 M NaCl)	0.13 (158)	>102 (155)	1.7 (121)	2.0 (121)	3.8 (86)	$<10^{-2}$ (101) fast (187)
Biphenyl	5.0 (116)	0.13 (116)			3.8 (121)	4.0 (121)		0.005 (101)
Pyrene-carboxaldehyde	16 (101)	2.5 (101)						
Aminopyrene	14 (101)	1.1 (101)						
Benzophenone	10 (5)	1.5 (5)		20 (5)		2.0 (5)		
Nitroanthracene	10	1.5 (204)		90 (204)				

Noncharged Micelles. The rates of reaction of e_{aq}^- and simple anions (e.g., I^-) with probe molecules located in nonionic micelles (i.e., Igepal, Triton X–100, or lysolecithin) are slower than in homogeneous solution (64). The retardation in rate may be due to the reluctance of the anions to leave the aqueous phase and enter the less polar regions of the micelle. The solutes are also screened by the micelle from reaction with e_{aq}^-, an effect that could reduce the rate by a factor of about 10, depending on the chosen molecular description of the reaction. The observed rates are, for the most part, only reduced by a factor of 10, or even less in many instances. Hence, the effect of nonionic micelles in reducing the rates of reaction of e_{aq}^- with many solutes may be readily explained in terms of simple, well-understood features of the reactants and the micelle.

In at least one instance, the observed e_{aq}^- + solute is greatly reduced by a nonionic micelle, namely the reaction of e_{aq}^- with β-carotene in Triton X-100 micelles (65). Pulse-radiolysis experiments show that e_{aq}^- does not react with β-carotene when this solute is solubilized in nonionic micelles. This result may be due to the deep solvation of the molecule in the micelle, away from the polar aqueous phase containing e_{aq}^-. However, inclusion in the system of biphenyl ϕ_2, which locates more towards the micelle surface, leads to the initial formation of ϕ_2^- anion,

$$e_{aq}^- + \phi_2 \text{ (micelle)} \rightarrow \phi_2^- \text{ (micelle)}$$

followed by electron transfer from ϕ_2^- to β-carotene.

$$\phi_2^- + \text{β-carotene (micelle)} \rightarrow \phi_2 + \text{(β-carotene)}^- \text{ (micelle)}$$

The biphenyl acts as mediator in this e^- transfer reaction; it enables e^- to penetrate the micelle by providing a more hydrophobic site for e^- (i.e., ϕ_2 acts as a carrier to move e^- from the aqueous phase to the hydrophobic micelle interior).

Anionic Micelles. The rates of reaction of e_{aq}^- with solutes in anionic micelles, such as NaLS or sodium taurocholate, are quite pronouncedly decreased from the rates observed in homogeneous solution. In some instances, such as, Cu^{2+}, Cd^{2+} (66) bipyridyl ruthenium (67), and methyl viologen (68), the larger decrease in reaction rate in NaLS micelles can be used to ascertain the degree of binding of the solute to the micelle. If the solute is located in the aqueous phase, it reacts 100 to 3000 times faster than when bound to NaLS micelles; hence, the observed rate of reaction of e_{aq}^- with the solute can be interpreted as the rate of e_{aq}^-

+ solute in the aqueous phase. The [solute] in the aqueous phase is then calculated from the observed rate R_1 and the established rate constant k in homogeneous solution by $[solute]_{aq} = k/R$.

The decrease in rate is again due to several effects introduced by the micelle; a steric effect due to incorporation of the molecule on or at the micelle surface, the tendency of e_{aq}^- to stay in the aqueous phase, and the strong repulsion of e_{aq}^- away from the micelle because of its strongly negative charge.

The last effect is dominant in controlling the rate of reaction. The first effect could contribute a 10-fold decrease in rate, but the last effect can easily contribute a 100- or 1000-fold decrease. The effect of charge is readily demonstrated by adding an inert solute to the system when the reaction rate of $e_{aq}^- + (solute)_{mic}$ increases dramatically. The effect of inert electrolyte has been discussed previously, and amounts to a compression of the Gouy–Chapman layer, which thereby increases the probability of penetration of e_{aq}^- to the micellar surface.

Many of the solutes investigated are quite large compared to the dimensions of the micelle (i.e., the micelle radius is about 20 Å and the solute may be about 10 Å). The exact location of the solute in the micelle is a matter of some consequence with regard to its reactivity with e_{aq}^- because the electrostatic repulsion increases sharply over the last few ångströms up to the micelle surface. The rate constant of reaction of e_{aq}^- and pyrene in NaLS is $k < 10^7$ M^{-1}s^{-1} compared to $k = 10^{10}$ M^{-1} s^{-1} in homogeneous solutions. However, a derivative such as pyrenecarboxaldehyde or aminopyrene reacts much more rapidly ($k = 10^8$ M^{-1}s^{-1}). These solutes are located on the micelle surface and are closer to the aqueous phase than pyrene. However, the pyrene fluorescence data indicate that pyrene is also in a polar environment and close to the micelle surface. Hence, e_{aq}^- does not necessarily have to penetrate into the hydrocarbon region of the micelle for reaction to occur. The similarity of rate data for $e_{aq}^- +$ solutes and I$^- +$ excited solutes suggests a similar mechanism for retardation of both reactions by the negative micelle surface. The electron may "tunnel" from the aqueous phase to the solute (69), and efficient reactions are observed when the energy levels of e_{aq}^- and the solute anion match closely. The micellar potential can alter those levels and, hence, alter the reaction rate. The main incentive for this theory is the premise that the solute is located in a hydrophobic core of the micelle, a situation that necessitates some sort of e^- hopping or tunneling mechanism. The more recent viewpoint (that solutes are located at the micelle surface) removes the dire necessity of the tunneling theory. Because the more conventional electrostatic picture describes the reactions well, it will be used in the present discussion.

Cationic Micelles. Positively charged micelles, either pure cationic micelles such as CTAB (70) or neutral micelles with added CTAB (71), markedly enhance the rate of reaction of e_{aq}^- with solutes located in these micelles. The effect is readily explained by an attraction force between the cationic micelle and e_{aq}^-, which draws e_{aq}^- to the micelle surface or to the location of the solute. Cetylpyridinium chloride is very reactive with e_{aq}^- in the monomeric state ($k = 10^{10}$ M^{-1}s^{-1}). The rate constant in the micelle form is ~2 × 10^{12} M^{-1}s^{-1} (72). This value provides a standard rate of reaction for e_{aq}^- with cationic micelles of radius 20 Å it is very nearly the product of the rate constant for the monomeric form and the aggregation number of the micelle. A similar rate constant can, therefore, be safely assumed for e_{aq}^- encounters with a CTAB micelle. However, because this micelle does not contain reactive groups, no actual rate of reaction is observed. The rate constant for reaction e_{aq}^- with a pyrene in CTAB micelles is 1.5 × 10^{11} M^{-1}s^{-1}, 10 times faster than that observed in a homogeneous solution. The rate constant decreases at higher pyrene/micelle ratios. This result is partly due to the nature of the Poisson distribution of pyrene among the micelles and also due to a decreased efficiency of reaction for micelles containing more than one pyrene molecule. The rate of reaction with micelles containing one pyrene molecule is quite high and indicates a 10% efficiency of reaction on every e_{aq}^-–micelle encounter. A greater efficiency of reaction is not achieved because of limited diffusion of e_{aq}^- in the micelle surface prior to reaction with pyrene or escape of e_{aq}^- to the aqueous bulk.

Photoionization

Phase or medium has a marked effect on the yield and threshold for photoionization of aromatic molecules (73) and micellar systems show the most pronounced effects. Laser photolysis of pyrene in NaLS micelles was first seen to produce photoionization in a two-photon process (74). The products of photolysis, which were identified by absorption spectroscopy, were the pyrene cation P$^+$ (λ_{max}, 455 nm), the hydrated electron e_{aq}^- (λ_{max}, 720 nm), the triplet (λ_{max}, 415 nm), and the excited singlet (λ_{max}, 470 nm); the excited states were a minor portion of the observed spectra. Completely different data were obtained in the laser photolysis in pyrene in alkanes, where excited states are the only transitory species observed spectroscopically (75). However, a small yield of two-proton photoionization of pyrene is observed in low dielectric liquids by sensitive electrical conduction methods (76). Photoionization of pyrene is observed both by optical and conduction methods in alcohols, although the yield is much smaller than that observed in many micellar systems. In detailed studies of the pyrene system (77), the initially formed geminate ion pair of P^+ and e^- recombine rapidly (10^{-10} s) in homogeneous solvents, a process that competes with escape of the ions to the bulk of the solution. Only these latter ions can be subsequently observed at longer time periods. The more polar liquids, such as alcohols, increase the extent of ion escape; and the negatively charged surface of a NaLS micelle repels e_{aq}^- while retaining P^+ and promotes charge separation. In the latter case, electrons are ejected into the Gouy–Chapman layer approximately 10–15 Å from the micelle surface (78) and, subsequently, diffuse into the aqueous bulk. The importance of the micelle surface in the photoionization process is also evident in the effect of additives such as CH_2I_2 and benzoic acid on the yield of e_{aq}^-. Such additives, which are located in the micelle surface, reduce the yield of e_{aq}^- but not the yield of P^+ (78). This result is reminiscent of the effect of high concentrations of solutes on the photoionization yield of pyrene in alcohols (79) and is attributed to a reaction of the solute with a precursor of the hydrated electron.

Several other molecules, such as phenothiazine (80), tetramethylbenzidine (81), and aminopyrene (73), also exhibit photoionization in micellar solutions. By contrast with the pyrene/NaLS system, photoionization of these solutes in NaLS micelles occurs with one photon of 347-nm light in micelle systems. Threshold studies indicate an onset of photoionization at about 3.0 eV, or more than 400 nm (73). These data show that the ionization potentials of these molecules are $\leq$ 3.0 eV in NaLS micelles, compared to $>$ 6.3 eV in the gas phase. An explanation for this effect has been couched in terms of e^- tunneling from the micelle lipid phase to the aqueous bulk (80), an apt explanation if the molecules are remote from the aqueous phase and in the micelle core. However, the molecules are now realigned to be on the micelle surface and in direct contact with the aqueous phase. Thus, the photoionization process is explained as a simple electron transfer from solute to solvent without any tunneling. In the process $P \rightarrow P^+ + e^+$, the ionization potential in solution I_s is related to that in the gas phase I_g by the expression:

$$I_s = I_g + P_+ + V_o$$

where P_+ is the polarization energy of the cation, and V_o is the energy state of e^- in water, which is -1.5 eV (82). The cation polarization energy P_+ may be calculated from the Born charging expression:

$$P_+ = \frac{e^2}{2r}\left(1 - \frac{1}{\varepsilon}\right)$$

where r is the interaction distance of the ion and the surrounding solvent molecules; e is the electronic charge; and ε is the fast or optical dielectric constant of the medium, approximately 2.0 for most media. If ε is much greater than 2, then the value of P_+ may double in the above expression. Normally, P_+ is -1.5 to -2.0 eV; therefore, $I_g - I_s$ is ~ 3.0 to 3.5 eV or larger, if ε is $>> 2$ for NaLS. Thus, photoionization of molecules is possible at much lower energies in the condensed phase than the gas phase.

Aminoperylene can be photoionized with one photon of green light (λ, 530 nm) in NaLS (83), ~ 4.0 eV below the gas phase ionization potential of this molecule. Micelles thus provide an ideal medium to reduce the ionization potential of a molecule and to increase the yield of long-lived ions.

Electron Transfer from Excited States

Both excited singlet and triplet states undergo electron transfer reactions to suitable acceptors. Many systems have been studied successfully in homogeneous solution (84–86). Several systems have been studied in micellar media; and, for the sake of discussion, we divide the work into systems involving excited singlet states and those involving excited triplet states.

Electron Transfer for Excited Singlet States. Excitation of many systems containing arenes and amines gives rise to either excited complexes, exciplexes, or, if the surroundings are suitable, to ions (87). A typical and well-studied system is pyrene (P) and dimethylaniline (D). Excitation of either the pyrene or the amine produces excited complexes, in nonpolar media such as benzene and cyclohexane, and to pyrene anions and amine cations, in polar media such as methanol and acetonitrile.

$$P \xrightarrow{h\nu} P^* + D \rightarrow (PD)^* \quad \text{(nonpolar media)}$$
$$\downarrow$$
$$P^- + D^+ \quad \text{(polar media)}$$

The various intermediates such as $(PD)^*$, P^*, P^-, D^+, etc. are readily identified by their established absorption or emission spectra. The yield of ions versus other photoproducts increases with an increase in solvent polarity. The lifetime of the ionic products is usually quite short (~ 10 μs) because the [ion] is high as a result of the high laser power used in the pulsed experiments and because the ion neutralization reaction, $P^- + D^+ \rightarrow P + D$, is diffusion controlled, $k \sim 10^{10}$ M^{-1}s^{-1}.

Excitation of this system in micelles produces results that depend on the nature of the micelle. The initial yield of electron transfer is large in anionic (NaLS), cationic (CTAB), and neutral (Igepal or Triton X-100) micelles, and approaches a quantum yield of unity. However, the subsequent fate of the ions is critically dependent on the micelle charge. The lifetime of the photoproduced ion is is short (μs) in neutral micelles, very short ($< 10^{-8}$ s) in NaLS, and extremely long (> 10 ms) in CTAB. The origin of this behavior is sought in the micelle charge and in the ease with which P^- and D^+ escape the micelle. Probably, P^- and D^+ escape the neutral micelle fairly rapidly, $< 10^{-6}$s (D^+, even faster), and subsequently react in the aqueous bulk phase. However, P^-, and certainly D^+, do not escape rapidly from NaLS because D^+ is held by the attractive field of the anionic micelle. Hence, P^- and D^+, which are held in close proximity, recombine rapidly. With CTAB, P^- is held by hydrophobic and electrostatic attractive forces to the micelle, and D^+ is expelled by this very potential and kept from further contact with P^-. This situation leads to the extremely long ion lifetimes observed.

Micelles do not have desirable effects on all exciplex systems, and ion lifetimes and yields may be low depending on the exact system used.

Nature of the Ion Separation in Cationic Micelles. The mechanism of ion separation in CTAB micelles is important. Excitation of ground state exciplexes, or small reactant systems, often does not lead to significant yields of ions in CTAB micelles. The former data indicate that contact of the ions is undesirable; the latter data indicate that delocalization of e^- in the anion may be important.

Two basic patterns of e^- transfer may be identified: electron transfer on collision contact, which could give rise to problems of eventual ion separation; and electron transfer over a distance, or e^- tunneling, where the ions are already produced in a separated condition.

Experiments have been carried out to identify the relative importance of these two mechanisms (90). In the pyrene/DMA system, e^- transfer from DMA to pyrene occurs at a diffusion-controlled rate on the micelle surface and no indication of the second process was forthcoming. The rate constant for the quenching of P* by DMA and the concomitant formation of P^- and D^+ was $k = 2 \times 10^7$ M^{-1}s^{-1}, which compares very favorably with rate constants for other diffusion controlled reactions in CTAB, such as pyrene excimer formation. Now the mechanism for subsequent charge separation has to be sought. The yield of ions was largest for systems operating on the surfaces of assemblies (e.g., pyrene/DMA on CTAB, pyrenesulfonic acid/DMA on CTAB/hexanol/dodecane microemulsion). The yield of ions in the pyrene/DMA system decreased progressively from 1.0 to 0.24 as the assembly was changed from CTAB micelle, swollen micelle, to microemulsion (i.e., as pyrene was pressed further into the assembly and away from the surface). Furthermore, in a CTAB reversed micelle, or water-in-oil microemulsion, the ion yield was zero for the pyrene and DMA system (90, 92). In this system the pyrene/DMA exciplex is formed in the oil phase. However, on collision with the water droplets the exciplex is efficiently quenched without formation of ions. The exciplex has a sandwich form; this configuration is undesirable for large yields of escaped ions. Perhaps, e^- is transferred from DMA to pyrene from a number of configurations, but back e^- transfer is only efficient in the sandwich configuration. This result is due to delocalization of e^- over the pyrene ring, and extensive overlap of DMA^+ and P^- (e.g., sandwich complex) is essential for efficient back e^- transfer. Other configurations slow down the back transfer and rapid diffusion of DMA^+ from the micelle surface produces ion separation. Surface solubilization does not promote the sandwich structure of P and DMA. This situation also contains features which explain the lack of large ion yields in other systems containing small molecules or ground state charge transfer complexes (93). The delocalization of e^- in small molecules is not sufficient to hinder back e^- transfer from many configurations, but the geometry of ground state complexes is suitable for rapid back e^- transfer.

Electron Tunneling. Data (94, 95) indicate that e^- jumping or tunneling may occur over significant distances (e.g., 10 Å) if the physical conditions of the aggregate and reactants are suitable. For the DMA/pyrene system, increasing the rigidity of the assembly by using a "polymerized" microemulsion leads to inefficient quenching of P* by DMA. The quenching occurs staticly and large yields of ions are produced. The physical nature of the system (i.e., close proximity and P* and DMA) and immobilization of reactants suggest that quenching occurs by e^- tunneling from DMA to P*.

Electron Transfer from Excited Triplet States

A wide variety of examples exist of electron transfer from excited triplet states to suitable acceptors. The long lifetime of the excited triplet state assists in efficient e^- transfer, which leads to significant yields of ionic products. A few important examples will be given here mainly as guidelines to what can be achieved; the number of possible systems could be legion!

The triplet states of both phenothiazine (PH) and tetramethylbenzidene (TMB), have a relatively high energy (>2.5 eV). This energy coupled with the low ionization potential of the donors leads to efficient e^- transfer from the excited triplet state to suitable acceptors such as quinones and metal ions (e.g., Cu^{2+} and Eu^{3+}). In a homogeneous solution (methanol), the rate constant for the e^- transfer reaction producing ions

$$\text{TMB}^\text{T} + \text{Eu}^{3+} \underset{\text{back}}{\overset{\text{forward}}{\rightleftharpoons}} \text{Eu}^{2+} + \text{TMB}^+$$

is $6.4 \times 10^9 \text{ M}^{-1} \text{ s}^-$, and the rate constant for the back e^- transfer reaction is $1.4 \times 10^7 \text{ M}^{-1} \text{s}^{-1}$. Excitation of this system in NaLS micelles, where the organic entity is close to the micelle surface and the metal ion is in the Stern layer, rapidly leads to the ionic products Eu^{2+} and TMB^+ [e.g., in 0.1 M NaLS and 3×10^{-3} M Eu^{3+}, the TMB^T decays rapidly, $t^{1/2} \sim 30$ ns, to give ions (80, 81); the TMB^T lifetime in absence of Eu^{3+} is > 800 μs]. In homogeneous solution the ions TMB^+ and Eu^{2+} rapidly back react, but some protection is provided in anionic micellar solutions. Quinones have also been used as the e^- acceptor, and other molecules, such as, aminopyrene (73), and chlorophyll (96), as the e^- donors. Excitation of a hydrophobic quinone, such as duroquinone, in micelles leads to a long-lived triplet state that is a very strong e^- acceptor. For example, excited duroquinone triplet (DQ^T) readily reacts with carbonate ions to produce long-lived DQ^- and CO_3^- (97)

$$\text{DQ}^\text{T} + \text{CO}_3^{2-} \rightarrow \text{DQ}^- + \text{CO}_3^-$$

A redox system of current importance in solar energy is tris-(bipyridyl)ruthenium (RuII) and methyl viologen (MV^{2+}) (98–102). The RuII complex may be excited in the visible part of the spectrum ($\lambda <$ 500 nm) to give the excited state, RuII*, which has a long lifetime (> 500 ns), and may be observed via its characteristic luminescence at $\lambda = 610$ nm. The excited RuII* reacts with MV^{2+} and other acceptors via e^- transfer,

$$\text{RuII}^* + \text{MV}^{2+} \rightarrow \text{RuIII} + \text{MV}^+$$

Micellar systems and other aggregates protect the products from back reaction and the presence of a sacrificial agent such as ethylenediamine-tetraacetate (EDTA), or ascorbic acid repairs the metal ion

$$\text{RuIII} + \text{SA} \rightarrow \text{RuII} + \text{SA}^+$$

and produces the original RuII and an unreactive SA^+ radical. The net result is a transfer of e^- from the sacrificial agent to MV^{2+} to give stable products; in other words, a net storage of energy is achieved. The MV^+ may be used to produce further products of interest (e.g., H_2).

Design of Surfactant. A functionalized surfactant is a modified surfactant that takes part in the photoinduced event and is not just a passive partner of the system. Such a surfactant is readily produced by replacing the inert counterion of the surfactant (Na^+ or Br^-) with a reactive ion (Eu^{3+} or Cu^{2+}) or anthraquinonesulfonate. A good example is provided by the system copper lauryl sulfate $[Cu(LS)_2]$ and the guest molecule N,N'-dimethyl-5,1,11-dihydroindolo[3,2]carbazole (DI) (103). DI shows an intense fluorescence in NaLS micelles with a lifetime of 144 ns. However, in $Cu(LS)_2$ micelles the fluorescence yield and lifetime are decreased some 300-fold as a result of a rapid e^- transfer from excited DI to Cu^{2+} ions

$$DI^* + Cu^{2+} \longrightarrow DI^+ + Cu^+$$

The rapid e^- transfer is due to the close proximity of reactants, that is, the locally high concentrations of Cu^{2+} and DI in the micelle surface. The Cu^+ ions formed are rapidly exchanged with Cu^{2+} ions of the water bulk; this result produces efficient charge separation of Cu^+ and DI^+. The Cu^+ can react further in the aqueous bulk with molecules such as ferricyanide, to produce Cu^{2+} and ferrocyanide. The back reaction of the products with DI^+ in the micelle is prevented by repulsion of $Fe(CN)_6^{4-}$ by the negative charge of micellar surface.

More complex functionalized surfactants have been constructed. Derivatives of methyl viologen with one CH_3 group replaced by a long chain hydrocarbon are cationic surfactants.

$$C_{14}H_{29} + N^+ \underset{}{\bigcirc} - \underset{}{\bigcirc} N^+ - CH_3$$

In homogeneous solution viologen VII will accept e^- from many excited molecules, in particular metalloporphyrins, to form reduced viologen MV^+ and the oxidized metalloporphyrin MP^+. However, the lifetime of the ions formed is not particularly long because of recombination in the aqueous bulk. The kinetics are modified in the presence of CTAC micelles where the long-chain viologen is incorporated into the micelle. Following e^- transfer to the viologen, the cationic porphyrin is repelled away from the micelle containing the viologen, and the back reaction is considerably retarded ($\sim$ 1000-fold).

Surfactants containing long hydrocarbon chains and polar macrocylic head groups have been constructed (105), and metal ions may be incorporated into the head group structure.

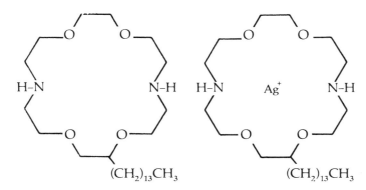

Excitation of various dyes (RuII, pyrene, etc.) in these systems leads to the formation of reduced Ag^+ (i.e., Ag) and the oxidized donor (106). The silver atoms are stable in the crown ether head group, and no back reaction of the products is observed. The micelle provides an environment of close proximity for Ag^+ and the excited reducing agent. This closeness leads to efficient e^- transfer and also protects the photoproducts.

Long-chain derivatives of phenothiazine have been constructed and, in the micellar form, they protect the phenothiazine cation formed in photolysis of the system from back reaction with the reduced reaction product. The normally unstable phenothiazine cation is stabilized for many days in these systems (107).

Literature Cited

1. Duynstee, E. F. J.; Grunwald, E. *J. Am. Chem. Soc.* **1959**, *81*, 4540.
2. Cordes, E. H. "Reaction Kinetics in Micelles"; Plenum Press: New York, 1973.
3. Oakenfull, D. G. *Chem. Soc. Rev.* **1977**, *6*, 25.
4. Fendler, J. H.; Fendler, E. J. "Catalysis in Micellar and Macromolecular Systems." Acad. Press: New York, 1975.
5. Kalyanasundaram, K. *Chem. Soc. Rev.* **1978**, *7*, 453.
6. Brown, J. M.; Bunton, C. A. *J. Chem. Soc., Chem. Commun.* **1974**, 969.
7. Thomas, J. K. *Chem. Rev.* **1980**, *87*, 283.
8. Bakalik, D. P.; Thomas, J. K. *J. Phys. Chem.* **1977**, *81*, 1905.
9. Henglein, A.; Proske, Th. *J. Am. Chem. Soc.* **1978**, *100*, 3706.
10. Dorfman, L. M.; Taub, I. A.; Bühler, R. E. *J. Chem. Phys.* **1962**, *36*, 3051.
11. Henglein, A.; Proske, Th. *Makromol. Chem.* **1978**, *179*, 2279.
12. Florence, A. T. In "Micellization, Solubilization, and Microemulsions"; Mittal, K. L., Ed.; Plenum Press: New York, 1977; p. 55.
13. Chen, T. Ph.D. Thesis, University of Notre Dame, Notre Dame, IN, 1978.
14. Almgren, M.; Greiser, F.; Thomas, J. K. *J. Chem. Soc., Faraday Trans.* 1 **1979**, *75*, 1674.
15. Matheson, M. S.; Mulac, W. A.; Weeks, J. L.; Rabani, J. *J. Phys. Chem.* **1966**, *70*, 2092.
16. Thomas, J. K. *Trans. Faraday Soc.* **1965**, *61*, 702.
17. Anbar, M.; Thomas, J. K. *J. Phys. Chem.* **1964**, 3829.
18. Frank, A. J.; Grätzel, M.; Kozak, J. J. *J. Am. Chem. Soc.* **1976**, *98*, 3317.

19. Proske, T. L.; Henglein, A. *Ber. Bunsenges Phys. Chem.*, **1978**, *82*, 711.
20. Breslow, R.; Kotabatake, S.; Rothbard, J. *J. Am. Chem. Soc.* **1978**, *100*, 8156; Breslow, R.; Rothbard, J.; Herman, F.; Rodrigues, M. L. Ibid. **1978**, *100*, 1213.
21. Mitani, M.; Suzuki, T.; Takenchi, H.; Koyama, K. *Tetrahedron Lett.* **1979**, 803.
22. Wagner, P. J. *Acc. Chem. Res.* **1971**, *4*, 168.
23. Turro, N. J.; Liu, K. C.; Chow, M. F. *Photochem. Photobiol.*, **1977**, *26*, 413.
24. Worsham, P. R.; Eaker, D. W.; Whitten, D. G. *J. Am. Chem. Soc.* **1978**, *100*, 7091.
25. Turro, N. J.; Kraeutler, B. *J. Am. Chem. Soc.* **1978**, *100*, 7432.
26. Turro, N. J.; Cherry, W. R. *J. Am. Chem. Soc.* **1978**, *100*, 7431.
27. Rico, I.; Maurette, M. T.; Oliveros, E.; Rivere, M.; Lattes, A. *Tetrahedron Lett.* **1978**, 4795.
28. Lee, K. H.; deMayo, P. *J. Chem. Soc., Chem. Commun.* **1979**, 493.
29. Nakamura, Y.; Imakuna, Y.; Monta, *J. Chem. Lett.* **1978**, 965.
30. Nakamura, Y.; Imakura, Y.; Kato, T.; Monta, Y. *J. Chem. Soc., Chem. Commun.* **1977**, 887.
31. Gorman, A. A.; Lovering, G.; Rodgers, M. A. J. *Photochem. Photobiol.* **1976**, *23*, 399.
32. Usui, Y.; Tsukada, M.; Nakamura, H. *Bull. Chem. Soc. Jpn* **1978**, *51*, 379.
33. Kalyanasundaram, K.; Thomas, J. K. "Micellization, Solubilization, and Microemulsion", Mittal, K. L., Ed.; Plenum Press: New York; 1977, Vol. 2, p. 569.
34. Teale, F. W. J. *Nature.* **1958**, *181*, 415.
35. Zenkevitch, E. I.; Losey, A. D.; Guirnovich, G. P. *Mol. Biol. Moscow.* **1972**, *6*, 824; Lehoczki, E.; Csatorday, K. *Biochim. Biophys. Acta* **1975**, *396*, 86; Csatorday, K.; Lehoczki, E.; Szalay, L. *Biochim. Biophys. Acta* **1975**, *376*, 268.
36. Singhal, G. S.; Rabinovitch, E.; Hevesi, J.; Srinivasan, V. *Photochem. Photobiol.* **1970**, *11*, 531.
37. Ghosh, A. K. *Indian. J. Chem.* **1973**, *11*, 1014.
38. Escabi-Perez, J. R.; Nome, F.; Fendler, J. H. *J. Am. Chem. Soc.* **1977**, *99*, 7749.
39. Almgren, M.; Grieser, F.; Thomas, J. K. *J. Am. Chem. Soc.* **1979**, *101*, 2021.
40. Aniansson, E. A. G. Ref. 15 of Ref. 39.
41. Porter, G.; Wright, M. R. *J. Chim. Phys.* **1958**, *55*, 705; *Faraday Discuss. Chem. Soc.* **1959**, *27*, 18.
42. Dexter, D. L. *J. Chem. Phys.* **1953**, *21*, 836.
43. Wagner, P. J.; Schott, H. N. *J. Phys. Chem.* **1968**, *72*, 3702.
44. Lamola, A. A.; Eisinger, J. In "Molecular Luminescence"; Lim, E. C., Ed.; Benjamin, New York, 1969.
45a. Ermolaev, V. L.; Tachin, V. S. *Opt. Spectra.* **1969**, *27*, 546.
45b. Rothenburgher, G.; Infelta, P. P.; Grätzel, M. *J. Phys. Chem.* **1979**, *83*, 1871; *J. Phys. Chem.* **1981**, *85*, 1850.
45c. Grieser, F.; Tausch-Treml, R. *J. Am. Chem. Soc.* **1980**, *102*, 7258.
45d. Moroi, Y.; Infelta, P. P.; Grätzel, M. *J. Am. Chem. Soc.* **1979**, *101*, 573.
46. Birks, J. B. *Photophysics of Aromatic Molecules.* London, 1970.
47. Förster, T.; Selinger, B. K. *Z. Naturforsch A.* **1964**, *19*, 38.
48. Pownall, H. J.; Smith, L. C. *J. Am. Chem. Soc.* **1973**, *95*, 3136.
49. Shinitzky, M.; Dianoux, A. C.; Gitler, G.; Weber, G. *Biochemistry* **1971**, *10*, 2106.
50. Loeb, A. L.; Overbeeck, J. T.; Wiersema, P. H. "The Electrical Double Layer Around a Spherical Colloid Particle"; MIT Press: Cambridge, MA, 1961.
51. Henglein, A.; Scheerer, R. *Ber. Bunsenges Phys. Chem.* **1978**, *82*, 1107.
52. Grätzel, M.; Infelta, P. *J. Chem. Phys.* **1979**, *70*, 179.
53. Atik, S. S.; Singer, L. A. *Chem. Phys. Lett.* **1978**, *59*, 519.

54. Atik, S. S.; Thomas, J. K. *J. Am. Chem. Soc.* **1981**, *103*, 3550.
55. Turro, N. T.; Yekta, L. A. *J. Am. Chem. Soc.* **1978**, *100*, 5951.
56. Infelta, P. P. *Chem. Phys. Lett.* **1979**, *61*, 88.
57. Rothenberger, G.; Infelta, P.; Grätzel, M. *J. Phys. Chem.* **1981**, *85*, 1850.
58. Infelta, P. P.; Grätzel, M.; Thomas, J. K. *J. Phys. Chem.* **1974**, *78*, 190.
59. Rothenberger, G.; Infelta, P.; Grätzel, M. *J. Phys. Chem.* **1979**, *83*, 1871.
60. Grätzel, M.; Thomas, J. L. *J. Am. Chem. Soc.* **1973**, *95*, 6885.
61. Chen, M.; Grätzel, M.; Thomas, J. K. *J. Am. Chem. Soc.* **1975**, *97*, 2052.
62. Grätzel, M.; Thomas, J. K. In "Modern Fluorescence Spectroscopy"; Wehry, E., Ed.; Plenum Press: New York, 1976.
63. Small, D. M. *Adv. Chem. Ser.* **1968**, *84*, 31.
64. Thomas, J. K. *Acc. Chem. Res.* **1977**, *10*, 133.
65. Almgren, M.; Thomas, J. K. *Photochem. Photobiol.* **1980**, *31*, 329.
66. Grätzel, M.; Thomas, J. K. *J. Phys. Chem.* **1974**, *78*, 2248.
67. Meisel, D.; Matheson, M. S.; Rabani, J. J. *J. Am. Chem. Soc.* **1978**, *69*, 1522; *J. Phys. Chem.* **1978**, *82*, 985.
68. Rodgers, M. A. J.; daSilva, M. F.; Wheeler, E. *Chem. Phys. Lett.* **1978**, *53*, 165.
69. Henglein, A.; Grätzel, M. In "Solar Power and Fuels"; Nyggard; Adler, Eds.; Acad. Press: New York, 1977, p. 53.
70. Grätzel, M.; Cooper, M.; Thomas, J. K. "Rad. Res."; Nyggard; Adler, Eds.; Acad. Press: New York, 1975; p. 511.
71. Henglein, A.; Proske, T.; Schnecke, W. *Ber. Bunsenges Phys. Chem.* **1978**, *82*, 956.
72. Thomas, J. K.; Patterson, L.; Grätzel, M. *Chem. Phys. Lett.* **1974**, *29*, 393.
73. Picuilo, P.; Thomas, J. K. *Adv. Chem. Ser.* **1980**, *184*, 97.
74. Wallace, S. C.; Grätzel, M.; Thomas, J. K. *Chem. Phys. Lett.* **1973**, *23*, 359.
75. Richards, J. T.; West, G.; Thomas, J. K. *J. Phys. Chem.* **1970**, *74*, 4137.
76. Beck, G.; Thomas, J. K. *J. Chem. Phys.* **1972**, *57*, 3649.
77. Picuilo, P.; Thomas, J. K. *J. Chem. Phys.* **1978**, *68*, 3260.
78. Grätzel, M.; Thomas, J. K. *J. Phys. Chem.* **1974**, *78*, 2248.
79. Bromberg, A.; Thomas, J. K. *J. Chem. Phys.* **1975**, *63*, 2124.
80. Alkaitis, S. A.; Beck, G.; Grätzel, M. *J. Am. Chem. Soc.* **1975**, *97*, 5723; Alkaitis, S. A.; Grätzel, M.; Henglein, A. *Ber. Bunsenges* **1975**, *79*, 541.
81. Alkaitis, S. A.; Grätzel, M. *J. Am. Chem. Soc.* **1976**, *98*, 3549.
82. Barker, G. C.; Bottura, G.; Cloke, G.; Gardner, A. W.; Williams, M. J. *Electroanal. Chem. Interfac. Electrochem.* **1974**, *50*, 323.
83. Picuilo, P.; Thomas, J. K. *J. Am. Chem. Soc.* **1978**, *100*, 3239.
84. Mataga, N. "Molecular Association", Forster, R., Ed.; Acad. Press: New York, 1979, Vol. II, p. 2.
85. Weller, A. *Pure Appl. Chem.* **1968**, *16*, 115.
86. Ottolenghi, M. *Acc. Chem. Res.* **1973**, *6*, 153.
87. Berlman, I. B. "Handbook of Fluorescence"; Acad. Press: New York, 1965.
88. Râzem, B.; Wong, M.; Thomas, J. K. *J. Am. Chem. Soc.* **1978**, *100*, 1679.
89. Waka, Y.; Hamamoto, K.; Mataga, N. *Chem. Phys. Lett.* **1978**, *53*, 242.
90. Atik, S. S.; Thomas, J. K. *J. Am. Chem. Soc.* **1981**, *103*, 3550.
91. Ibid. *J. Am. Chem. Soc.* **1981**, *103*, 4367.
92. McNeil, R.; Thomas, J. K. *J. Colloid Interface Sci.* **1981**, *83*, 57.
93. Matsuo, T.; Nagamura, T.; Itoh, K.; Nishijima, T. *Mem. Fac. Eng.; Kyushu Univ.* **1980**, *40*, 25.
94. Atik, S. S.; Thomas, J. K. *J. Am. Chem. Soc.* **1981**, *103*, 7403.
95. Atik, S. S., University of Notre Dame, unpublished data.
96. Wolff, C.; Grätzel, M. *Chem. Phys. Lett.* **1977**, *52*, 542.
97. Scheerer, R.; Grätzel, M. *J. Am. Chem. Soc.* **1977**, *99*, 865.
98. Grätzel, M.; Kiwi, J. *J. Am. Chem. Soc.* **1979**, *101*, 7214.
99. Lehn, J. M.; Sauvage, J. P. *Nov. J. Chem.* **1977**, *1*, 449.
100. Lehn, J. M.; Sauvage, J. P. *Nov. J. Chem.* **1979**, *3*, 423.
101. Shilov, A. E. *Dokl. Akad. Nauk SSSR* **1977**, *238*, 620.

THE CHEMISTRY OF EXCITATION AT INTERFACES

102. Kalyanasundaram, K.; Kiwi, J.; Grätzel, M. *Helv. Chim. Acta* **1978**, *61*, 2720.
103. Moroi, Y.; Braun, A. M.; Grätzel, M. *J. Am. Chem. Soc.* **1979**, *101*, 567.
104. Krieg, M.; Braun, A. M.; Humphry-Baker, R.; Grätzel, M. In press.
105. Cinquini, M.; Montanani, E.; Tundo, P. *J. Chem. Soc., Chem. Commun.* **1975**, 393.
106. Humphry-Baker, R.; Grätzel, M.; Tundo, P.; Pelizzetti, E. *Angew. Chem.* **1979**, *91*, 630.
107. Krieg, M.; Braun, A. M.; Humphry-Baker, R.; Grätzel, M. To be published.
108. Thomas, J. K.; Almgren, M. In "Solution Chemistry of Surfactants"; Mittal, K. L., Ed.; Plenum Press: New York, 1979; Vol. II, p. 599.

8

Reversed Micelles

IN MANY COLLOIDAL MICELLAR SYSTEMS, THE BULK PHASE is water, and the percentage of hydrocarbon is a minor, although important, constituent of the system. However, systems in which the bulk phase is hydrocarbon and the minor constituent is water are also quite common. For convenience these systems may be divided into two classes: (a) systems with only one surfactant, an oil, and water, called "reversed micelles"; and (b) systems with many surfactant components in addition to water and oil, called emulsion or microemulsion systems.

A stable colloidal solution of water in oil is formed in both systems. In photochemistry transparent solutions are desirable. Although this restriction usually removes emulsions from serious consideration, microemulsions are clear enough for optical studies.

Some Structural Details of Reversed Micelles

Three well-defined reversed micellar systems are described according to the surfactant used: 1, sodium diisooctylsulfosuccinate (AOT) and other derivatives with different hydrocarbon chains (1–6); 2, benzylhexadecyldimethylammonium chloride (BHDC), suitable only in aromatic solvents such as benzene (7, 8); and 3, surfactants of the dodecylamine propionate type with variations of the fatty acid (2–9); reversed micelles of lecithin in benzene also belong to this group (10, 11).

Groups 1 and 2, the ionic surfactants, may be distinguished from group 3, and nonionic surfactants, by greater water-solubilizing power. The ratio of water to surfactant molecules in group 3 is usually small; in groups 1 and 2 the ratio is very large and can be greater than 40:1. The AOT–alkane–water and BHDC–benzene–water systems will be a model for all three groups. Whether surfactants form reversed micelles in the absence of water is debatable because water is necessary as a focus for precise aggregation of the surfactant (12).

AOT–Alkane–Water Reversed Micelles. Large concentrations of AOT may be dissolved in many liquid hydrocarbons; thus, large amounts of water are solubilized in the system. In many spectroscopic measurements, the water is in small clusters or pools rather than monodispersed

0065-7719/84/0181-0191$06.00/1
© 1984 American Chemical Society

through the system. The size of the water pools depends on temperature and, most critically, on the water-to-surfactant ratio, W_o (5). Figure 1 shows typical data for the dependence of the water-pool radii (r) on these physical parameters. The nature of the hydrocarbon and the alkane chains of the AOT also play a minor but important role in establishing the size of the water pools. The micelle is thought to be a spherical shape with a water core separated from the hydrocarbon bulk by the surfactant. The polar head groups of the surfactant are in the water pools, and the hydrophobic tails are in the hydrocarbon. Penetration of the surfactant head group area by the solvent must take place, but little is known of this effect.

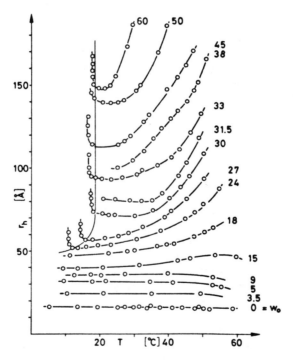

Figure 1. Measurements of the Stokes radii (r_v) of microemulsions of isooctane (AOT) and water as a function of the temperature, T. AOT concentrations were between 5.5 and 8.1 × 10⁻² M; the scattering angles varied between 8° and 90°. The vertical line at 18 °C indicates a stability boundary. To the left of this boundary, spontaneous growth of the aggregates occurs and will ultimately lead to phase separation. The data points shown in this region indicate successive measurements in 30-s time intervals (59).

Nature of the Water Pools. The following physical properties, which reflect on the nature of the water pool, show quite marked changes as the W_o is varied: a, NMR spectroscopy, both ¹H of water and ²³Na of AOT counterion (8, 14); b, IR spectroscopy (13, 14) of water pools; c, visible spectroscopy of coordinated ions such as Co^{2+} (8, 11, 15, 16) and charge-transfer spectra such as I^- (14); d, rigidity of the water pool as measured by fluorescence spectroscopy (4); and e, fluorescence spectral changes as monitored by sodium anilinonaphthalenesulfonate (4) or 8-hydroxy-1,3,6-pyrenetrisulfonate (pyranine) (17).

Figure 2 shows the effect of water content in the BHDC–benzene system on the $(CoCl_4)^{2-}$ absorption spectrum at λ_{max} = 694 nm. NMR data showing the spin–lattice relaxation times of the water peak are included; similar data are also obtained with the spin–lattice relaxation time of the N-methyl peak (8). Both absorption spectroscopy and NMR data of $(CoCl_4)^{2-}$ show two distinct breaks at about 0.1 and 0.5% water content. These breaks are interpreted as follows in terms of two states of water in the system: Up to a water content of 0.1%, water associates

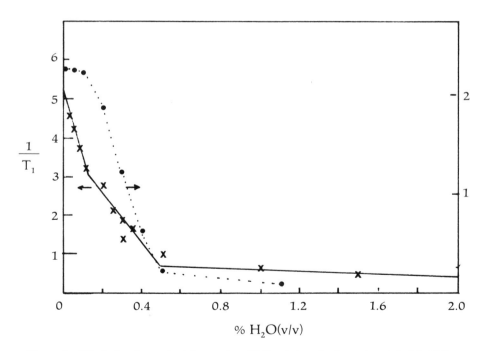

Figure 2. Effect of water content (percentage H_2O, v/v) on the intensity of the $[CoCl_4]^{2-}$ at λ_{max} 694 nm, in a BHDC–benzene–H_2O reversed micelle (●). Effect of water content on reciprocal spin–lattice relaxation times of the water peaks, $1/T_1$ (×) (8).

with the charged group of the surfactant; this effect saturates at 0.1% water content. Beyond 0.1%, water is available to interact with added solutes, and beyond 0.5%, the availability of water to added solutes approaches the properties of bulk water.

Co^{2+} in Cl^- solution exists in two forms: as $(CoCl_4)^{2-}$, which exhibits a strong absorption with λ_{max} 694 nm; and as $Co(H_2O)_6^{2+}$, which exhibits a weak absorption at λ_{max} 625 nm. The data in Figure 2 show a strong $(CoCl_4)^{2-}$ peak (i.e., 694 nm) at low water content up to about 0.1% water. However, the 694-nm absorption decreases at higher water content as progressively more water becomes available for complexation with Co^{2+} to form the weaker $[Co(H_2)_6]^{2+}$, which absorbs at 625 nm.

The NMR spectrum of the BHDC–benzene–water systems shows only one water peak; the chemical shift of this peak increases with an increase in water content. This behavior is explained by the increased covalent character of the water as well as by an increase in hydrogen bonding (18–20). The change in the reciprocals of the spin–lattice relaxation time of the water pools also exhibits this behavior. The data in Figure 2 show two peaks with behavior indicative of bulk water at water contents > 0.5%.

The IR spectra of the BDHC–benzene–water solution show absorption bands at 3620 and 3430 cm^{-1} with widths at half-heights of 55 and 120 cm^{-1}. These absorptions are assigned to bound and free water, respectively, from similar data observed in other systems (13). The 3620-cm^{-1} absorption did not increase in intensity with an increase in water content above 0.1%. This observation is in agreement with the suggestion that, up to this limit, water is merely associated with the polar head group of the surfactant. However, the 3430-cm^{-1} absorption increased with further addition of water above 0.1%, in agreement with the suggestion of progressive introduction of free water into the water pool.

Interestingly, NMR studies indicate that the benzyl group is placed away from the water pool and is probably located between the surfactant head groups. This condition (i.e., the introduction of a second chain or spacing group) is necessary for the stable formation of a reversed micelle.

AOT–Alkane–Water System. Figure 3 and Table I show typical data for the influence of water content on the following properties of the AOT–heptane–water system: I^- absorption spectra (14); ^1H NMR relaxation of H_2O (14); ^{23}Na NMR spectra (14); and anilinonaphthalene-sulfonate (ANS) fluorescence spectrum (4).

Table I shows that the ^1H-NMR spectrum of water exhibits a down-field chemical shift with an increase in water content of the system that

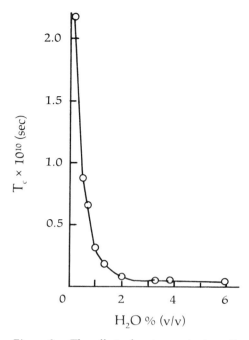

Figure 3a. The effect of water content on the rotational correlation times (τ_c) in 3% AOT–H_2O–C_7D_{16} reversed micelles (14).

gradually approaches ordinary water. As the water content increases, both the spin–lattice relaxation rates ($1/T_1$) and the spin–spin relaxation rates ($1/T_2$) of ^1H of water decrease significantly up to 1% H_2O; after this point, a much slower decrease is observed. Calculations of the rotational correlational time (τ_c) and T_1 show that the water molecules are highly immobilized in small water pools because of strong ion–dipole interactions with Na^+ counterions. When the ratio $H_2O/Na^+ \simeq 6$ (i.e., completion of the Na^+ hydration shell) the rigidity of the water pool is greatly reduced. Sodium-23 NMR measurements show an analogous decrease in $1/T_2$ with an increase in H_2O content.

The absorption spectrum of I^- in the water pool shows large differences from that observed in bulk water. The effect is associated mainly with the extinction coefficient of I^-, which is much smaller than bulk water ($\sim 1/100$) even in the largest water pool.

The spectral maximum shows a small red shift at lower water contents. The UV absorption spectrum of I^- involves formation of a spherically symmetrical excited state that involves charge-transfer character with the surrounding water (21–24). The decreased extinction coefficient of I^- in the water pools is explained as the lack of sufficient

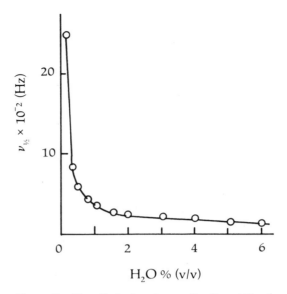

Figure 3b. The effect of water on the linewidth of
^{23}Na-NMR absorption lines (14).

Table I.
The Spectral Shift and Extinction Coefficient of 5 × 10^{-4} M I$^-$ in the Reversed Micelles and Water

Percent Water	λ_{max}, nm	ε, M^{-1} cm^{-1}
0.55	236.6	7.4
0.8	236.8	14.9
1.0	237.0	19.9
1.5	237.4	35.2
2.0	237.7	48.4
3.0	237.8	79.5
4.0	238.0	105.2
6.0	238.0	160.8
100	226.0	1270

available water to solvate I$^-$ and thus to produce the necessary environment for charge transfer to solvent.

In the water-free AOT solution, anilinonaphthalenesulfonate (ANS) displays a very intense fluorescence, which appears as a broad structureless band, λ_{max} 450 nm. The fluorescence is rapidly quenched by addition of water, which produces an accompanying red shift in the spectrum. The fluorescent properties of ANS are quite solvent dependent; the excited state is quenched in environments of high polarity (25, 26). The ANS data may be explained with the same model used for the I$^-$ spectrum and NMR data (i.e., increasing the water content of the pools leads to a more polar micelle core). This effect is less pronounced after H$_2$O/Na$^+$ $\simeq$ 6. However, the fluorescence yield and spectral maximum obtained with the longest water cluster (ϕ, 0.018; λ_{max}, 484 nm) are still considerably different from the respective parameters in bulk water (ϕ, 0.0038; λ_{max}, 510 nm). Thus, the effective polarity of a water pool of radius 70 Å is smaller than that of bulk water.

These four physical measurements indicate the same behavior for AOT–heptane–water pools as the water content is increased: Below a ratio H$_2$O/Na$^+$ $\simeq$ 6, little free water is available to participate in hydrating added solutes; the water pool is quite rigid because of a lack of free water; and beyond a ratio H$_2$O/Na$^+$ $\simeq$ 6, physical properties reminiscent of bulk water are approached. Decreasing the temperature of the system produces a decrease in λ_{max} and an increase in ϕ for ANS fluorescence. This effect is particularly marked below H$_2$O/Na$^+$ $\simeq$ 6, and again indicates a restriction of available H$_2$O to participate in solvation of excited ANS.

Fluorescence polarization measurements also reflect on the rigidity of the micelle core (14). Rhodamine B fluorescence is polarized in AOT–heptane–water micelles at low H$_2$O content, which indicates a restriction to movement of the probe by the micelle environment. Increasing the water content produces a decrease in fluorescence polarization because of the large water-pool size, which indicates a greater movement of the probe in these pools. The fluorescence polarization is particularly marked below H$_2$O/Na$^+$ $\simeq$ 6, but much less marked beyond a ratio of 6. This behavior is quite similar to that observed with other physical measurements in Figure 3.

The use of probes to reflect on the nature of micelles depends heavily on the location of the probe in the assembly. Some information is available on probe location in reversed micelles.

Location of Solutes

A small, hydrophilic anion such as I$^-$ is assumed to be located in the water region of a reversed micelle, away from the anion head group. A cation such as Cu^{2+} could be located in the water pool and be bound to the anionic head groups. The reverse is true for a cationic surfactant such as BHDC. The position of charged organic molecules, such as ANS (anionic) or rhodamine B (cationic), may also be guessed. For example, rhodamine B should be located at the head groups in AOT micelles, but ANS should be repelled. The reverse is true for BHDC micelles. However, the organic hydrophobic moiety of these molecules also provides some affinity for the organic region of the surfactant interface.

The following observations can be surmised for the location of solutes in reversed micelles: If the solute has a charge that is opposite to the charge of the surfactant head group, then the solute will be bound to some extent to the head group. A counterion, such as Na^+, Br^-, or Cl^-, is replaced in this process. However, solutes normally used (e.g., trisbipyridylruthenium ion in AOT or ANS in BHDC) are much more strongly bound than the surfactant counterion. If solute and surfactant head groups have like charges, then the solute associates mainly with the water pool. With noncharged solutes, partitioning occurs between the water pool, the head group region, and the oil phase to an extent that depends on the hydrophobicity of the solute.

Nonpolar Surfactants

Although strong evidence exists for the formation of reversed micelles by several ionic surfactants, the same situation is by no means true for nonionic surfactants (*12*). However, evidence has been presented for the formation of reversed micelles with alkylammoniumcarboxy-lates (*27, 30*) and poly(oxyethylene)nonylphenols (*2*). Typical ^{1}H-NMR data are shown in Figure 4 for the chemical shift of the $CH_3CH_2CO_2^-$ protons as a function of surfactant concentration for several alkylam-monium propionates in CCl_4 at 33 °C. The ^{1}H-NMR spectra consist of single weight-averaged resonances for the discrete protons of the mon-omeric and aggregated surfactant; therefore, equilibrium of the two forms is rapid on the NMR time scale. When a physical property of the system is plotted versus surfactant concentrates, quite often breaks in the plots are observed (Figure 4). These breaks are indicative of micellar behavior because they occur at the critical micelle concentration (CMC). Although this behavior is reminiscent of aqueous micellar systems, identical behavior may not occur in nonaqueous systems. Indeed, multiple equilibrium similar to that purported for alcohol–alkane systems has been suggested for the present surfactant system (*31*). For poly(oxyethylene)nonylphenol, the fluorescence of the phenol group, which has a lifetime of 21 ns in the monomeric form, changes at the CMC into two exponential decays with lifetimes of 21 and 9 ns, respectively (*30*). The shorter lifetime of 9 ns is associated with surfactant aggregation.

The spectroscopy of probe molecules such as 1-ethyl-4-carbome-thoxypyridinium iodide (*32*), vitamin B_{12} (*33*), and hemin (*34*) has been used to study water pools in reversed micelles formed by nonionic surfactants. Results similar to the AOT systems are found; an increase in water content increases the effective polarity of the water pool from a nonpolar condition until it approaches the polarity of bulk water.

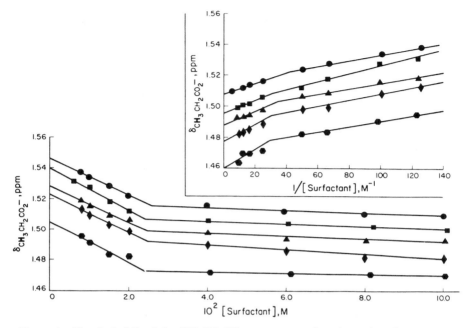

Figure 4. Chemical shift of the $CH_3CH_2CO_2{}^-$ protons as functions of surfactant concentration in carbon tetrachloride at 33 °C. ●, *butylammonium propionate;* ■, *hexylammonium propionate;* ▲, *octylammonium propionate;* ◆, *decylammonium propionate; and* ⬤, *dodecylammonium propionate (2).*

Studies with acid–base indicators, such as bromophenol blue, thymol blue, methyl orange, and malachite green, in Igepal CO-530 reversed micelles show that the pK_a values of these dyes are as much as 7 units lower in the micelles compared to the values in bulk water (35).

Catalysis of Thermal Reactions

Reversed micelles catalyze many thermal reactions; the most spectacular is the 5×10^6 increase in rate for the aquation of tris(oxalato)chromium(III) ion in octadecyltrimethylammonium tetradecanoate micelles over the rate in water (36).

The nature of the catalysis can take several forms. The rate of reaction can tend to a plateau with respect to both surfactant and substrate concentration. Typical examples of this form are the mutarotation of 2,3,4,6-tetramethyl-α-D-glucose (37), the *trans–cis* isomerization of sodium bis(oxalato)diaquachromate(III) (38), and the decomposition of meisen heimer complexes (39). Rate maxima have been observed for bimolecular reactions such as ligand-exchange reactions of vitamin B_{12} (40). Finally, the surfactant may enter into the reaction and cause a linear increase in rate constant with an increase in surfactant concentration. An example is the solvolysis of 2,4-dinitrophenyl sulfate, where the alkylammonium carboxylate surfactant acts as a general acid or general base catalyst.

Hence, reversed micelles catalyze bimolecular reactions by a proximity effect, as with normal micelles. For some reactions, however, bond breaking, which is involved in the kinetic process, is assisted by concerted proton transfer from the surfactant. Surfactants that cannot transfer protons often do not catalyze this type of reaction. However, assignment of catalytic properties to the solubilized water is necessary to explain catalysis with some systems because changing the water content drastically changes the catalysis effect. Spectacular catalysis of thermal reactions has been observed in many thermally reactive systems (2). The exact nature of the catalysis may still be debatable. Although proximity effect of micelles (local crowding of reactants) is important, we must also consider the unusual properties of the water pools, which do not exhibit physical properties normally associated with bulk water.

Fast Thermal Reactions

Several stop-flow studies have been carried out in reversed micelles, in particular metal–ligand substitution reactions (41–43). Because the activation energies of these reactions are unchanged, the main effect is the concentrating influence of the micelle surface on the reactants. The aquated nickel ion–murexide ion reaction is conveniently followed by stop-flow methods because the product of the reaction, the nickel murexide complex, is highly colored (44). Both reactants are confined to the water pools of AOT reversed micelles, and reaction proceeds by migration of the reactants from one water pool to another. This latter process is faster than the actual complexation process, which is rate limiting and decreases as the water pool size decreases. The rate-limiting step in this phase of the reaction is suggested to be the rate of loss of the first water molecule from the first coordination shell of the metal ion. The subsequent complexing of the Ni^{2+} with murexide is very fast and diffusion controlled.

The reaction is modified by reacting $Ni^{2+}(aq)$ with the bidentate ligand pyridine-2-azo-p-phenol (PAP). Unlike murexide, which is located in the water pool, PAP is located at the surfactant–water interface, where the $Ni^{2+}(aq)$ + PAP reaction takes place (45). Again, the rate-limiting step is the $Ni^{2+}(aq)$ + PAP reaction because exchange of reactants between water pools is rapid. The energetics of the reaction in reversed AOT micelles are similar to those observed for the reaction in NaLS–water micelles, where reaction occurs at the anionic surface.

Radiation-Induced Reactions: Capture of Ions by Water Pools

Capture of Electrons. Probably the simplest example of ion capture by a water pool of a reversed micelle is the capture of e^- by water pools of AOT in alkane. Radiolysis of these systems leads to the formation of e^- in the alkane phase of the system. Electrons are also produced in the micellar phase, but the yield is small compared to the yield produced in the bulk alkane phase because it is proportional, to the electron fraction of the phase.

$$\text{alkane} \rightarrow (\text{alkane})^+ + e^-$$

The e^- liberated in the alkane phase is captured by the water pools and forms e_{aq}^-, identified by characteristic absorption spectrum and kinetic behavior (*46, 47*).

$$e^- + (\text{micelle}) \rightarrow (e_{aq}^-)\ \text{micelle}$$

The yield of captured electrons, as observed by absorption spectroscopy, increased with water content, or water-pool radius.

Subsequent picosecond pulse radiolysis studies (*48*) showed that the rate constants for capture of e^- by the water pools were very rapid. Although the e^- attachment rates to the pools in the AOT–water–isooctane system were rapid, they were less than the diffusion-controlled rate at $W_o = [H_2O]/(AOT)(<12)$. However, the rates increased to the diffusion-controlled rate of $\sim 10^{15}$ M^{-1}s^{-1} at $W = 37$, where the micelle radius is ~ 100 Å. The transition from nondiffusion-controlled rates to diffusion-controlled electron attachment implies that free, or non-AOT bound, water in the micelle is required for efficient electron attachment. This observation is consistent with the previously mentioned NMR, fluorescence, and polarization studies, which indicated properties indicative of bulk water at $[H_2O]/(AOT)$ of ~ 30. The data shows the significance of free water for the formation of e_{aq}^-.

Photoionization studies of several solutes in AOT reversed micelles have been carried out (*48, 49*). The photoejected e^- is again captured by the water pool if the excited solute is close to the micelle. This situation requires a polar solute (e.g., tetramethylbenzidene or pyrenesulfonic acid), which is located at the micelle surface.

Capture of Anions by Reversed Micelles. Pulse radiolysis of alkane solutions of biphenyl ($C_{12}H_{10}$) produces the biphenyl anion ($C_{12}H_{10}^-$) via electron capture of e^- liberated in the alkane phase. The $C_{12}H_{10}^-$ is readily identified via its strong absorption spectrum with λ_{max} at 408 nm. In pure alkane solutions, a part of $C_{12}H_{10}^-$ disappears rapidly

($\sim$ 100 ns) via geminate recombination with the countercation, and a small portion survives for $\sim 10^{-6}$ s or longer. In the presence of AOT–water reversed micelles, the excited $C_{12}H_{10}^{-}$ spectrum decays rapidly to produce a modified $C_{12}H_{10}^{-}$ spectrum with λ_{max} at 395 nm. This decay is indicative of a very polar environment for the anion (50), and the kinetics indicate capture of $C_{12}H_{10}^{-}$ by the polar reversed micelle. This process is similar to e^- capture but the rate constant is much lower because of the much lower mobility of $C_{12}H_{10}^{-}$ compared to e^- in alkanes ($\sim 10^{-3}$ to 10^{-4} less).

Various ionic species that are solubilized in reversed micelle, such as H_3O^+, Cu^{2+}, and pyrenesulfonic acid (PSA), react with $C_{12}H_{10}^{-}$ via electron transfer. For PSA, the PSA anion (PSA$^-$) is readily observed (λ_{max} 493 nm). Table II shows the rate constants for reaction at several W_o (i.e., different size water pools). These rate constants are calculated from the quotient of the rate of reaction and the solute concentration. The latter may be bulk concentration or the concentration of the solute present only in the micelle. Both types of calculated rate constants are shown in Table II. The rate constant for reaction of $C_{12}H_{10}^{-}$ and PSA was found to be independent of [PSA] for 2×10^{-4} to 10^{-3} M (bulk concentration). For Cu^{2+}, the rate constant decreased at higher [Cu^{2+}]. The rate constants for reaction of $C_{12}H_{10}^{-}$ and PSA are independent of water content if the [PSA] in the micelle is used to calculate the rate constant. This observation is not true for Cu^{2+} or H^+, where the rate constant decreases with decrease in water content. The rate constants are lower than those observed for these systems in water. The lower rate constants and the effect of water indicate the association of Cu^{2+} and H^+ with the micellar anion head groups. The effect increases as the water content decreases; the rate of the e^- transfer reaction from $C_{12}H_{10}^{-}$ is correspondingly decreased.

Reactions of Excited States

Excited states of aromatic solutes, such as biphenyl, are produced on radiolysis of alkane solutions of these solutes. Both singlet and triplet excited states are formed; but, with biphenyl, rapid intersystem crossing ($\sim 10^{-8}$ s) produces the triplet state, $C_{12}H_{10}^{T}$. The biphenyl triplet energy is efficiently transferred to other solutes (such as pyrene), which have lower energies. The rate constant, k, is close to the diffusion-controlled rate constant, $k = 2.7 \times 10^{10}$ M^{-1}s^{-1} (Table II) for the following process:

$$\phi_2^* \xrightarrow{\sim 10^{-8} \text{ s}} \phi_2^T + P \rightarrow P^T + \phi_2$$

However, in a reversed micellar system, where the pyrene chromophore

Table II.
Bimolecular Rate Constants of Some Reactions Taking Place in the Reversed Micellar System 3% AOT–H_2O–Heptane

Reactions H_2O% (v/v)	K ($M^{-1}s^{-1}$)			
	6	3	2	1
$C_{12}H_{10}^{-}$ + PSA^a → $C_{12}H_{10}$ + PSA^-	$1.05 \times 10^{10}(6.3 \times 10^8)$	$1.84 \times 10^{10}(5.5 \times 10^8)$	$3.7 \times 10^{10}(7.4 \times 10^8)$	$6.5 \times 10^9(6.5 \times 10^7)$
$C_{12}H_{10}^{T}$ + PSA^a → $C_{12}H_{10}$ + PSA^T	$6 \times 10^9(3.6 \times 10^8)$	$7.5 \times 10^9(2.25 \times 10^8)$		
$C_{12}H_{10}^{T}$ + PBA^b → $C_{12}H_{10}$ + PBA^T	$6.4 \times 10^9(3.8 \times 10^8)$			
$C_{12}H_{10}^{T}$ + Py^c → $C_{12}H_{10}$ + Py^T	2.7×10^{10}			
$C_{12}H_{10}^{T}$ + $PTSA^d$ → $C_{12}H_{10}$ + $PTSA^T$	$<2 \times 10^9$			
H_3O^{+e} + $C_{12}H_{10}^{-}$ → $C_{12}H_{10}H$ + H_2O	$>5 \times 10^9(3 \times 10^8)$	$5.8 \times 10^9(1.78 \times 10^8)$	$4.6 \times 10^9(9.2 \times 10^7)$	$4.7 \times 10^9(4.7 \times 10^7)$
Cu^{2+f} + $C_{12}H_{10}^{-}$ → Cu^+ + $C_{12}H_{10}$	$<6.3 \times 10^9(3.78 \times 10^8)$	$3.0 \times 10^9(1.0 \times 10^8)$	$3.4 \times 10^9(6.8 \times 10^7)$	$1.7 \times 10^9(1.7 \times 10^7)$

NOTE: Rate constants were calculated from the bulk solute concentration over the whole solution. The rate constants in the parentheses were calculated from the local solute concentration in the micelle. In all cases, [$C_{12}H_{10}$] was 2×10^{-2} M.
[a] [PSA] varied from 10^{-4} to 10^{-3} M.
[b] [PBA] varied from 5×10^{-5} to 5×10^{-4} M.
[c] [Py] varied from 5×10^{-5} to 10^{-4} M.
[d] [PTSA] varied from 5×10^{-5} to 2×10^{-3} M.
[e] [H^+] varied from 10^{-3} to 10^{-2} M.
[f] [Cu^{2+}] varied from 10^{-4} to 2×10^{-3} M.

is attached to the micelle by use of a polar acidic group such as pyrene-butyric acid (PBA), or pyrenesulfonic acid (PSA), the rate constant is smaller. This fact indicates the inhibiting effect of the micelle surface on the approach of $C_{12}H_{10}{}^T$ to PSA or PBA. For pyrenetetrasulfonic acid (PTSA), the rate constant is considerably smaller because this solute is located in the water pool, an unattractive region for hydrophobic $C_{12}H_{10}{}^T$.

Excited states of pyrene may be formed by direct excitation of pyrene or its derivatives in a reversed micelle system. Pyrene is located in the alkane phase, and the presence of a reversed micelle does not affect the lifetime of the excited singlet state. However, polar quenchers such as Cu^{2+} and I^-, which are in the water pools, do quench the excited pyrene singlet state. The rate constants for quenching are less than 10 times smaller than those found in homogeneous solution. Addition of inert sales (e.g., Mg^{2+}) or an increase in the water-pool size reduces the rate constants to 1/100 of those found in homogeneous solution. Again, the explanation centers on the inability of hydrophobic pyrene to enter the polar core of the micelle. Cu^{2+} is located, to some extent, in the head group region as well as in the water pool. Cu^{2+} in the head group region can quench excited pyrene; however, the Cu^{2+} or I^- in the water core cannot quench excited pyrene efficiently. An increase in the water-pool size or the addition of inert cations that bind to the anionic head groups increases the extent to which Cu^{2+} is located in the water pool and decreases the extent of binding to the anionic micelle surface. These effects lead to decreased probability of encounter of the ionic quencher and excited pyrene.

The situation may be changed by situating the pyrene chromophore at the micelle surface (i.e., by using PSA). Figure 5 shows the effect of water content on the quenching of excited PSA by various quenchers in a AOT–water–heptane reversed micelle system. The quenching rate constants are calculated from local quencher concentrations for very polar quenchers, which are located in the micelle, and from bulk concentrations for nonpolar quenchers, which are located in the heptane phase but which also move to the micelle surface. All quenchers, apart from O_2 and CH_2I_2, show increased rate constants with increase in water content. Oxygen shows a weak maximum at ~1% water, and the quenching rate constant of CH_2I_2 shows a definite decrease as the water content increases.

The data show that increasing micellar size (i.e., increasing water-pool radius) increases the rates of reactions when the two reactants are hydrophilic and are located in the micelle. A few examples are the quenching of PSA* by I^-, Cu^{2+}, Tl^+, and also nitromethane and the electron transfer from $C_{12}H_{10}{}^-$ to Cu^{2+}. This effect is probably due to the enhanced mobility of solutes in the micellar water phase with increasing water content.

If one reactant is hydrophobic (located in the alkane phase) and one is hydrophilic (located in the micelle), then increasing water content produces a decrease in the rate of reaction. This result may be due to a decrease in the probability of reactant encounters with the increase in water content. This probability decrease in turn is due to the lack of penetration of the hydrophobic reactants into the micelle. Typical examples are observed in the quenching of P^* with I^- and Cu^{2+}, and PSA^* with CH_2I_2.

Oxygen is solubilized in both the water and the liquid alkane; the solubility is larger in the latter phase by a factor of about five. The quenching of PSA^* by O_2 in AOT reversed micelles is larger than in water. This quenching is interpreted by the observation that the PSA^* in the micelle is effectively exposed to the O_2 available in the alkane phase (4). Experiments with reaction rates of e_{aq}^- with O_2 in the water

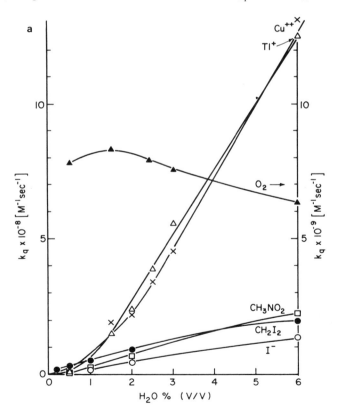

Figure 5a. The effect of water content on the quenching of excited PSA by various quenchers in 3% AOT–H_2O–heptane reversed micelles. The quenching rate constants are calculated from the local or micelle solute concentrations.

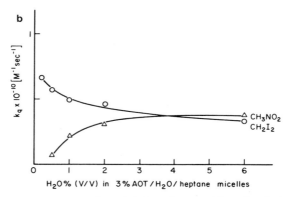

Figure 5b. The effect of H_2O on the quenching of excited PSA by CH_3NO_2 and CH_2I_2. The quenching rate constants are calculated from the bulk solute concentration (50).

pool also indicate a high O_2 concentration of 5.5×10^{-4} M/L in AOT–water–heptane micelles. The quenching rates of PSA* with O_2 are almost independent of water content. Unlike other quenchers such as I^-, Ca^{2+}, O_2 moves freely in this system.

Cationic Reversed Micelles

The data are sparse for photoinduced reactions in cationic reversed micelles; the only system studied was BHDC–water–benzene. The data are, for the most part, similar to what has been observed in AOT systems. PSA* in BHDC is quenched by Br^- ions with little effect of added water. This observation confirms that both anions (Br^- and PSA*) are bound to the cationic surface. The magnitude of the quenching is similar to normal CTAB micelles (51). Data similar to CTAB systems have been observed for PSA* quenching by Tl^+, dimethylaniline (DMA), nitromethane, and I^- (8). Quenching of P* by DMA in these systems produces the exiplex (P, DMA)*, which is identified by its characteristic spectroscopic properties. However, the water pools of the reversed micelle rapidly quench the exiplex ($k = 6 \times 10^{11}$ $M^{-1}s^{-1}$). No ions are observed, unlike the CTAB–Hartley micelle system (52, 53). However, large yields of ions are observed in the PSA*–DMA system in BHDC reversed micelles. This result leads to speculation on the nature of the geometric form of reactants that lead to large yields of long-lived ion formation. The sandwich exiplex form is apparently unattractive as back electron transfer is rapid (53).

Transport of Ions in AOT Reversed Micelles

An important feature of reaction kinetics in reversed micelles is the movement of ionic or polar species from one water pool to another.

Measurement of the rate of this process is possible for ionic species that are exclusively confined to water pools. The concept of the experiment is to set up a system of reactants, A and B, where a portion of the reaction requires movement of A or B from one pool to another pool containing B or A. Some initial rapid reaction arises when A and B are initially contained in the same water pool and has to be eliminated for the movement from pool to pool to be observed. Two sets of experiments via flash photolysis (54, 55) and stop flow (56) have been successful for observing the ion-exchange process.

Pulsed laser photolysis has been used to investigate the reaction kinetics of the following processes in AOT–water–alkane reversed micelles:

$$RuII \xrightarrow{h\nu} (RuII)^* + Fe(CN)_6^{3-} \rightarrow RuIII + Fe(CN)_6^{4-}$$

$$(RuII)^* + MV^{2+} \rightarrow RuIII + MV^+$$

$$PTS \xrightarrow{h\nu} PTS^T + FS \rightarrow PTS$$

$$PTS^T + Cu^{2+} \rightarrow PTS^+ + Cu^+$$

where RuII is tris(bipyridyl)ruthenium, MV is methyl viologen, PTS is pyrenetetrasulfonic acid, and FS is Fremy's salt. One of the reactants, RuII or PTS, is excited directly by a short laser pulse. The quenching of the excited state by the quencher, Cu^{2+}, $Fe(CN)_6^{3-}$, MV^{2+}, or Fremy's salt, is followed spectrophotometrically. An initial rapid decay as a result of micelles containing both reactants is observed. This decay is followed by a slower decay as a result of the migration of the quencher ion from a water pool, other than that which contains the excited $(RuII)^*$ or PTS^T, to the excited probe (54, 55).

According to the data, the reactants distribute themselves among the micelles according to a Poisson rather than a geometric distribution. From the kinetics of quenching, ion exchange only occurs on collision of water pools, and the effectiveness of exchange is 1% or less. Various additives affect the exchange rate as shown in Table III.

The data, typical values of which are shown in Table III, show that both cationic Cu^{2+} and anionic Fremy's salt behave similarly in all systems. The rate constants for ion transfer are 10^{-1} to 10^{-2} smaller than the rate constants for the collision of water pools, which are calculated from the diffusion equation. Perhaps, ions are exchanged with low efficiency in water-pool collision. Modification of the micelle surface by various additives markedly affects the efficiency of ion exchange. Benzyl alcohol, which sits in the head group region, shows the largest effect and increases the rate of ion exchange. Most other additives either do not significantly affect the rate or, as with toluene and benzene, cause a significant decrease in ion-exchange efficiency. Benzene is also located in the vicinity of the micelle head group region

Table III.
Effect of Additives on the Exit Rate Constant, k_e, of Fremy's Salt

Additive	$k_e \times 10^{-6}\ M^{-1}s^{-1}$
None	13
0.02 M Sodium lauryl sulfate	22
0.02 M Sodium laurate	15
0.02 M Sodium p-tolyl sulfate	18
0.02 M CTAC	3.3
0.02 M Tetramethylammonium chloride	6.8
0.3 M Hexanol	7.5
0.3 M Benzyl alcohol	330
40% Benzene	2.5
40% Toluene	1.6
None, but dodecane in place of heptane	50

NOTE: Solution consists of 2.0×10^{-5} M PTS, 0.2 M AOT–heptane ($W_o = 11$), and 1.0×10^{-3} M Fremy's salt.

where it blocks efficient ion transfer. The situation is quite reminiscent of ion transport across membranes. The exact mechanism of ion transport is not known, but two ideas immediately spring to mind: (a) the complete fusion of two micelles thereby mixing their contents, or (b) the formation of a temporary bridge between two micelles. Additives, which are situated in the micelle–hydrocarbon boundary should readily affect both mechanisms. Further work is required to place a more precise mechanism on this important process.

Electron Transfer

Reversed micelles are effective in promoting photoinduced charge separation. The photoinduced e^- transfer from RuII to benzylnicotinamide (BN$^+$) in DAP–benzene–water reversed micelles produces reduced nicotinamide (BH). Because BH is more hydrophobic than BN$^+$ and escapes into the bulk hydrocarbon phase, efficient charge separation and a large yield of photoinduced e^- transfer is produced (57). Reversed micelles promote efficient e^- transfer from tetraphenylporphyrinzinc (ZnTP) to either MV^{2+}, to give MV$^+$, or anthraquinone sulfonate, AQS (58). The charge of the micelle is important for maximum ion yield and efficient transfer. The ZnTP–MV^{2+} system works well in anionic AOT, and the ZnTP–AQS system works in cationic BHDC. ZnTP is solubilized at the micelle surface and/or in the alkane phase, and e^- transfer to MV^{2+} or AQS occurs from the long-lived triplet, ZnTP. The electron acceptors have to approach the surface for e^- transfer to occur from excited ZnTP; MV^{2+} is partly located at the AOT surface but repelled from the cationic BHDC surface. The anionic AQS is repelled from the

AOT surface and attracted to the BHDC surface. Hence, simple electrostatic features explain the marked effect of micelle surface charge on the efficiency of the e^- transfer reaction.

Conclusion

An understanding of much of the photochemistry of systems in reversed micelles is possible in terms of what is known in homogeneous solutions and in terms of what is known in normal, conventional Hartley micelles. An additional feature of interaction in reversed micelles is in their ability to produce various water pools where physical properties other than those of bulk water exist. This feature can lead to modified properties of many photochemical systems, and, coupled with catalysis, provides systems of great interest to enzyme chemists. Reversed micelles provide systems that in many ways are intermediate between Hartley micelles and microemulsions.

Literature Cited

1. Fendler, J. H.; Fendler, E. J. "Catalysis in Micellar and Macromolecular Systems"; Acad. Press: New York, 1975.
2. Fendler, J. H. *Acc. Chem. Res.* **1976**, *9*, 153.
3. Menger, F. *Acc. Chem. Res.* **1979**, *12*, 111.
4. Wong, M.; Grätzel, M.; Thomas, J. K. *J. Am. Chem. Soc.* **1976**, *98*, 2391.
5. Eiche, H. F.; Shepherd, J. C.; Steinermann, A. *J. Coll. Interface Sci.* **1976**, *56*, 168.
6. Robinson, B. H. *J. Chem. Soc., Faraday Trans. 1* **1979**, *75*, 481. Day, R. A.; Robinson, B. H.; Clarke, J. H. R.; Doherty, J. V. *J. Chem. Soc., Faraday Trans. 1* **1979**, *75*, 132.
7. Miller, D. J.; Klein, U. K. A.; Hauser, M. *J. Chem. Soc., Faraday Trans. 1* **1977**, *73*, 1654.
8. McNeil, R.; Thomas, J. K. *J. Coll. Interface Sci.* **1981**, *83*, 57.
9. Winsor, P. *Chem. Rev.* **1968**, *68*, 1.
10. Walter, N. V.; Hayes, R. G. *Biochim. Biophys. Acta*, **1971**, *249*, 528.
11. Wells, M. A. *Biochem.*, **1974**, *13*, 2943.
12. Eiche, H. F.; Christen, H. *Helv. Chim. Acta* **1978**, *61*, 2258.
13. Kertes, A. S. In "Micellization, Solubilization, and Microemulsions"; Mittal, K. L., Ed.; Plenum Press; New York, 1977; Vol. 1, p. 445.
14. Wong, M.; Thomas, J. K.; Nowak, T. *J. Am. Chem. Soc.* **1977**, *99*, 4730.
15. Cotton, F. A.; Goodgame, D. M. L.; Goodgame, M. *J. Am. Chem. Soc.* **1961**, *83*, 4690.
16. Sunamoto, J. Hamada, T. *Bull. Chem. Soc., Jpn* **1978**, *51*, 3130.
17. Correll, G. D.; Cheser, R. N.; Nome, F.; Fendler, J. H. *J. Am. Chem. Soc.* **1978**, *100*, 1254.
18. Hague, R.; Tinsley, I. J.; Schenedding, D. *J. Chem. Soc.* **1977**, *247*, 157.
19. Hansen, J. R. *J. Phys. Chem.* **1974**, *78*, 256.
20. Clifford, J.; Pethica, B. H. *Trans. Faraday Soc.* **1964**, *60*, 1483.
21. Franck, T.; Platzman, R. L. *Z. Phys. Chem.* **1954**, *138*, 411.
22. Jortner, J.; Ray, B.; Stein, G. *Trans. Faraday Soc.* **1960**, *56*, 1273; *J. Chem. Phys.* **1961**, *34*, 1455.
23. Stein, G.; Trienin, A. *Trans. Faraday Soc.* **1958**, *54*, 338.

24. Smith, M.; Symons, M. C. R. *Trans. Faraday Soc.* **1958**, *54*, 338; *J. Chem. Phys.* **1956**, *25*, 1074.
25. Weber, G.; Laurence, D. J. R. *Biochem.* **1954**, *56*, 31.
26. Kosower, E. M.; Doderik, H.; Tanigawa, K.; Ottolenghi, M.; Orbach. *J. Am. Chem. Soc.* **1975**, *97*, 2167.
27. Fendler, J. H.; Fendler, E. J.; Medary, R. T.; Elseoud, O. A. *J. Chem. Soc., Faraday Trans. 1* **1973**, *69*, 280.
28. Fendler, J. H.; Fendler, E. J.; Medary, R. T.; Elseoux, O. A. *J. Phys. Chem.* **1973**, *77*, 1432.
29. Ibid., *J. Phys. Chem.* **1973**, *77*, 1876.
30. Fendler, E. J; Constion, V. G.; Fendler, J. H. *J. Phys. Chem.* **1975**, *79*, 917.
31. Muller, N. *J. Phys. Chem.* **1975**, *79*, 287.
32. Fendler, J. H.; Lui, L. J. *J. Am. Chem. Soc.* **1975**, *97*, 999.
33. Fendler, J. H.; Nome, F.; Van Woert, H. C. *J. Am. Chem. Soc.* **1974**, *96*, 6745.
34. Hinze, W.; Fendler, J. H. *J. Chem. Soc., Dalton Trans.* **1975**, *238*.
35. Nome, F.; Chang, S. A.; Fendler, J. H. *J. Chem. Soc., Faraday Trans. 1* **1976**, *72*, 296.
36. O'Connor, C. J.; Fendler, E. J.; Fendler, J. H. *J. Am. Chem. Soc.* **1973**, *95*, 600; and *J. Chem. Soc., Dalton Trans.* **1974**, *625*.
37. Fendler, J. H.; Fendler, E. J.; Medary, R. T.; Woods, V. A. *J. Am. Chem. Soc.* **1972**, *94*, 7288.
38. O'Connor, C. J.; Fendler, E. J.; Fendler, J. H. *J. Am. Chem. Soc.* **1973**, *96*, 370.
39. Fendler, J. H.; Fendler, E. J.; Chang, S. A. *J. Am. Chem. Soc.* **1973**, *95*, 3273.
40. Fendler, J. H.; Nome, F.; Van Woert, H. C. *J. Am. Chem. Soc.* **1974**, *96*, 6745.
41. James, A. D.; Robinson, B. H. *J. Chem. Soc., Faraday Trans. 1* **1978**, *74*, 10.
42. Holzwarth, J.; Knocke, W.; Robinson, B. H. *Ber. Bunsenges. Phys. Chem.* **1978**, *82*, 1001.
43. Reinsborough, V.; Robinson, B. H. *J. Chem. Soc., Faraday Trans. 1* **1978**, *74*, 10; and *J. Chem. Soc., Faraday Trans. 1* **1979**, *75*, 2395.
44. Robinson, B. H.; Steytler, D. C.; Jack, R. D. *J. Chem. Soc., Faraday Trans. 1* **1979**, *75*, 481.
45. Fletcher, P. D. I; Robinson, B. H. In "Further Applications of Relaxation Spectrometry", Getins, W. J.; Wyn-Imers, E., Eds.; Holland Press, 1979.
46. Wong, M.; Grätzel, M.; Thomas, J. K. *Chem. Phys. Lett.* **1975**, *30*, 329.
47. Thomas, J. K.; Grieser, F.; Wong M. *Ber. Bunsenges. Phys. Chem.*, **1978**, *82*, 937.
48. Bakale, G.; Beck, G.; Thomas, J. K. *J. Phys. Chem.* **1981**, *85*, 1062.
49. Calvo-Perez, V.; Beddard, G. S.; Fendler, J. H. *J. Phys. Chem.* **1981**, *85*, 2316.
50. Wong, M.; Thomas, J. K. "Micellization, Solubilization, and Microemulsions", Mittal, K. L., Ed.; Plenum Press: New York, 1977; Vol. II, p. 647.
51. Miller, D. J.; Klein, U. K. A.; Hauser, M. *J. Chem. Soc., Faraday Trans. 1* **1977**, *73*, 1654.
52. Razem, B. K.; Wong, M.; Thomas, J. K. *J. Am. Chem. Soc.* **1978**, *100*, 1679.
53. Atik, S. S.; Thomas, J. K. *J. Am. Chem. Soc.* **1981**, *103*, 3550.
54. Atik, S. S.; Thomas, J. K. *Chem. Phys. Lett.* **1981**, *79*, 351.
55. Atik, S. S.; Thomas, J. K. *J. Am. Chem. Soc.* **1981**, *103*, 3543.
56. Robinson, B. H. Private Communication.
57. Ford, W. E.; Otvos, J. W.; Calvin, M.; Wilner, I. *Nature*, **1978**, *274*, 507; Ibid. **1979**, *280*, 823.
58. Costa, S.; Thomas, J. K. Proceeding 3rd International Conference, Solar Energy Meeting: Boulder, CO, 1980.
59. Zularuf, M.; Eiche, H. F. *J. Phys. Chem.* **1979**, *83*, 480.

9

Microemulsions and Liquid Crystals

THE SOLUBILIZATION OF AN OIL IN WATER or water in an oil by a simple surface-active agent leads to emulsions. These entities, although of great interest in industry, are problematical in photochemistry because of their optical opaqueness. However, smaller colloidal particles may be formed by using selected surfactants such as sodium diisooctylsulfosuccinate (AOT), which forms reversed micelles, or by the introduction of a cosurfactant to form microemulsions. Microemulsions were first formed by Schulman (1, 2) and have been extensively developed (3–6). The cosurfactant, by interacting with the head group region, introduces into the system a geometric character that is conducive to the formation of small particles. In other words, the geometric shape of the cosurfactant and surfactant resembles that of the specialized surfactants such as AOT and benzylhexadecyldimethylammonium chloride (BHDC), that is, two hydrocarbon chains in the vicinity of the head group.

The process may be approached more succinctly in thermodynamic language. For total miscibility of oil and water, the interfacial tension must be zero. For example, the interfacial tension of hexadecane and water is 53 dynes/cm, and it is reduced to 10 dynes/cm in the presence of potassium oleate. Addition of hexanol molecules leads to an interfacial film of potassium oleate and hexanol with a spreading pressure π given by $\pi = \gamma_o - \gamma_f$, where γ_o is the interfacial tension of a potassium oleate film (10 dynes/cm), and γ_f is the interfacial tension of this film with hexanol also incorporated. If π becomes larger than γ_o, or 10 dynes/cm, then the resulting interfacial tension, γ_f, will become negative and will lead to breakdown of the interface until curvature of the microemulsion drop attains equilibrium with its surroundings. Hence the hexanol, or cosurfactant, breaks up the large droplets of an *emulsion* system into the smaller droplets of a *microemulsion* system.

0065-7719/84/0181-0211$09.00/1
© 1984 American Chemical Society

Hexanol–Hexadecyltrimethylammonium Bromide–Water System

The hexanol–hexadecyltrimethylammonium bromide–water (H–CTAB–W) system is one good example of the many different assemblies that can be formed by varying the components of a surfactant system (7, 8). Figure 1 shows the composition diagram with indications of the micelle, microemulsion, and liquid crystal regions (9, 10). At very low or zero concentrations of hexanol, CTAB forms conventional "Hartley type" micelles in water, where the core of the micelle is hydrophobic. On the other hand reversed micelles that contain polar cores can be formed with CTAB in hexanol. Under the correct conditions, indicated in Figure 1, water can be solubilized in the cores. By suitably choosing the ratios of CTAB, hexanol, and water, bilayer-type liquid crystals or neat phases or cylindrical aggregates or middle phases can be formed. The reversed micelle phase is really a water-in-oil type microemulsion.

The influence of the various phases of the system on the rate of hydrolysis of p-nitrophenyl laurate has been studied (9, 10). The product of the reaction, nitrophenolate, is readily determined by its strong absorption spectrum at 400 nm (11). Figure 2 shows the variation in the rate constant for Reaction I in the liquid crystal neat phase as a function of water content and at several hexanol/CTAB ratios. The pseudo first-order reaction rate constants were determined from the rate of increase of the 400-nm absorption with time, and the second-order rate constants were calculated by dividing these values by the mole fraction of water. The rate constants show a pronounced increase when the water content is increased beyond 45%. NMR data show changes in the water mobility and counterion binding in this range. Above 45% water the [81]Br-NMR

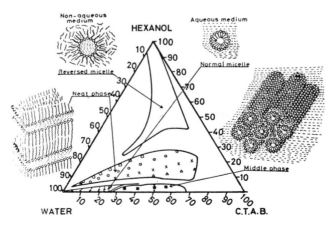

Figure 1. Regions and structure for the phases of the H–CTAB–W system. The compositions of the samples are marked (9).

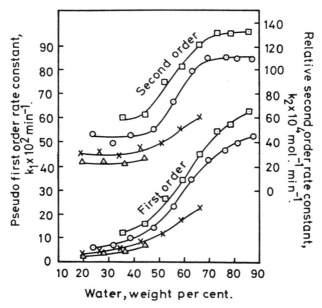

Figure 2. Reaction rates in the liquid crystal neat phase. The hexanol/CTAB ratios (w/w) were 0.77 (□), 0.55 (○), 0.34 (×), and 0.23 (△) (9).

$$NO_2 - \langle \rangle - O - \overset{\overset{O}{\|}}{C} - C_{12}H_{23} + OH^- \longrightarrow NO_2 - \langle \rangle - O^-$$

$$+ C_{12}H_{23}COOH$$

Reaction I

linewidth narrows, a result that indicates increased hydration of the Br^- ions accompanied by a decreased binding to the cationic group of the CTAB (12). Such a structural change may facilitate the formation of charged activation complexes in the vicinity of the polar layer. The hydrolysis rates also increase with increasing hexanol content, an effect that also gives rise to a decrease in counterion binding. The increase in rate constant appears to be fairly independent of phase type and to depend quite critically on the water content or degree of head group ionization.

Oleate–Alcohol–Alkane System

A commonly used microemulsion system is formed from the components potassium oleate, hexanol or pentanol, hexadecane, and water

(13–15). Figures 3a and 3b show the variation in the aggregate type and the electrical conduction on increasing the water-to-oil ratio. Up to a water-to-oil (volume/volume) ratio, r, of about 0.6, the system is optically clear, and is thought to consist of isotropic spheres containing water in their cores, surrounded by hexanol and oleate, with the oil as the bulk phase. These systems may be thought of as water-in-oil microemulsions. When $r > 0.7$, the system is turbid and exhibits birefringence; the viscosity of the system is greatly increased; and the electrical conduction also increases sharply. Possibly, the structure of the system changes from isotropic water spheres to water cylinder to water lamellae. Above $r = 1.6$, a continuous water phase is formed with spherical aggregates possessing oil centers surrounded by hexanol

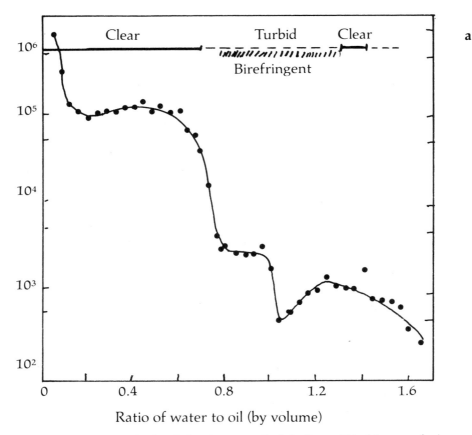

Figure 3a. Variation in electrical resistance, optical clarity, and birefringence of microemulsions as the water content increases. The microemulsion contains 0.20 g of potassium oleate in 1 mL of oil; the ratio of hexanol to oil is 0.40 (by volume) (5, 14, 15).

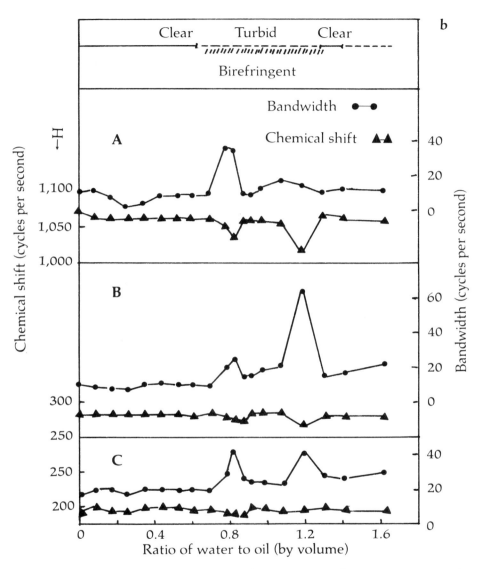

Figure 3b. Variation in the bandwidth at half-light and chemical shift of (A) water, (B) methylene, and (C) methyl protons in high-resolution NMR (220-MHz) spectra is the same microemulsion as that in Figure 3a, as the water constant increases. The upper part of the diagram shows corresponding data in optical clarity and birefringence (5, 14, 15).

and oleate. These systems are correctly called oil-in-water microemulsions.

The continual decrease in the electrical resistance of the system on increasing the water content is indicative of the formation of a more continuous water structure, such as water cylinders and water lamellae. The optical birefringence is typical of the liquid crystalline nature of the latter two structures, and the continuous nature of the aggregates gives rise to the high viscosity in this part of the system. Further evidence for these assignments is obtained from NMR studies. The NMR measurements of the bandwidth at half-height and chemical shift of the protons of water and hexanol (–OH), methylene (–CH$_2$–), and methyl (–CH$_3$) show distinct changes occurring in the birefringent region; the chemical shift of water protons is more pronounced than that of (–CH$_2$–) or (–CH$_3$). Two distinct molecular environments for water exist in this region. In the first, the chemical shift of water protons is moved up field by 25 Hz, and in the second by 50 Hz, when compared to water spheres. The water proton bandwidth at half-height is much greater in the first compared to that in the second environment. A greater bandwidth is associated with a lower molecular mobility; that is, water is more restricted in the first environment. The bandwidth of the methyl protons suggests that the hydrocarbon moieties are less mobile in the second environment. The measurements are consistent with increasing water content first forming water cylinders dispersed in oil, and the second environment consists of water and oil lamellae.

Many microemulsion and liquid crystalline systems can be engineered from surfactant–cosurfactant–water–oil systems of the correct composition. Several excellent reviews outline other systems of interest (16).

Sites of Solubilization and Environments of Reactants

As indicated earlier, much has been written about the nature of microemulsion systems. It is probably safe to assign the following structures to microemulsions:

Oil-in-Water Systems. The bulk phase is water with small quantities of surfactant and cosurfactant in equilibrium with the microemulsion particle surface. This projection comes from ultrasound studies (21) that demonstrated two relaxation processes due to the two surfactant components in different surroundings, micelle and water bulk. The surface consists of a single layer of surfactant and cosurfactant separating the oil drop center from the aqueous phase. The core of the particle consists primarily of oil with some cosurfactant. This statement is partly a projection of the known solubility properties of systems and observed NMR properties of these systems, which indicate significant cosurfactant concentrations in the oil phase (22). The particles are spherical and vary in size from radii of 100 Å upwards, depending on the components of the system. The last statement is taken from the earliest work of Schulman, who prepared electron microscope pictures of these systems, and also from numerous light scatter studies.

Water-in-Oil Systems. The bulk phase is oil that contains some cosurfactant and little if any surfactant. Ultrasound velocity measurements indicate only one relaxation process due to a single component in equilibrium with the microemulsion particle (22). The long-chain alcohol or cosurfactant is relatively soluble in oil, but the ionic surfactant is insoluble. The microemulsion particle surface consists of cosurfactant and surfactant, and the core consists of water. The particles are spherical, and the radii can range from about 100 Å upwards.

In both systems the geometry of the particles is stated to be spherical. However, under stated conditions, the spheres can coalesce to form cylinders with oil or water cores. These cores often pack to form organized systems or liquid crystals; layered liquid crystal structures can also be formed. Most of the work reported has been carried out in spherical systems.

Location of Guest Molecules

The foregoing description of microemulsion systems predicts several sites of solubilization for guest molecules: (1) the microemulsion core, (2) the surface, and (3) bulk phase.

Because the systems are dynamic, only average location sites for guest molecules can be allocated, and the surfaces of the structures themselves dictate preferred regions of solubilization. Ionic molecules (e.g., counterions such as Br^-, I^-, or Na^+ and trisbipyridylruthenium) locate both at the interface and in the aqueous regions; nonpolar molecules primarily locate in the oil phase but also move into the surface region; molecules of intermediate hydrophobicity (e.g., dimethylaniline) locate primarily at the interface with penetration into the oil phase and into the aqueous bulk. Increasing the hydrophobicity (e.g., dibutylaniline) draws the molecule away from the aqueous regions and more into the oil phase (23). The fluorescences of these guest molecules are quite solvent dependent and show red shifts with increasing solvent polarity. In oil and water microemulsions, the fluorescence is indicative of an environment less polar than water, more polar than the oil, and similar to that of the long-chain alcohol or cosurfactant, Table I.

Very hydrophobic molecules are located in the oil phase, but do approach the particle surface, and have a very transitory existence in

Table I.
Wavelength of Maximum Fluorescence of DMA
and DBA in Various Media

Medium	λ_{max}, nm	
	DMA	DBA
Cyclohexane	332	337
Hexadecane		338
Ethanol	345	345
Water	361	341
0.05 M CTAB	353	345
0.01 M DDAB	354	346

the water phase. This point is well illustrated by the guest molecule pyrene.

Photoinduced Reactions in Microemulsion Systems

Radical Reactions. On photolysis, tetra-*tert*-butylphthalocyanine, an oil-soluble derivative in an oil-in-water microemulsion, abstracts an H atom from the 1-pentanol cosurfactant used in the system (26). The pentanol radical produced reacts with oxygen and produces a peroxy radical. Singlet oxygen is also generated and is detected via trapping with a spin trap to produce a diamagnetic species. Oxygen is consumed during photolysis and the phthalocyanine dye is destroyed. Successful attempts were made to generate the pentanol radical by photolysis of $S_2O_8^{2-}$ in the system, which produces OH and SO_4^- radicals. These radicals react with the pentanol to produce a radical similar in character to that produced by H atom abstraction by excited phthalocyanine, a result that supports the proposed mechanism.

Electron-Transfer Reaction. Photoinduced electron-transfer reactions have been studied extensively in micellar media, and many of the same reactions have been investigated in oil-in-water microemulsions. In particular, the systems duroquinone (DQ)–diphenylamine (DPA) and methyl viologen (MV^{2+})–N-methylphenothiazine (MPTH), where the first component of the couple, the electron acceptor, and the second component, the e^- donor, show significant photoinduced charge separation in microemulsions (27). With DQ–DPA both components are associated with the oil drop, but the exact location of the solutes is uncertain. Photoexcitation of the system leads to electron transfer from DPA to DQ. Laser flash photolysis studies indicate that the process occurs in two well-defined time regions, a rapid (10^{-9} s) region involving singlet excited DQ and DPA,

$$DQ^s + DPA \rightarrow DQ^- + DPA^+$$

and a slower (10^{-6} s) transfer involving triplet excited DQ,

$$DQ^T + DPA \rightarrow DQ^- + DPA^+$$

The cationic DPA$^+$ associates with the negatively charged microemulsion surface and reacts to form multicomponent complexes, and DQ$^-$ escapes to the water bulk.

For the system MV^{2+}–MPTH, MV^{2+} is adsorbed at the particle surface in place of the cationic counterion, and MPTH is associated with the micelle anterior, probably both in the oil core and at the surface. Photoinduced electron transfer again occurs via two well-separated time-dependent processes, an initial very rapid followed by a slower e^- transfer from singlet excited MPTH to MV^{2+} to give reduced MV$^+$, which is observed readily via its characteristic absorption spectrum (λ_{max} = 390 and 510 nm). The rapid e^- transfer from singlet excited MPTH is in competition with intersystem crossing of the excited singlet to the excited triplet state, which subsequently slowly transfers e^- to MV^{2+} (10^{-6}-s range). This latter process probably occurs at a diffusion-controlled rate on the particle surface, which indicates that the much faster initial e^- transfer may occur via e^- tunneling or via ground state pairs of MV^{2+} and MPTH. The latter two processes do not require diffusional movement of the reactants, and may be very rapid.

Photoexcitation of Chlorophyll. Electron transfer from excited chlorophyll to duroquinone occurred readily in sodium lauryl sulfate (NaLS) micelles, and the micelle promoted the escape of the ionic products (28). Similar data are observed in oil-in-water microemulsions using sodium cetyl sulfate, hexadecane, and pentanol (29). Excitation of chlorophyll at λ = 694 nm with a pulsed ruby laser gives rise to a short-lived transient with λ_{max} = 465 nm, which is associated with the triplet state of chlorophyll a. The triplet state decays rapidly in the presence of MV^{2+} to produce reduced methyl viologen MV$^+$ and the chlorophyll cation. At high (MV^{2+}) (5 × 10^{-3} M), the efficiency of the process approaches unity. The chlorophyll cation reacts with the repair agents ascorbate ion and NADH and gives chlorophyll and a radical state of the repair agents. In the presence of colloidal platinum the MV$^+$ can be induced to produce H$_2$ by water decomposition and MV^{2+}.

Steady-state photolysis studies have been carried out with chlorin, a mixture of chlorophyll a and pheophytin a (30) in several oil-in-water microemulsions. Excitation of the system in the presence of methyl red and ascorbate leads to a chlorophyll sensitized reduction of the methyl red dye by ascorbate. At a bulk pH 7 the quantum yields for the process decrease in the following order for different microemulsions: anionic > neutral > cationic. This result is attributed to the increased pH at the micelle surface, pH(cationic) > pH(neutral) > pH(anionic), an effect that decreases the quantum yield in homogeneous media. However, the reversed order is obtained for the variation in quantum yield with ascorbate concentration. Ascorbate probably is more strongly bound to the cationic microemulsion surface, and this binding leads to higher yields of photoreduction of the methyl red, which is also associated with micelles. The system is complicated, because the photoproducts such as N,N-dimethyl-p-phenylenediamine, even at millimolar concentrations, increase the photoreduction process by two- to three-fold.

Quenching of Excited States

The quenching of excited singlet states by oxygen is diffusion controlled in homogeneous solutions. A small diminution in rate constant for this process is observed in micellar solutions and is interpreted as being due to restriction of movement of reactants by the assembly. However, most likely the oxygen solubility is higher in micelles because of the hydrophobic alkane-like environment. Indeed, this higher solubility has been found both from solubility (31) and quenching (32) studies. The increased $[O_2]$ is more pronounced in microemulsion systems (25, 33) as seen from measurements of the O_2 solubility in various systems at 21 °C measured by a Van Slyke apparatus (25). $[O_2]$ of 1.3, 2.5, 8.0, and 1.4 mM were found for water–NaLS–pentanol, dodecane oil-in-water microemulsion, pentanol–dodecane mixture (2:1), and 0.2 M NaLS, respectively. The solubility in the microemulsion is about twice that in water. Pulse radiolysis experiments, utilizing the established reaction of e_{aq}^- with O_2, showed that the [oxygen] in the bulk aqueous phase was that of pure water. Thus the increased $[O_2]$ resides in the oil drops and is comparable to that in dodecane or pentanol. The rate of pyrene fluorescence quenching by O_2 is much larger in a microemulsion system than in micelles or water. Analysis of the data indicates a rate constant for entry of O_2 into the microemulsion k_+ of 6.4×10^{10} $M^{-1}s^{-1}$, and of exit, $k_- = 1.5 \times 10^8$ $M^{-1}s^{-1}$. Similar data are obtained in other microemulsion systems (33, 34); all data indicate that the oil core of the microemulsion can be regarded as small but simple homogeneous portion of the system. The viscosity of the oil core is similar to that of the oil constituting the system, a value determined from photokinetic studies such as excimer formation and fluorescence quenching.

Fluorescence polarization studies indicate that the rigidity of the microemulsion surface is much lower than that of a micelle made from the same surfactant. This condition also is true with a micelle swollen by the cosurfactant. At 20 °C the microviscosities obtained are 52, 0–2, 4–5, and 1.26 cp for 0.2 M NaLS, 0.2 M NaLS saturated with pentanol, NaLS–pentanol–dodecane microemulsion, and dodecane, respectively (25). However, quenching studies to be described later indicate a slightly different picture of the nature of solute mobility in the particle surface.

Quenching by Tl^+ Ions

The effect of Tl^+ ions on excited singlet states of pyrene, P^*, and pyrenebutyric acid, PBA^*, depends on the nature of the microemulsion. The quenching rate constants for PBA^* by Tl^+ ions are similar in 0.2 M NaLS, 0.2 M NaLS saturated with pentanol, and for an NaLS–pentanol–dodecane microemulsion (25), being $k_q = 5.0 \times 10^9$, 2.6×10^9, and 3.0×10^9 $M^{-1}s^{-1}$, respectively. Probably the pyrene chromophore is held close to the surface in all systems, and encounters with Tl^+ are comparable in each assembly. The slight decrease in k_q of the microemulsion and swollen micelle with respect to the micelle is probably due to the decreased binding of Tl^+ in the first two systems compared to the micelle.

A different picture emerges with excited pyrene where the quenching rate constant with Tl^+ is much smaller in the microemulsion compared to the micelle or swollen micelle, k_q being 0.8×10^9, 7.0×10^9, and 9.7×10^9 $M^{-1}s^{-1}$, respectively. As pyrene is situated further into the assembly core in the microemulsion compared to the micelle, or swollen micelle, encounters with Tl^+, which is bound to the surface, most likely are severely contained. Excited pyrene has to reside at the surface during the residence period of Tl^+ in this region. The quenching curves were always exponential, and indicate that Tl^+ exchanges readily between the various aggregates in all systems. Similar effects are also noted for I^- quenching of pyrene and PSA in cationic microemulsions (34). The data show that the quenching rate constant is similar for PSA in a micelle or microemulsion, and that for pyrene is much lower in the microemulsion compared to the micelle.

Oleate Microemulsion

Different kinetic features are observed in an oleate–hexanol–hexadecane microemulsion (15, 33). Table II shows the rate constants for quenching of excited pyrene and excited pyrenebutyric acid (PBA) by CH_2I_2, Tl^+, dimethylaniline (DMA), and CH_3NO_2 in an oleate micro-

Table II.

Results of Fluorescence Quenching Studies in 30% (w/w) Water (ω_o)/Oleate Microemulsions

Probe	$k_q \times 10^{10} \ M^{-1} \ s^{-1}$			
	Tl^+	DMA	$[O_2]^* \times 10^3$	I^-
P	0.4	4.0	2.0	<0.005
PSA	3.3	3.0	0.41	0.14
PBA	2.1	1.3	1.0	0.05

NOTE: Solutions of 1.0×10^{-5} M probe were outgassed by bubbling N_2. Calculated from the relationship $1/\tau = (1/\tau_o) + [O_2]k_q$ where k_q is taken as $10^{10} \ M^{-1} \ s^{-1}$ for air-equilibrated samples.

emulsion and oleate micelle. The simplest approach to explaining the data is to state that hydrophobic molecules (e.g., pyrene and CH_2I_2) are located in the interiors of the assemblies, and polar molecules (e.g., CH_3NO_2, PBA and DMA) are in the aqueous phase and in the surface region of the assembly; cations such as Tl^+ are bound to the $-COO^-$ head groups in the assembly surface. The cation binding to $-COO^-$ groups is considerably stronger than to $-OSO_3^-$ head groups of the NaLS assemblies (35). The rate constant for quenching of pyrene by CH_2I_2 (both in the assembly interior) is much larger than that for PBA and CH_2I_2, because PBA lies in the surface region. The opposite is true for CH_3NO_2 quenching, where the CH_3NO_2 is in rapid equilibrium between the aqueous phase and the water bulk. These latter data are similar to the Tl^+ experiments of the NaLS systems. An interesting situation occurs with strongly bound surface quenchers (e.g., Tl^+ and DMA) that apparently still quench excited pyrene more rapidly than excited PBA. This result indicates that pyrene approaches the surface quite readily to react with these molecules. However, the data indicate an apparent restriction of lateral movement along the surface for Tl^+ or DMA, quenching of excited PBA. These data are more in line with micelle kinetics where restricted surface movement is observed. Data to support the suggestion (33) that movement of molecules at the surface is restricted compared to that in bulk are forthcoming for the efficiency of excimer formation of pyrene and PBA in these systems. The efficiency of excimer formation is comparable for pyrene and PBA in homogeneous solution and in micelles, but is much smaller for PBA compared to pyrene in microemulsions. These data again indicated restricted lateral movement in the head group region for PBA, probably due to counterion binding, but movement perpendicular to the surface from the core occurs readily. Thus the fluorescence polarization data do not represent the true kinetic situation at the surface. Movement other than gross displacement, such as rotation, plays a big role in fluorescence polarization.

In all of the cases just described, the quenching kinetics exhibit a single exponential decay. However, under special circumstances, as in the case of micelles, the decay kinetics are not simple because of the distribution of the reactants among the particles, a distribution that is Poisson in nature.

Poisson Kinetics in Microemulsions

Following previous experience with micelles, solutes are assumed to be distributed among micelles according to a Poisson distribution. If the solute, such as pyrene, is distributed among assemblies according to Poisson statistics, then a reaction such as the decay of pyrene fluorescence giving rise to excimer formation is described by (37)

$$I_F(t) = I_F(0)e^{-\{k_1 t + \bar{n}(1 - e - k_e t)\}} \tag{1}$$

where $I_F(t)$ and $I_F(0)$ are the fluorescence intensities of excited pyrene at time t and 0, respectively; $\bar{n}$ is the average pyrene occupancy of a microemulsion particle given by $[P]/[\mu E]$; k_e is a first-order intramicroemulsion excimer formation rate constant for a particle containing two pyrene molecules, one of which is in its excited singlet state; and k_1 is the first-order rate of decay of P^* solubilized in a particle that does not contain other pyrene molecules (i.e., the lifetime of P^* under conditions where no excimer emission is observed at $\bar{n} \leq 0.1$).

For a certain $\bar{n}$ value ($\bar{n} \geq 0.2$) where Poisson law ($\alpha_n = \bar{n}^{-n}e^{-\bar{n}}/n!$), ($\alpha_n$ is the probability of finding a microemulsion particle containing n probe molecules) predicts the existence of particles with two or more solubilized pyrene molecules, the fluorescence decay curve shows two components (as predicted by Equation 1): an initial fast one attributed to particles having solubilized two or more P molecules, and a limiting slow one corresponding to particles that contain only one P molecule.

At sufficiently long times when all the fluorescence emitting from multiple occupied particles has decayed completely (i.e., $e^{-k_e t} = 0$), the remaining fluorescence intensity can be ascribed to the quenched P or singly occupied particles and will be given by

$$I_F(t) = I_F(0)e^{-(\bar{n} + k_1 t)}$$

Therefore, the limiting linear slope of the log plot of $I_F(t)$ versus t will be given by $(-k_1)$, and the extrapolated intercept will be equal to

$(-\bar{n})$. From this experimentally determined value of $\bar{n}$ and the molar concentration of P, the concentration of microemulsion particles in solution can be calculated.

The described case is not always realized and can only be observed for systems where $k_e \gg k_1$. In cases where k_e is comparable to k_1, the fluorescence decay will no longer exhibit two distinct components because the contribution of excimer fluorescence to the total fluorescence intensity from multiple occupied particles will always be significant even at very long times $t \gg 1/k_e$. Nonetheless, in such cases the observed nonexponential fluorescence decay can be fitted to the equation by a two-parameter ($\bar{n}$ and k_e) computer program, and therefore an estimate of $\bar{n}$ can still be achieved.

Examples of the two cases just discussed are given in Figure 4. The continuously nonexponential fluorescence decay is actually observed for the CTAB micelle. However, upon addition of hexanol, the decay exhibits two easily discernible components resulting from an increase in the rate of excimer formation (k_e), which may be associated with an increase in the microfluidity of the mixed micelle. The two-component fluorescence decay is also noted for the microemulsion particle. The solid curves in Figure 4 show the best-fit simulations according to the equation. The results of such a time-resolved fluorescence-decay analysis are detailed in Table III.

The surfactant aggregation number (N) for the various molecular assemblies is calculated from the derived values of $\bar{n}$, which may be expressed as $N[P]/([CTAB] - CMC)$, with the literature critical micellar concentration (CMC) value of 9.6×10^{-4} M for the micelle and the estimated value of 5.0×10^{-4} M for the CTAB–hexanol mixed micelle and for the microemulsion particle. Under the experimental condition of $[CTAB] \gg CMC$, the calculated values of N should not be affected by errors in the approximated values of the CMC.

Table III also shows the numbers of hexanol and dodecane molecules per aggregate, calculated on the basis of the assumption that they are completely incorporated into the aggregates. Such an assumption is justified because neither hexanol nor dodecane is very soluble in water. Simple calculations using the density of dodecane and the concentration of the microemulsion particles in solution yields a radius of 40 Å for the oil droplet.

Photoionization in Microemulsions

As discussed, the threshold and yield of photoionization were decreased and increased, respectively, in micelles compared to homogeneous solution (38). Similar effects were observed in small microemulsion systems [e.g., NaLS, pentanol–dodecane (25)] where the yields of

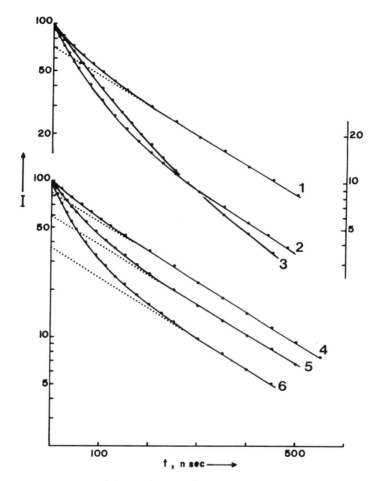

Figure 4. Rate of decay of pyrene fluorescence, I, at various [pyrene] and conditions: 1, [pyrene] = 1.0 × 10⁻⁴ M (CTAB); 2, [pyrene] = 1.0 × 10⁻⁴ M (0.01 M CTAB + 0.05 M hexanol); 3, [pyrene] = 1.0 × 10⁻⁴ M (CTAB microemulsion); 4, [pyrene] = 5.0 × 10⁻⁵ M (0.01 M CTAB + 0.05 M hexanol); 5, [pyrene] = 1.5 × 10⁻⁴ M (0.01 M CTAB + 0.05 M hexanol); and 6, [pyrene] = 3.0 × 10⁻⁴ M (0.01 CTAB M + 0.05 M hexanol) (23).

photoionization for pyrene and N,N'-tetramethylbenzidene (TMB) are similar to those obtained in micelles. Although threshold values were not measured, the photoionization of TMB was one photon with 3471-Å light, as in micelles. Larger systems show different behavior; for example, the yield of photoionization of pyrene in oleate microemulsions was only 38% of that in oleate or NaLS micelles, but that of PBA was similar in all systems (38). Pyrene is situated further from the

Table III.
Kinetic Parameters Obtained from Intraaggregate Excimers
Fluorescence Quenching Decay Analysis

[P], M	$\overline{n}$	$k_1 \times 10^{-6}$, s^{-1}	$k_e \times 10^{-7}$, s^{-1}	CTAB	Hex-anol	Dodec-ane
0.01 M CTAB						
5.0×10^{-5}	0.40	5.88	0.50	75	—	—
1.0×10^{-4}	0.82	5.85	0.60	—	—	—
0.01 M CTAB + 0.05 M Hexanol						
5.0×10^{-5}	0.17	4.35	1.30	33	170	—
1.5×10^{-4}	0.53	4.30	1.40	—	—	—
3.0×10^{-4}	1.00	4.40	1.35	—	—	—
CTAB-μE (O/W)						
1.0×10^{-4}	1.35	4.00	0.80	135	670	800

aqueous phase than either pyrene in micelles or PBA in any system. Hence, the probability of e^- escape from the pyrene–oleate microemulsion system is lower than that from the other systems. Similar effects of site or depth of solubilization of molecules on the photoionization yield have been observed previously for micelles of various sizes (39).

Photoinduced Electron-Transfer Reactions

Photoexcitation of the charge-transfer systems pyrene–dimethylaniline and pyrene–dibutylaniline has been studied in several organized assemblies, micelles, microemulsions, and vesicles. Both steady-state measurements and pulse laser photolysis data show that the quenching of excited pyrene on the surface by the anilines is rapid but can be described by diffusional-type processes. The main products of the quenching are pyrene anions and dialkylaniline cations. Increasing the size of the assemblies (i.e., micelle to microemulsion) or increasing the rigidity of the reactants' environment (i.e., micelle to vesicle) led to decreased yields of ions. The yields of ions in these latter systems can be restored if polar derivatives of pyrene are used in place of pyrene, thus the pyrene chromophore is located in the region of the assembly surface.

Effect of Structure on the Yield of Photoelectron Transfer

Table IV shows the yield of P^-, measured at $\lambda = 4930$ Å, in the photoassisted electron transfer from DMA or DBA to pyrene, pyrene-butyric acid, pyrenesulfonic acid (PSA), or pyrenetetrasulfonic acid (PTSA)

Table IV.
Relative Yield of P⁻ (493 nm)

Probe	CTAB	Medium		
		A	B	C
P	1.0 (0.40)	0.50 (0.32)	0.24 (0.12)	0.85 (0.67)
PBA	1.0	0.90	0.77	0.80
PSA	1.0	0.87	0.66	0.96
PTSA	1.0	0.95	0.90	—

NOTE: 5.0×10^{-5} M P (PBA, PSA, PTS)/$5-10 \times 10^{-3}$ M DMA. Better than 95% of the fluorescence is quenched: Medium A, 0.05 M CTAB + 0.1 M hexanol; Medium B, 0.05 M CTAB + 0.25 M hexanol + 0.3 M dodecane (1.83% (w/w) CTAB, 2.5% hexanol, 5.0% dodecane); and Medium C, 0.01 M DDAB. Values in parentheses were obtained for DBA (0.005 M).

in various organized assemblies constructed as follows: (1) spherical micelle of CTAB in water (5×10^{-2} M); (2) rod micelle, 0.1 M CTAB in water with 0.25 M KBr; (3) vesicle, didodecyldimethylammonium bromide in water following sonication; and (4) microemulsion, oil-in-water type of CTAB, hexanol, and dodecane in water.

The following additional observations on ion yields in the P/DMA system are helpful in interpreting the data in Table IV.

1. The yield of ions from the exciplex increases with increasing dielectric constant or medium polarity.

2. The yield of ions tends to decrease with increasing viscosity; for example, the yield of ions in ethylene glycol is 10% of that in acetonitrile. The dielectric constants of both media are similar, but the viscosity of ethylene glycol is 30 times that of acetonitrile. The lower yield in glycol is probably due to a decreased probability of escape of ions from the solvent cage in more viscous media.

The introduction of the P–DMA into more viscous media by changing the system, spherical micelle, rod-micelle, or vesicle, leads to a lower ion yield. Viscosity or rigidity plays a similar role when DBA is used in place of DMA. The larger butyl group leads to a slower diffusion of DBA compared to DMA, which, in turn, leads to a smaller ion yield. For a polar form of pyrene such as PSA, which resides at the micelle surface, the restriction of movement or viscosity experienced by the ions on the P/DMA or PSA/DMA systems is similar, and similar ion yields are observed.

The medium of pyrene in microemulsions is less rigid than that in a micelle, but lower yields of ions are observed. This result is due to polarity effect because pyrene is located primarily in the oil center of the microemulsion, and photoexcitation leads to excited pyrene that encounters DMA in the vicinity of the microemulsion surface (15, 33). However, the environment of the excited encounter complex is not that of the surface, and ion formation is reduced. However, polar derivatives of pyrene are located at the microemulsion surface, and the encounter complex experiences a polar environment as in a micelle.

Addition of a cosurfactant to the system leads to a decrease in the ion yield. Previous studies (25) show that the micellar environment of pyrene becomes less viscous on addition of a cosurfactant, but also less polar. The decreased ion yield in these systems suggests that the overall effect is due to the decreased polarity of the encounter complex.

Other e^- transfer systems (43) show similar effects; the yield of ion products on excitation is highest in micellar systems and decreases in larger systems such as vesicles.

In reversed micellar systems (44) and water-in-oil microemulsions (34), the exciplex of pyrene and dimethylaniline is quenched by the water pool, and no ions are formed. This finding contrasts with derivatives associated with the water pools such as PSA, which do give ions on excitation in the presence of amines. The data suggest that the initial electron transfer to pyrene takes place without restriction with regard to orientation of pyrene and amine. The back electron transfer, however, appears to be inefficient unless a significant overlap of pyrene anion and amine cation occurs, such as in an exciplex where a sandwich structure is formed. Delocalization of e^- over the pyrene ring structure hinders the back electron transfer in other more random arrangements of the ions, such as those occurring in micelles. Similar photo-"diode" effects are seen in systems where the donor and acceptor are joined by short methylene chains (45).

Reactions of P^- in Micelles

Table V shows the rate constants for reaction of P^- in CTAB micelles with CO_2, O_2, Eu^{3+}, methyl viologen (MV^{2+}), and cetylpyridinium chloride (CP^+); the rate constants in homogeneous solutions of acetonitrile are also shown for comparison. In all cases the rates in homogeneous solution approach diffusion control and are 10^{10} $M^{-1}s^{-1}$. The rate constant for reaction of P^- with O_2 is significantly decreased in CTAB micelles, an effect that is similar to the decreased rate of quenching of pyrene fluorescence by O_2, which is diffusion controlled in methanol, $k = 2 \times 10^{10}$ $M^{-1}s^{-1}$, but only 5×10^9 $M^{-1}s^{-1}$ in CTAB micelles. This decreased rate constant is interpreted as a decreased motion of O_2 due to the rigidity of the micellar environment.

Both Eu^{3+} and MV^{2+} show significantly decreased rates of reaction with P^- in CTAB micelles. This result is not unexpected because these

Table V.
Reaction Rate Constants for Pyrene Anions, P^-, with Acceptors

Solute	Homogeneous Soln, $M^{-1} s^{-1}$	Cationic Micelle, $M^{-1} s^{-1}$
O_2	2.0×10^{10}	5.0×10^9
Eu^{3+}	2.7×10^9	1.3×10^7
		(10^{-2} M CTAB)
MV^{2+}	2.6×10^{10}	3.3×10^7
		(6×10^{-2} M CTAB)
		1.8×10^7
		(10^{-2} M CTAB)
Cetylpyridinium chloride	2.6×10^{10}	10^7 (s^{-1}) [1st order decay]
CO_2	$\sim 10^7$	$\sim 10^7$

cations should be repelled by the positive potential of the micelle. The rates of reaction increased linearly with [cation] and are independent of [micelle], a result that indicates that these ions approach the micelle surface to react with P^-, rather than P^- exiting into the aqueous phase followed by an encounter with the cation.

CP$^+$ is located in the micelle, and the kinetic motion leading to a decreased yield of P^- is that of P^- and CP^+ on the micelle surface. The rate constant k_q is similar to those found for CP^+ quenching of excited pyrene in CTAB micelles ($k_q = 1.4 \times 10^7$ s^{-1}) (11) and for pyrene excimer formation ($k = 5.0 \times 10^6$ s^{-1}). These data suggest that P^- and CP^+ must diffuse together on the micelle surface for the reverse e^- transfer process to occur. This process takes 10^{-7} s for e^- transfer from P^- to a cation; the DMA^+ exit rate from the CTAB micelles probably is much faster than 10^{-7} s. However, the rate of back reaction is not as rapid as that observed for P^- and DMA^+ in anionic NaLS micelles, which occurs within 20 ns. This effect may be due to the higher mobility of DMA^+ on the micelle surface compared to CP^+ or to a tunneling of e^- from P^- back to DMA^+, a process that does not occur with CP^+ and P^-. Some evidence to this effect was found for a rapid 1-ns back reaction of P^- and DMA^+ when linked together by a propyl chain (45).

The reactions just described are probably electron transfer in nature. This suggestion was substantiated in the case of MV^{2+}, where the enhanced decay of P^- was matched by a concomitant rise of the reduced MV^{2+} observed at 6000 Å.

MV^{2+} quenches excited pyrene P* with $k_q = 10^{10}$ M^{-1}s^{-1} in methanol and 5.0×10^7 M^{-1}s^{-1} in CTAB micelles. The decreased rate constant observed in CTAB compared to acetonitrile, 1/200, is close to that observed in similar experiments for the reaction of P^- and MV^{2+}. This observation also suggests that for reaction to occur, MV^{2+} approaches either P^- or P^* in the micelle surface. No evidence suggests that e^- tunnels to MV^{2+}. Data in microemulsions are similar, but with higher O_2 rates due to the higher $[O_2]$, and reaction rates of cations are slower.

Water-in-Oil Microemulsions

For the most part, the few photochemical studies of water-in-oil microemulsions tend to follow what has been found fruitful in reversed micelles. A typical example is the quenching of excited tris(bipyridyl)ruthenium (RuII) by methyl viologen (MV^{2+}). Both salts are soluble exclusively in the water pools of a water-in-oil cationic microemulsion (34).

$$Ru(II)^* + MV^{2+} \rightarrow Ru(III) + MV^+ \qquad (2)$$

The existence of water pools in the system, in which RuII and MV^{2+} are statistically distributed, is confirmed by the observed upward curvature in the Stern–Volmer plot, which can be described by the relationship

$$\frac{I_F}{I_F^0} = \sum_{n=0}^{\infty} \frac{\alpha_n}{1 + n(k_q/k_1)}$$

where I_F and I_F^0 are the luminescence yields of RuII in the presence and absence of MV^{2+}, respectively; α_n is the Poisson probability for the existence of a water pool that contains n quenchers and is given by $(\bar{n}^{-n} e^{-\bar{n}}/n!)$ where $\bar{n} = [MV^{2+}]/[WP]$; and k_1 is a first-order rate of decay of excited RuII in a water pool that does not contain any quencher (Q) (equivalent to the luminescence lifetime of (RuII)* in the absence of Q).

·The presence of water pools is further substantiated by the observed lower quenching efficiency of heptyl viologen (HV^{2+}) as compared to MV^{2+}, which is due to its decreased mobility as a result of increased binding at the water–hydrocarbon interface by the heptyl chains.

The adjustable parameters in Equation 2 are [WP] and k_q. The size of the water pool may be determined from the value of [WP] that produces the best fit to the experimental Stern–Volmer curve. The luminescence decay of (RuII)* in the presence of MV^{2+} does not show a single exponential but can be analyzed according to the equation:

$$I_F(t) = I_F(0)e^{-\{k_1 t + \bar{n}(1 - e^{-k_q t})\}}$$

where $I_F(t)$ and $I_F(0)$ are the fluorescence intensities of (RuII)* at time t and zero, respectively; $\bar{n}$ is the average MV^{2+} occupancy of the microemulsion [μE] given by [(MV^{2+}]/[μE]; k_q is the first-order quenching rate constant for a water pool with one (RuII)* and one MV^{2+}; and k_1 is the first-order decay constant for a pool where [MV^{2+}] = 0. The [water pool] can be determined from the derived ratio of $\bar{n}$ and the known [MV^{2+}], and hence the water pool radius can be derived from the known water content of the system.

The fluorescence quenching technique can also be used to comment on the locations of the molecules in the system.

Excited P, PSA, PBA, and PDA are quenched with various quenchers that are expected to exhibit different modes of interaction in the μE system. Typical quenchers used are DMA, which probably has some affinity to the water pool and also moves indiscriminately throughout the system; N-methyl-N-dodecylaniline (MDA), which favors the hydrocarbon phase rather than the water pool structure; cetylpyridinium chloride (CPC), which should be completely associated with the water pool with the pyridinium quenching head group located at the interface in contact with water; and NaI, which would be exclusively dissolved in the water pool.

The results of the quenching experiments are given in Table VI. These quenching data seem to indicate that P, which resides in the hydrocarbon medium, is more efficiently quenched by quenchers that partition strongly into the hydrocarbon phase. On the other hand, PBA and PDA, which are strongly bound to the water pool structure with their carboxylate groups anchored at the interface, strongly interact with the trimethylammonium groups of the surfactant molecules and are more strongly quenched than P by quenchers that are associated with the water pool aggregate. However, PSA, which is likely to be asso-

Table VI.
Fluorescence Quenching Data of 5.0 × 10^{-5} M P or P Derivatives at 0.02 M Quencher in CTAB Microemulsion (W_o = 20.3) at Room Temperature (22 °C)

	$\tau_F°$, ns		[O_2] ×	$k_q × 10^{-9}$, M^{-1} s^{-1}			
Probe	N_2	Air	10^3, M^a	DMA	MDA	CPC	I^-
P	380	22	4.2	2.7	2.0	0.42	0.40
PSA	65	35	1.2	2.3	0.8	5.3	7.4
PBA	195	41	2.0	2.0	0.8	2.1	0.75
PDA	200	28	3.1	2.0	0.85	1.0	0.50

^{a} Calculated by using the fluorescence lifetimes under N_2 bubbled and nondegassed conditions and assuming $k_q(O_2)$ = 10^{10} M^{-1} s^{-1}.

ciated with the head groups of the surfactant molecules in the Stern-layer region, is most efficiently quenched by I^- and CPC, which are localized within the water pool structure.

Photoinduced Electron-Transfer Reactions

P/DMA. In the CTAB–hexanol–dodecane microemulsion system with the water-to-surfactant ratio $(W_o) = 20.3$, DMA efficiently quenches the fluorescence of pyrene P, as well as that of PSA, PBA, and PDA (Table VI). This quenching leads to very weak exciplex emission (lifetime 60 ns) for P, whereas no exciplex emission is observed for the P derivatives. However, in a mixture of dodecane and hexanol having the same composition as in the microemulsion system, DMA quenching of the fluorescence of P as well as its derivatives produces a strong exciplex emission with a measured lifetime of 100 to 120 ns. The much shorter lifetime of the P/DMA exciplex in the microemulsion system seems to indicate that it is strongly quenched by the water pools. On the other hand, PSA, PBA, and PDA are expected to be bound to the water pool structure, and therefore the absence of exciplex emission can be ascribed to static quenching by the water pools.

Laser flash photolysis studies on pyrene and its derivatives showed a very low yield of P^-, but a very high yield of anions for the pyrenyl derivatives. This result is in accord with the explanation given earlier for the low yield of P^- in the oil-in-water CTAB microemulsion system, that is, that high yields of ions in photoinduced electron-transfer (ET) reactions can only be achieved when the reactants are solubilized close to the micelle–water interface, where escape of the photoproduced ions is most favorable. Therefore, quenching of the P/DMA exciplex by collision with the water pools probably does not lead to a separation of its constituent radical ions.

In CTAB microemulsion systems containing small water pools $(W_o = 20.3)$, the decay of the absorption of the anion of the pyrene derivatives (monitored at 493 nm) manifests itself in two fractions: a fast decay component (half-life 100 ns) attributed to first-order intra-water pool back ET reaction, involving pairs of (P^-, DMA^+) residing in isolated water pools, and a slow second-order decay (half-life 2.3 μs). In microemulsion systems that contain larger water pools, the intrawater pool back ET process is projected to occur within a few microseconds and therefore would be mixed in with the second-order inter-water pool reaction that occurs within a comparable time period.

RuII/MV^{2+}/HV^{2+}. The laser flash photolysis technique has also been applied to investigate the ET from photoexcited RuII to MV^{2+}, a reaction that is expected to take place inside the water pool, and to HV^{2+}, which, as indicated earlier, is probably located at the interface boundary as a result of its amphiphilic structure. Results showed no yield of MV^+ (monitored at $\lambda = 395$ nm) ions, presumably because of a very fast back ET from MV^+ to RuIII, but a high yield of HV^+ was observed. The transient absorption of HV^+ decayed by two processes (similar to the case of P/DMA discussed) intra- and inter-water pool back ET reaction occurring in a time range of ~ 10 μs.

Attempts to intercept the back transfer of the electron from MV^+ to RuIII by addition of EDTA, which would react with RuIII to regenerate RuII, were not successful. However, with RuII/HV^{2+}, addition of EDTA increased the lifetime of HV^+ by more than an order of magnitude.

Oleate Water-in-Oil Microemulsion. The quenching of (RuII)* by MV^{2+} in an anionic oleate water-in-oil microemulsion is much less efficient than in the cationic CTAB systems. The reason is that both (RuII)* and MV^{2+} are bound to the COO$-^-$ head groups thus restricting movement of the reactants. (46). However, (RuII)* is rapidly quenched by $Fe(CN)_6{}^{3-}$ because this ion moves readily in the water pool to the bound (RuII)*. Poisson statistics are again used to explain the nonlinear quenching behavior. The quenching rate constant k_q for one of each reactant in the water pool depends on the water pool radius. The variation in the (RuII)* lifetime at fixed $Fe(CN)_6{}^{3-}$ as a function of water content is shown in Figure 5, and the data are limited to other physical measurements that monitor behavior of the water pools in the system. Figure 5 shows the variation in the spectroscopic properties of anilinonaphthalenesulfonate (ANS), pyrenecarboxaldehyde (PCHO), and diphenylhexatriene (DPH) at various water contents. The variation in the lifetime of (RuII)* in an undegassed system containing 10^{-4} M RuII and 5.0×10^{-3} M $Fe(CN)_6{}^{3-}$ is also shown. Under these conditions where $\bar{n} > 1$ and $k_q < \tau_o{}^{-1}$, the luminescence decay of (RuII)* is nearly exponential.

Increasing the water content of the system leads to a decrease in the ANS fluorescence intensity and an increase in the PCHO fluorescence intensity, together with a concomitant increase in λ_{max} for this probe; the effects appear to level out above 15% water. The trends in the spectroscopic data show that the environments of the probes show increasing polarity with increasing water and approach that of bulk water at 15% water.

The RuII/$Fe(CN)_6{}^{3-}$ lifetime data and the DPH fluorescence polarization data both monitor the rigidity of the environments of these systems. The RuII/$Fe(CN)_6{}^{3-}$ system monitors the water pool, and the DPH monitors the hydrocarbon phase. The lifetime of (RuII)*(τ) decreases with water content up to 15%; thus, the efficiency of quenching

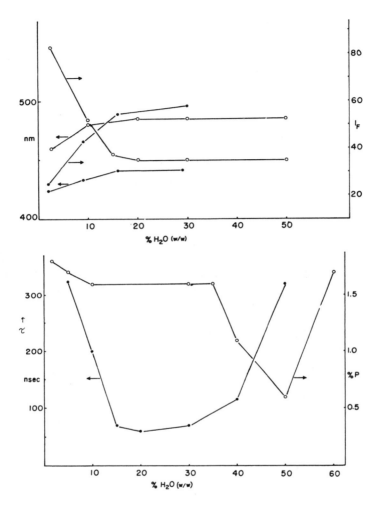

Figure 5. Top: Variation in the spectroscopic properties λ_{max} *and flu-orescence intensity of pyrenecarboxaldehyde (●) and anilinonaphthal-enesulfonate (○) with percentage water. Bottom: Variation with per-centage water of (●) the lifetime of RuII* fluorescence with* 5.0×10^{-3} *M Fe(CN)*$_6^{3-}$ *and (○) with the degree of fluorescence polarization of diphenylhexatriene (46).*

of (RuII)* by Fe(CN)$_6^{3-}$ increases as the reactants become more mobile. However, the τ increases again; thus, a lower mobility of reactants occurs above 40% water. The DPH fluorescence polarization decreases slightly up to 15% water, but shows a much larger decrease above 38% water with an increase above 50% water. The fluorescence polarization of this probe varies with the degree of rigidity of the probe environment (i.e., the hydrocarbon).

The lifetime of (RuII)* is almost independent of W_o (780 ns at W_o = 16, and 750 ns for W_o = 53), and longer than the τ observed in water (τ = 650 ns). Moreover, the wavelength of maximum emission of (RuII)* at λ = 630 nm is found also to be invariant with respect to change in W_o, and is ~20 nm red-shifted from that observed in pure water. These results could probably be interpreted to mean that RuII stays electrostatically bound to the anionic surface of the water pool during its excited state lifetime.

Photochemical Reactions

Table VI lists the rate constants for several photochemical reactions induced in the microemulsion system. In each case the pyrene chromophore is excited and the quenching of the fluorescence by Tl^+, DMA, O_2, and I^- is monitored. Pyrene is located in the hydrocarbon phase, and PBA and PSA are located at the water bubble–hydrocarbon interface. Tl^+ and I^- are confined to the water pools; Tl^+ is bound to the negatively charged surface; and I^- is in the pool itself. Both DMA and O_2 move in both hydrocarbon and water phases. Excited pyrene is poorly quenched by Tl^+, as P* must approach the aqueous phase for reaction to occur. The I^- quenching of P* is almost nonexistent because P* must enter the pool for reaction to occur. Tl^+ readily quenches both PSA and PBA because all reactants are located at the interface; again I^- quenching is lower because it is repelled from the interface where PSA and PBA reside. DMA efficiently quenches all these molecules, a result that indicates no significant barrier to its movement throughout the system. The same applies to O_2 although PSA and PBA, the more polar derivatives of pyrene, are slightly lower in reactivity than P. This condition is most probably due to the lower $[O_2]$ in the water bubble compared to the bulk hydrocarbon phase.

The data show that the polarity of the water pools, as monitored by the fluorescence probes ANS and PCHO, increases with increasing W_o. Similar data have been observed for AOT–heptane systems. The properties reminescent of bulk water are approached beyond 15% water.

Figure 5 shows that the lifetime of (RuII)* in the presence of $Fe(CN)_6^{3-}$ decreases; that is, the reaction rate increases with addition of water up to 15% water in a similar fashion to the fluorescence probe data. Increasing the radius of the water pool tends to decrease the rigidity of this region of the system and leads to more rapid reaction. Further increase in water content, after 40%, leads to a sudden increase in the (RuII)* lifetime, or decrease in reaction rate. This result is due to a change in the system from spherical water pools or reversed micelles to rod-like structures, as noted previously. The (RuII)* and $Fe(CN)_6^{3-}$ are no longer confined to a small part of the system, a situation that encourages rapid reaction, but are now spread throughout larger structures. The reaction rate drops, and the lifetimes of (RuII)* increases.

The polarization of diphenylhexatriene decreases abruptly at 40% water when the large structures start to be formed. This probe is located in the oil part of the system and indicates a disruption of the oil as the larger structures are formed; the disruption leads to a higher mobility of the probe. At larger water contents the polarization increases again as the system forms a more organized assembly.

Reaction rate constants given in Table V for several pyrene chromophores and quenchers indicate the compartmentalized nature of the system. The probes PBA and PSA, and quenchers Tl^+ and I^-, are located in the water pools. Tl^+ is probably in the vicinity of the head group region near the carboxylate groups, while I^- is repelled from the surface toward the pool. Pyrene is located only in the alkane, but may approach the pool, and DMA and O_2 move fairly freely through the system. The quenching rate constants of DMA and O_2 with pyrene, PSA, and PBA are similar in keeping with the model given. Iodide ion quenches both PSA and PBA, but not pyrene because this probe cannot enter the water pool interior. Thallous ion quenches all three probes but pyrene least efficiently, although more efficiently than I^-. This result is in keeping with Tl^+ being located at the interface and an inefficient approach of pyrene to this region of the system.

If a k of $10^{10} M^{-1} s^{-1}$ is assumed for O_2 reacting with the probes, then the $[O_2]$ can be calculated in the bulk hydrocarbon and in the water pool. The $[O_2]$ is larger in the hydrocarbon than in the water, again in keeping with the established solubilities of O_2 in the bulk liquids.

These studies establish the kinetic patterns that take place with photoinduced reactions in oleate microemulsions. The kinetics follow structural changes in the system, which are also monitored by other physical measurements.

Exchange of Ions Between Water Pools

The exchange of ions between water pools in AOT reversed micelles was found to be an inefficient process that varied with the additives in the system (47). Similar data were found for oleate microemulsions with hexanol as a cosurfactant (46). However, the exchange is much faster if pentanol is used as a cosurfactant (48). Earlier conductance studies (14, 15) indicated a large electrical conductivity in pentanol–oleate microemulsions compared to hexanol–oleate microemulsions. Discrete water pools were thought to exist in the hexanol system but not in the pentanol system. Positronium annihilation studies agreed with this interpretation (49). However, more recent photochemical studies, similar to those described earlier, showed that discrete water pools exist in both the pentanol and hexanol systems. The larger conductance of the former could be explained by the increased ion mobility observed in the photochemical data.

Polymerized Microemulsions

The oil at the core of a microemulsion may be replaced by a polymerizable monomer, such as styrene or methyl methacrylate, and subsequently polymerized to form what might be called a "permanent" microemulsion. An excellent system is made from CTAB, hexanol, and a 50% mixture of styrene and divinylbenzene. The divinylbenzene is required to cross-link the polymer and to form a stable core (50). The system is polymerized by high-energy radiation or by heating with the initiator azobisisobutyronitrile (AIBN). Both dynamic light scatter and electron microscopy indicate particles of fairly uniform scope of radii in the 200- to 400-Å region.

The stability of the polymerized particles, PμE, is readily demonstrated by the use of probes such as pyrene or pyrene carboxaldehyde that are solubilized in hydrophobic particles. These probes show specific fluorescence indicative of their environment. No spectral changes are observed on dilution of a PμE system containing a probe molecule; the data show that a hydrophobic probe environment is always maintained. However, dilution of micelles or unpolymerized microemulsions containing probes shows that these systems dissociate at high dilution (beyond the CMC) and that the probe molecule then exhibits spectroscopic properties indicative of an aqueous environment. The relative ease of formation of pyrene excimers from excited and ground state pyrene is often used as a measure of the mobility of pyrene in a system. Excimers are readily formed in micelles and microemulsions, but under similiar conditions no excimer formation is observed in PμE. This result indicates a very low mobility of pyrene in PμE and is expected from the cross-linked, polymerized nature of the system. However, pyrenebutyric acid, which is located in the nonpolymerized surfactant coating of the PμE, exhibits considerable excimer formation and is comparable to a simple unpolymerized microemulsion system. Hence, the two sites of solubilization in PμE are (1) the rigid nonpolar polymerized core, and (2) the fluid surfactant coating that contains the polar surface of the particle. The binding constant of pyrene to the PμE is larger by a factor of 6.8×10^3 than that to a CTAB micelle.

The two sites of solubilization of pyrene in the PμE are readily demonstrated by kinetic observations on the fluorescence of pyrene. Two decays of pyrene fluorescence are observed with lifetime τ of 180 and 425 ns, corresponding to solubilization in the surfactant coating and the polymerized core of the PμE, respectively. About 75% of the pyrene is solubilized in the core and 25% in the surfactant coating. The fluorescence of pyrene in the core is unaffected by O_2 and surface bound quenchers such as I^- or cetylpyridinum ion, which, however, quench excited pyrene in the surfactant coating. Nonpolar molecules such as dimethyl terephthalate quench core bound pyrene fluorescence.

The PμE systems lend further variation to organized assemblies both by providing rigid media in the particles to investigate effects of diffusion on photoinduced reactions, and by enabling the construction of multicomponent systems. Micelles can coexist independently in the presence of PμE; thus different reactions can be reduced in the two organized assemblies, with subsequent mixing of the products. The rigid nature of the system promotes e^- tunneling reactions, because diffusion is low in these systems.

Liquid Crystals

Liquid crystals are presently used quite extensively in many electronic devices, a situation that is in direct contrast to chemistry, where they are only of academic interest. Figure 6 gives a diagrammatic representation of the three different classes of liquid crystals. Although all three types are liquids, they have considerably more order than is usually found in fluid systems, but still lack the order of crystalline solids. Simple surfactant molecules readily form smectic and nematic structures, and organic molecules of the cholesterol type with bulky, long chains form cholesteric type structures. The method of preparation of these systems and their salient properties have been reviewed extensively (51–55).

The diagrammatic representation of liquid crystals in Figure 6 immediately suggested that guest molecules will tend to "line up" with the ordered structures of these systems. Furthermore, the liquid crystals themselves may be aligned relative to a convenient laboratory axis. This effect may be achieved by the use of external magnetic fields, or electrostatically by enclosing the liquid crystal system between two microscope slides followed by unidirectional stroking of the slides with a soft cloth. Alternatively, the liquid crystal system can be wiped onto an oriented polymer, an action that organizes the direction of the liquid crystal (56). In spite of these unique properties, little chemistry has been reported in these systems; a few examples of what has been attempted are given as follows.

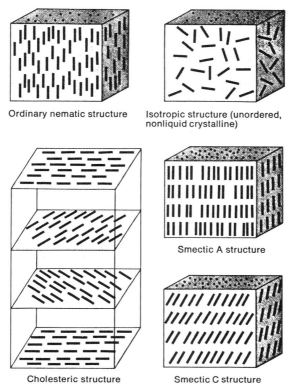

Ordinary nematic structure   Isotropic structure (unordered, nonliquid crystalline)

Smectic A structure

Cholesteric structure   Smectic C structure

Figure 6. Representation of various types of liquid crystals (70).

The CTAB–hexanol–water system forms a liquid crystal phase that catalyzes the hydrolysis more efficiently than any other phases of the system (9).

Sodium undecanoate forms a lyotropic crystal in water at 60 °C (57, 78). Optical microscopy and low-angle X-ray diffraction indicate that the structure is an array of hexagonal closely packed cylinders with the hydrophobic part of the soap in the center of the cylinders. Polymerization of the system leads to an isotropic structure at 60 °C, which is changed into a lamellar liquid crystal on cooling to 20 °C. The transition is reversible. The lamellar phase is stated to have a polyethylene backbone with decanoate chains extending to the aqueous phase. This intriguing example of a planar polymerization is induced by the structure configuration of the starting material.

Studies in the depolarization of the fluorescence of 1, 6-diphenylhexatriene (DPH) in methylcyclohexane, paraffin oil, and the nematic liquid crystal, cholesteryl laurate chloride (CLC) show that rotational diffusion is 1 ns in the simple hydrocarbon, several nanoseconds in paraffin oil, and is beyond the limit of measurement in the liquid crystal (50, 60). The guest molecule DPH lines up with the long axis of the CLC, and its mobility is severely restricted. This situation provides a method of conveniently orienting molecules with respect to a laboratory axis to determine the polarization of a selected excited state.

Liquid crystals may be oriented in electric (3–4 $\times$ 10^4 V/cm) and magnetic fields. In some instances the fields may produce a phase change, as in the case of p,p'-diactive amyloxyazoxybenzene in p,p'-dihexyloxyazoxybenzene (61) where NMR studies show that magnetic fields produce a cholesteric to nematic phase change. However, the sample still maintains orientation properties of use in photochemistry; usually the long axis of the molecule lies parallel to the applied electric field (62–64). Long molecules such as $CH_3NC_6H_4N:NNC_6H_4NO_2$ align with the long axis parallel to the long axis of a liquid crystal solvent. A typical system would be a mixture of cholesteryl chloride and cholesteryl myristate. The systems align perpendicular to an applied external magnetic field (62). β-Carotene and benzoquinone align with the polarization of the IR visible absorptions in the direction of the long axis of a liquid crystal solvent. Spectroscopic studies of octatetrabutylphenoquinone and 2,3-benzanthracene in liquid crystals show that the first allowed transitions of these molecules are polarized along the short molecular axis. For convenience a sample may be frozen after orienting in a magnetic field, and the ordered state maintained indefinitely (63). Charge-transfer complexes can also be investigated in liquid crystals. The charge-transfer complex, tetrachlorophthalic anhydride and chloranil, and the complexes of dimethylaniline and coronene, naphthalene, or pyrene align with the molecular planes of the complex parallel to the axis of the solvent, and the charge-transfer absorption band is polarized perpendicular to the solvent axis (65). The fluorescences of pyrene and diphenylhexatriene are polarized parallel to the long axis of a cholesteryl liquid crystal, while the excimer emission of pyrene and tetracene, and the exciplex emission of pyrene and p-methyldimethylaniline are polarized perpendicular to the long axis of the solvent (66).

Several studies have concerned photochemical reactions in liquid crystals (67–69). Significant effects on the photodimerization of acenaphthalene, the isomerization of azobenzene, the quenching of pyrene fluorescence, and on Norrish Type II photoprocesses have been observed. Studies in liquid crystals indicate that the reactant geometry for quenching of pyrene fluorescence of p-methoxydimethylaniline probably resembles that found in exciplexes.

These early studies hold much promise for the use of liquid crystals to aid in the determination of spectroscopic properties of excited states and in the estimation of reactant geometry.

Literature Cited

1. Stoechenius, W.; Schulman, J. H.; Prince, L. M.; *Kolloid Z.* **1960**, *169*, 170.
2. Schulman, J. H.; Stoeckenius, W.; Prince, L. M. *J. Phys. Chem* **1959**, *63*, 1677.
3. Shinoda, K.; Friberg, S., *Adv. Coll. Interface Sci.* **1975**, *4*, 281.
4. Friberg, S.; Buraczewska, I. In "Micellization, Solubilization and Microemulsions"; Mittal, K. L., Ed.; Plenum Press: New York, 1977; Vol. II, p. 791.
5. Shah, D. O.; Walker, R. D.; Hsieh, W. C.; Stal, N. J.; Dwivedi, S.; Pepensky, R.; Deamer, D. W., *S.P.E. Tech. Pap.* **1976**, 5815.
6. Hermansky, C.; MacKay, R. A. In "Solution Chemistry of Surfactants"; Mittal, K. L., Ed.; Plenum Press: New York, 1979; p. 723.
7. Ekwall, P.; Mandell, L.; Fontell, K. *J. Coll. Interface Sci.* **1969**, *29*, 639.
8. Ekwall, P.; Mandell, L.; Solyom, P. *J. Coll. Interface Sci.* **1971**, *35*, 519.
9. Ahmad, S. I.; Friberg, S. *J. Am. Chem. Soc.* **1972**, *94*, 5196.
10. Friberg, S.; Ahmad, S. I. *J. Phys. Chem.* **1971**, *75*, 2001.
11. Cordes, E. H.; Dunlap, R. B. *Acc. Chem. Res.* **1969**, *2*, 329.
12. Waugh, J. S.; Wang, C. H. *Phys. Rev.* **1967**, *162*, 209.
13. Shah, D. O.; Hamlin, R. M. *Science* **1971**, *171*, 483.
14. Falco, J. W.; Walker, R. D.; Shah, D. O. *A.I.C.h.E. J.* **1974**, *20*, 510.
15. Shah, D. O.; Tamjeedi, A.; Falco, J. W.; Walker, R. D. *A.I.C.h.E. J.* **1972**, *18*, 1116.
16. Winsor, P. A. *Chem. Rev.* **1968**, *68*, 1.
17. Shinoda, K.; Friberg, S. *Adv. Coll. Interface Sci.* **1975**, *4*, 281.
18. MacKay, R. A.; Letts, K.; Jones, C. In "Micellization, Solubilization and Microemulsions"; Mittal, K. L. Ed.; Plenum Press: New York, 1977; Vol. II, p. 801.
19. Prince, L. M. in Ref. 18, p. 45.
20. Forrest, B. J.; Reeves, L. W. *Chem. Rev.* **1981**, *81*, 1.
21. Lang, J.; Djavanbakht, A.; Zana, R. *J. Phys. Chem.* **1980**, *84*, 1541.
22. Tricot, Y.; Kiwi, J.; Niederberger, W.; Grätzel, M. *J. Phys. Chem.* **1981**, *85*, 862.
23. Atik, S. S.; Thomas, J. K. *J. Am. Chem. Soc.* **1981**, *103*, 3550.
24. Gregoritch, S., M.Sc. Thesis, University of Notre Dame, 1979.
25. Almgren, M.; Greiser, F.; Thomas, J. K. *J. Am. Chem. Soc.* **1980**, *102*, 3188.
26. Harbour, J. T.; Hair, M. L. *J. Phys. Chem.* **1980**, *84*, 1500.
27. Kiwi, J.; Grätzel, M. *J. Am. Chem. Soc.* **1978**, *100*, 6314.
28. Wolff, C.; Grätzel, M. *Chem. Phys. Lett.* **1977**, *52*, 542.
29. Kiwi, J.; Grätzel, M. *J. Phys. Chem.* **1980**, *84*, 1503.
30. Jones, C. A.; Weaner, L. E.; MacKay, R. A. *J. Phys. Chem.* **1980**, *84*, 1495.
31. Matheson, I. U. C.; King, A. D. *J. Coll. Interface Sci.* **1978**, *66*, 464.
32. Turro, N. J.; Aikawa, M.; Yekta, A. *Chem. Phys. Lett.* **1979**, *64*, 473.
33. Gregoritch, S.; Thomas, J. K. *J. Phys. Chem.* **1980**, *84*, 1491.
34. Atik, S. S.; Thomas, J. K. *J. Am. Chem. Soc.* **1981**, *103*, 4367.
35. Thomas, J. K. Unpublished data.
36. Turro, N. J.; Yekta, A. *J. Am. Chem Soc.* **1978**, *100*, 5951.
37. Atik, S. S.; Singer, L. A. *Chem. Phys. Lett.* **1978**, *59*, 519; *Chem. Phys. Letts.* **1979**, *66*, 234.
38. Thomas, J. K.; Piculo, P. Am. Chem. Soc. Adv. Chem. Series *184*, 97, 1980.
39. Wallace, S. C.; Grätzel, M.; Thomas, J. K. *Chem. Phys. Lett.* **1973**, *23*, 359.
40. Waka, Y.; Hamamoto, K.; Mataga, N. *Chem. Phys. Lett.* **1978**, *53*, 242.
41. Katusin-Razem, B.; Wong, M., Thomas, J. K. *J. Am. Chem. Soc.* **1978**, *100*, 1679.
42. Atik, S. S.; Thomas, J. K. *J. Am. Chem. Soc.* **1981**, *103*, 3550.
43. Matsuo, T.; Nagamura, T.; Itoh, K.; Nishijima, T. *Mem. Fac. Eng. Kyushu Univ.* **1980**, *40*, 25.
44. McNeil, R.; Thomas, J. K. *J. Coll. Interface Sci.* **1980**, *73*, 517.
45. Eisenthal, K. *Acc. Chem. Res.* **1975**, *8*, 118.

46. Atik, S. S.; Thomas, J. K. *J. Am. Chem. Soc.* **1981**, *103*, 7403.
47. Atik, S. S.; Thomas, J. K. *Chem. Phys. Lett.* **1981**, *79*, 351.
48. Atik, S. S., Thomas, J. K. *J. Phys. Chem.* **1981**, *85*, 3921.
49. Boussaha, A.; Djermouni, B.; Fucuganchi, L. A.; Ache, H. J. *J. Am. Chem. Soc.* **1980**, *102*, 4654.
50. Atik, S. S.; Thomas, J. K. *J. Am. Chem. Soc.* **1981**, *103*, 4367.
51. Saeva, F. D. "Liquid Crystals"; Marcel Dekker; New York, 1979.
52. Gray, G. W. "Molecular Structure and Properties of Liquid Crystals"; London Academy Press: London, 1962.
53. Brown, G. H.; Shaw, W. G. *Chem. Rev.* **1957**, *57*, 1049.
54. Brown, G. H., Ed.; "Advances in Liquid Crystals"; Acad. Press: New York, 1974–1977; Vol. III.
55. de Gennes, P. G. "The Physics of Liquid Crystals"; Oxford Univ. Press: London, 1974.
56. Novak, T.; MacKay, R. A.; Poziomick, E. J. *Mol. Cryst. Liq. Cryst.* **1973**, *20*, 213.
57. Thundathil, R.; Stoffer, J. O.; Friberg, S. E. *J. Polym. Sci.* **1980**, *18*, 2629.
58. Friberg, S. E.; Thundathil, R.; Stoffer, J. O. *Science* **1979**, *205*, 607.
59. Cehelnik, E. D.; Cundall, R. B.; Lockwood, J. K., Palmer, T. F. *Trans. Faraday Soc.* **1974**, 244.
60. Cehelnik, E. D.; Cundull, R. B.; Timmons, C. J. *Proc. R. Soc. London Ser. A.* **1978**, *335*, 387.
61. Sackmann, E.; Meiboom, S.; Snyder, L. C. *J. Am. Chem. Soc.* **1967**, *89*, 5981.
62. Sackmann, E. *J. Am. Chem. Soc.* **1968**, *90*, 3569.
63. Sackmann, E.; Meiboom, S.; Snyder, L. C.; Meixner, A. E.; Dietz, R. E. *J. Am. Chem. Soc.* **1968**, *90*, 3567.
64. Sackmann, E. *Chem. Phys. Lett.* **1969**, *3*, 253.
65. Sackmann, E.; Krebs, P. *Chem. Phys. Lett.* **1969**, *4*, 65.
66. Sackmann, E.; Rehm, D. *Chem. Phys. Lett.* **1970**, *4*, 537.
67. Nerbonne, J. M.; Weiss, R. G. *J. Am. Chem. Soc.* **1978**, *100*, 2571.; *J. Am. Chem. Soc.* **1979**, *101*, 402; and *J. Am. Chem. Soc.* **1978**, *100*, 5953.
68. Nerbonne, J. M.; Weiss, R. G. *Isr. J. Chem.* **1979**, *18*, 266.
69. Anderson, V. C.; Craig, B. B.; Weiss, R. G. *J. Am. Chem. Soc.* **1981**, *103*, 7169.
70. Brown, Glen; Crooker, P. P. *C & EN* **1983**, 24.

Bilayer and Monolayer Systems

THE COLLOIDAL STRUCTURES DISCUSSED SO FAR have been formed by the general sequestering of a material (e.g., surfactant) to withdraw the hydrophobic part of the molecule away from the aqueous phase. The geometry of the system is such that an oil droplet may form in water (e.g., micelle and oil-in-water microemulsion) or a pool of water may form in an oil (e.g., a reversed micelle and water-in-oil microemulsion). Alternative structuring of the surface-active material is also possible; the same physical and chemical factors come into play. For example, a surfactant may be floated on water to form a monolayer, a system that, although not colloidal, is of importance and will be considered subsequently. If the geometry of the surfactant is of the correct form, then vesicles or closed bilayers may be formed. These systems are colloidal in nature, and are directly related to biological membranes.

Vesicles

The factors that promote vesicle or bilayer formation rather than micelle formation have been discussed (1, 3). In all cases, two long hydrocarbon chains have to be connected to the polar head groups for the surfactant to form vesicles because single-chain surfactants only form micelles. The geometry of the double-chain surfactants constrains the packing of the molecule, which attempts to minimize water contact, so that a small surface curvature is obtained. One possible structure is a curved bilayer that encloses an inner compartment containing water from the outer bulk-water phase.

Preparation. The preparation of vesicles is somewhat more complicated than the procedure to produce simple micelles or microemulsions. The basis of all techniques is to disperse the double-chain surfactant efficiently in the aqueous phase. Inefficient dispersion can lead to large multilayered structures of nonuniform shape and size.

0065-7719/84/0181-0243$06.75/1

Sonication. Sonication of a lipid, such as lecithin, or a surfactant, such as dioctadecyldimethylammonium bromide, in water gives rise to small vesicles. The surface-active material in dissolved in a volatile solvent and then a portion of this solution is dried on the walls of a flask; thus, a thin film of material is produced. Water is added and the flask and contents are sonicated until vesicles are formed (4). Typically, with a Bronson Model 200K probe sonicator, 1.5–2-h sonication time at temperatures above the phase transition temperature ($\sim$ 50 °C) is required. Sonication should be carried out under anaerobic conditions to eliminate oxidation. The final mixture is centrifuged to remove probe metal particles from the sonicator, and small vesicles are produced, $r \sim$ 200 Å.

Precipitation. The surface-active material is dissolved in a solvent that is miscible with water (e.g., ethanol). Small amounts of this solution are then injected into vigorously stirred water (6, 7). The surfactant is thus precipitated in a finely divided form because the solvent dissolves in the aqueous phase. The surface-active material may be solubilized in water by means of another surfactant; for example, lecithin is solubilized by sodium cholate. When this solution is then dialyzed to remove the simple surfactant, the surface-active vesicle material is precipitated into the water and vesicles are formed. All methods are successful; the alcohol technique is easiest and fastest to use, and the sonication method is more conventional and more widely used.

General Properties of Bilayers

The properties of current interest in vesicle chemistry are readily extrapolated from known micelle chemistry, provided some restraint is used. Unlike a micelle, equilibrium between vesicles and monomeric surfactant does not exist because the critical vesicle concentration is too low. Hence, the structures are usually considered as static colloidal particles. However, guest molecules located in the vesicle do move backward and forward between the vesicle and the bulk-water phase. A vesicle has an additional feature of interest, that is, a phase change. Typically, for distearyllecithin or dioctadecyldimethylammonium bromide, a sharp change in physical properties associated with the vesicle occurs around 50 °C. Several physical parameters show sharp changes at this temperature (e.g., light scatter, heat capacity, volume changes, fluorescence polarization, and other photophysical effects) (8, 9). The phase change is associated with a "melting" of the surfactant hydrocarbon chains, and the temperature at which the change occurs (T_c) decreases with decreasing chain length.

Solubilization of Guest Molecules. Four sites of solubilization may be identified as follows in an aqueous vesicle system:

1. The bulk water phase—molecules may move back and forth between this phase and the vesicle.

2. The inner water region enclosed by the lipid vesicles—molecules may be trapped in this region for large periods of time, often days. Leakage from the inner water region to the bulk and vice versa is an important property of the system and simulates transport in cell membranes.

3. The polar head group region of the vesicle, which may be the outer region, and is in contact with bulk water, or the inner region, which is in contact with the inner water region.

4. The nonpolar hydrocarbon phase.

The surface area of the inner head group region ($r \sim 100$ Å) is only 20% of that of the outer head group region ($r \sim 200$ Å). Thus with this ratio a guest molecule located in the head group region may be distributed between the inner and outer head groups. Furthermore, the inner region is more rigid than the outer region because of packing constraints. This rigidity may also favor solubilization in the outer region. Experimental evidence is sparse on these points.

Solubilization of Hydrophobic Molecules. By comparison with micelle solubilization, hydrophobic molecules (e.g., pyrene) are expected to be solubilized in the head group and hydrocarbon tail regions of the vesicle. Fluorescence studies indicate that pyrene is located in the vicinity of the head groups at temperatures below the phase change; but, at the phase change temperature, an abrupt increase in the hydrophobicity of the pyrene environment is observed (10). This increase is associated with the increased fluidity of the hydrocarbon chains above T_c; this fluidity, in turn, permits greater penetration of pyrene into this region. Fluorescence stopped-flow techniques have been used to study the exit rates of pyrene and perylene from soybean lecithin vesicles containing 10% dicetyl phosphate and 8 volume % ethanol (11, 12). At 18 °C, the exit rate constants are 3 ± 0.5 s^{-1} for perylene and 77 ± 65 s^{-1} for pyrene. Similar data are obtained for egg lecithin and dimyristoyl-phosphatidylcholine vesicles; however, no exit was detected in pure dicetyl phosphate vesicles. Analysis of the data indicates that earlier theories of micellar solubilization (13) may be applied to explain these data in vesicles.

Movement into and out of Inner Water Pools. Several techniques have been used to determine the rates of movement of molecules into and out of the interior water pool. Radioactive tracer studies and NMR measurements show that, unlike exit rates from the hydrocarbon phase, exit rates from the water pools are quite slow (hours or days) but can be increased by molecules that interact with the vesicle wall (14).

Movement in the Vesicle Wall. The earliest estimates of movement of guest molecules in vesicles come from fluorescence measurements (15). These measurements indicated that the vesicles severely restricted the motion of fluorescent probe molecules. Also, the data clearly showed the vesicle phase change at T_c. This result is expected because of what is known of vesicle structure; however, the exact nature of the molecular motion giving rise to the fluorescence depolarization is not clearly understood. The kineticist needs to know the gross displacement of a reactant molecule in a vesicle, rather than rotational movement, to explain the data. First attempts at this approach are available from excimer studies with pyrene (10, 16–19). In this technique, pyrene is excited with a pulse of light and its fluorescence is observed. Movement of excited pyrene gives rise to excimers of pyrene P_2^* by lateral movement in the membrane

$$P \xrightarrow{h\nu} P^* \xrightarrow{+P} P_2^*$$

The lifetime of P^* and the ratio of P_2^*/P^* can give information on the movement of P^* in the system. However, problems arise in vesicles at temperatures below the phase change (16, 20). Figure 1 shows the increase of the P_2^*/P^* ratio, measured as fluorescence of P_2^* and P^* versus temperature for pyrene in lecithin vesicles. The ratio at first rises with increasing temperature, as would be expected because of increased fluidity of the vesicle and, hence, increased movement of P^* and P to form P_2^*. However, at T_c an abrupt drop in P_2^* is observed. This observation is in agreement with an increase in hydrophobicity of the probe molecule as observed by fluorescent fine structure data where an opening of the hydrocarbon region will permit P^* and P to diffuse in a larger vesicle volume. Thus, the local concentration of both reactants decreases, and the decrease in turn leads to a decreased yield of P_2^*. Increasing the temperature further leads to an increase in P_2^* relative to P^* because of increasing mobility of P^* and P. Thus, a simple interpretation of the data that does not take into account the local concentrations of the reactants is very misleading. The interpretation of the rate of growth of P_2^* below the phase change is even more misleading. Figure 2 shows typical data for the rate of formation of P_2^* with time below and above T_c. Below T_c the rate of growth of P_2^* is affected very little by pyrene concentration, [P]. This result is in contradiction to the second-order mechanism of formation of P_2^* in which the total yield of P_2^* does increase with [P]. Above the phase change the rate of growth of P_2^* increases linearly with increasing [P] as re-

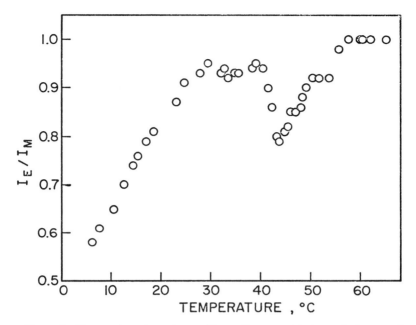

Figure 1. Temperature dependence of $I_{excimer}/I_{monomer}$ ratio for excited pyrene in DPL dispersions. [Pyrene] = 10^{-4} M; and [DPL] = 2×10^{-3} M.

quired by the suggested mechanism. Thus, above T_c the data are fully explained by simple mechanisms established in homogeneous solution. Below T_c, however, clusters of pyrene may rapidly form P_2^* on excitation (16). The extent of these clusters increases with [P]. At the phase change T_c, these pyrene clusters disperse in the larger vesicle volume that suddenly becomes available. This result is indeed a complication of the system and makes an interpretation of P_2^* formation in terms of guest molecule mobility susceptible to error. However, the data show that a full kinetic analysis of the system can describe the unique events in vesicle systems, such as the discussion of simpler micelles and microemulsions.

Addition of Quenchers. Table I shows the rate constants for quenching of pyrene fluorescence in micellar lysoplasmalogen and lysolecithin; vesicular dipalmitoyllecithin (DPL), distearoyllecithin, and dilauryllecithin; and homogeneous solutions, either water or methanol. Apart from iodoheptane, the rate constants for reactions are decreased compared to water for all systems, a result that reflects on the enclosed nature of the guest molecule pyrene in the assembly. For the most part the vesicles exhibit much lower rates of reaction than the micelle. This result is to be expected for the polar quenchers Tl^+, I^-, and CH_3NO_2,

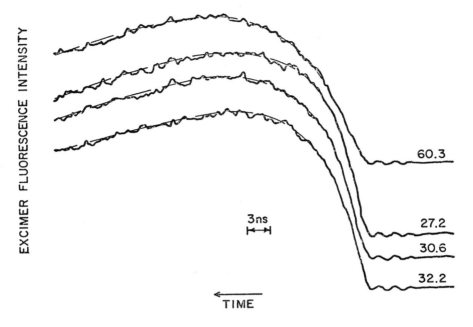

Figure 2. Growth curves of pyrene excimer fluorescence in DPL bilayers at different temperatures. [Pyrene] = 10^{-4} M; and [DPL] = 2×10^{-3} M.

Table I.

Rate Constants of Quenching of Pyrene in Phospholipid Dispersions at 23 °C, $k_2 \times 10^{-9}$ M^{-1}/s^{-1}

Quencher	LP	LL	DLL	DPL	DSL	Homogeneous Solution
O_2	6.4	7.2	14	2.8	2.2	20
CH_3NO_2	0.63	0.57	0.36	0.052	0.057	8.1
Tl^+	0.48	0.39	0.091	0.018	0.004	5.0
I^-	0.59	0.51	0.14	0.004	0.004	3.0
Iodoheptane	—	—	—	160	—	1.7

NOTE: LP = lysoplasmalogen; LL = lysolecithin; DPL = dipalmitoyllecithin; DSL = distearoyllecithin; and DLL = dilauryllecithin.

which reside mainly in the aqueous phase and attempt to penetrate the micelle or vesicle to quench pyrene fluorescence. The data indicate that either pyrene is secluded further from the aqueous phase in vesicles compared to micelles or penetration of polar and ion material is more difficult in the former systems. Oxygen shows high reactivity in all systems; the long-chain lecithin imposes more restriction on its movement to the site of the pyrene probe than the other systems. Hydrophobic iodoheptane is located in the assemblies with pyrene. In micelles, Poisson statistics are observed as with other quenchers. In DPL the 100-fold enhancement compared to homogeneous solution should be modified by the $\sim$ 1000-fold enhanced local iodoheptane concentration due to solubilization in the vesicle. Hence, the rate constant in DPL is actually 10 times less than in homogeneous solution. Therefore, vesicle microviscosity is about one-tenth that of methanol or water. This value is larger than estimates from fluorescence polarization; however, two different types of motion are depicted in the measurements: lateral motion in iodoheptane quenching, and rotational relaxation in fluorescence polarization studies.

Effect of Surfactants on Vesicles

Surfactants (e.g., CTAB, NaLS, sodium cholate, and lysolecithin) markedly affect the movement of guest molecules within and across the vesicle wall. Across the vesicle wall, transport is affected between the bulk water phase and the inner water pool (9, 14, 21–25). Movement of Eu^{3+} across the vesicle wall is readily observed by ^{31}P-NMR. Such studies show that distearyllecithin (DSL) vesicles are impermeable to Eu^{3+} for days, but addition of lysolecithin (LL) sharply increases vesicle permeability to times on the order of minutes. Other studies, such as fluorescence polarization, fluorescence fine structure, and excited state quenching, amplify the Eu^{3+} data.

The overall picture may be briefly recorded as follows: Addition of lysophosphatidylcholine to membrane bilayer causes increased fluidity of the membrane (pyrene dimer/monomer ratio, D/M, fluorescence depolarization of diphenylhexatriene); increased penetration of H_2O into membrane (pyrene fluorescence fine structure ratio I_3/I_1, pyrene D/M); increased penetration of polar species Eu^{3+}, H_2O, and CH_3NO_2 into and through the assembly; lysis of bilayer at greater than 40% lysophosphatidylcholine to give mixed micelles; and, up to 40% lysophosphatidylcholine, physical properties characteristic of bilayer are observed.

Changes in the photophysical properties of the excited pyrene (e.g., pyrene/I_3/I_1, τ_o, quenching rate) have been explained in terms of changes in the lipid structures that lead to greater penetration of polar molecules via H_2O (Eu^{3+} and CH_3NO_2) into and through the bilayer. Many aromatic molecules form weak complexes with quaternary ammonium compounds (25), and pyrene forms complexes with tetraethylammonium chloride in 1-pentanol solution; and I_3/I_1 ratio changes from 1.07 to 0.61 for concentrations of the ammonium compound from 0 to 1.4 M. Complex formation also reduces the yield of pyrene excimers. Perhaps similar complexes exist in the pyrene–phosphatidylcholine systems. For example, the decrease in τ_o of excited pyrene is due to complex formation with monomeric lysophosphatidylcholine. However, the formation of a single component micelle or bilayer structure hinders the formation of the complex, and much higher τ_o and I_3/I_1 values are observed.

Addition of lysophosphatidylcholine to phosphatidylcholine causes a disruption of the bilayer structure as witnessed by the increased permeability of the structure to polar molecules. The fluorescence polarization of diphenylhexatriene also decreases; this increase indicates that the structure is less ordered and less rigid. In disordered structures, pyrene has a more ready access to the head group structure of the phosphatidylcholine and forms a complex that decreases I_3/I_1.

The formation of a complex between pyrene and the partially charged nitrogen of the choline group leads to a diminution in the ease of formation of the excimer P_2^*. Increasing the temperature tends to dissociate the complex, and the P_2^* yield and I_3/I_1 increases. However, at approximately 40% the increases are very abrupt at the phase temperature of approximately 40 °C. At this temperature the phosphatidylcholine structure suddenly becomes less rigid and leads to greater access of pyrene to the lipid interior. The I_3/I_1 ratio increases as the pyrene–phosphatidylcholine complex breaks up; P_2^* increases because P^* is now free to move in the structure to form the excimer.

The data show quite clearly that addition of lysophosphatidylcholine to bilayers increases movement of polar molecules across these assemblies because of a break-up of the head group structure. Certain molecules, in particular aromatic (carcinogen-type) molecules, show additional interesting features; they have increased complexation by the bilayer head group. These data have a possible bearing on the penetration and solubilization of carcinogenic aromatic compounds in these structures of biological interest.

Reactions of Free Radicals with Vesicles

The reactivity of hydrated electrons (e_{aq}^-) and OH radicals, the primary products of the radiolysis of water, with vesicles and with guest molecules in vesicles is particularly important in radiobiology.

Hydrated electrons react with pyrene in the bilayer and produce the pyrene anion (P^-), which decays slowly via second-order kinetics. Because the anion is more hydrophilic than pyrene, it may exit into the water where it is annihilated (27). The rate of reactions of e_{aq}^- and pyrene increases above the temperature of the phase change in the vesicle when the bilayer becomes more fluid. However, increasing the pyrene concentration does not produce a corresponding increase in the rate of formation of P^-, but the rate tends to a maximum. The rate-controlling feature of the kinetics is the approach of e_{aq}^- to the vesicle; once at the vesicle, the reaction with pyrene is rapid. This type of behavior was indicated earlier for polymers (28) and is a feature of the clustering of the system. It also occurs with OH radical reactions, as will be discussed.

The hydrated electron does not readily react with lecithin vesicles. However, some reactivity is noted with the OH radical (29), and the point of attack may be toward the surface of the vesicle in the choline group rather than at the fatty acid region below the glycerol unit. With pyrene located in the bilayer, OH attack produces the spectrum of the OH–pyrene adduct. In homogeneous solution the rate constant for this reaction is 1.34×10^{10} M^{-1}s^{-1}; the rate varies linearly with pyrene concentration and is diffusion controlled. The rate of appearance of the OH–pyrene adduct and the yield of this product increase with increasing pyrene concentration in the bilayers; the rate of reaction increases slightly, and the yield of product approaches a maximum. The rate limiting step is that at which the OH radical reaches the vesicle.

The rate constant for a diffusion-controlled process can be written as $k = 4\pi r D \times 6 \times 10^{20}$ M^{-1}s^{-1}, where r is the sum of the reactant radii ($r_{OH} + r_{pyrene} = 6$ Å) and D is the sum of the diffusion constants (in the present system, $D = 4 \times 10^{-5}$ cm^{2}s^{-1}). Then k is calculated as 1.8×10^{10} M^{-1}s^{-1}. This k value is typical of many OH radical reactions and close to that measured for the reaction of OH and pyrene. The radius of a vesicle is about 250 Å, and the diffusion of the vesicle is slow; therefore, $D = 2 \times 10^{-5}$ cm^{-2} s^{-1} and the calculated rate constant is now 3.9×10^{11} M^{-1}s^{-1}. This k may be taken as a measure of the rate at which OH radicals encounter vesicles. There are some 1000 lecithin molecules in the vesicles, so k (OH ± lecithin) is 3.9×10^8 and close to that measured. Addition of pyrene introduces competition between the vesicle and pyrene for OH radicals. The yield of the OH–pyrene adduct increases with increasing pyrene concentration. However, the rate of reaction remains relatively unchanged because the rate dependent step is that at which OH encounters the vesicle. The rate tends to a limit of 5×10^5 s^{-1} at a vesicle concentration of 2×10^{-6} M. The rate constant for reaction of OH and a vesicle is then 2.5×10^{11} M^{-1}s^{-1}, in agreement with the previous discussion.

These aggregated systems, which are reminiscent of true biosystems, introduce new features into the convenient kinetic patterns. However, these features are now understood and may be used to explain more complicated systems. Reaction rate studies for reaction of e_{aq}^- with diphenylhexatriene (DPH) in didodecyldimethylammonium bromide (DDAB) vesicles show that the rate of reaction is limited by the rate of encounter of e_{aq}^- with the vesicle (30). The electron transfer to DPH was about 100% effective at a DPH/lipid monomer ratio of 1/50.

Electron transfer from the diphenylhexatriene radical anion in the positively charged DDAB vesicle to both a negatively and a positively charged ion was investigated. The Br^- counterions may be replaced with Fe(III)EDTA$^-$, which is an e^- acceptor; the lifetime of the hydrated electron and the yield of DPH$^-$ then decreased because of the direct reaction of e_{aq}^- with FeEDTA$^-$. The decrease in lifetime of DPH$^-$ indicates an electron transfer from DPH$^-$ to FeEDTA$^-$. The data suggest a linear correlation of FeEDTA$^-$ and rate of decay of DPH$^-$ from which a second-order rate constant of 1×10^{10} M^{-1}s^{-1} can be deduced. The surface counterion is thus a very efficient acceptor for electrons from DPH$^-$ in the vesicle. This observation is in marked contrast to a cation, such as Eu^{3+}. EuCl$_3$ acts as an efficient e^- acceptor in ethanol; the rate constant of DPH$^-$ and Eu^{3+} is close to the diffusion controlled $k = 1.1 \times 10^{10}$ M^{-1}s^{-1}. In DDAB vesicular solution, no evidence of electron transfer could be obtained up to 0.1 M EuCl$_3$.

Electron transfer from the radical anion of carotene in egg lecithin vesicles to Eu^{3+} was studied in an experiment that was designed to give evidence for an eventual transfer of electrons through the vesicle membrane. The vesicles were prepared by injection of an alcohol solution containing lecithin and β-carotene (molar ratio 1/100) into water to produce a concentration of 2 mM of lipids. One sample was injected into water containing 1 mM EuCl$_3$. The Eu^{3+} ions bind to these vesicle membranes, and in this procedure the vesicles are formed with Eu^{3+} at both the inside and the outside surfaces. The Eu^{3+} at the outside was subsequently removed by dialysis, first in EDTA for some hours and then overnight in water. A control sample without Eu^{3+} was dialyzed simultaneously in the same solvent baths.

On pulse radiolysis of both samples a long-lived absorption from the β-carotene anion (car$^-$) was observed at 750 nm. In the control sample the car$^-$ absorption amounted to about 40% of the initial e_{aq}^- absorption at the same wavelength. The latter decayed with a half-life of about 0.6 μs. In the sample with Eu^{3+}, the long-lived car$^-$ absorption was only 10% of the initial e_{aq}^-, but the portion of the signal between the 40 and 10% levels decayed more slowly ($\tau_{1/2} = 1.2$–2.0 μs) than the initial e_{aq}^- signal.

This observation indicates that car$^-$ was initially produced in approximately equal amounts in both samples, and that about two-thirds of the car$^-$ was deactivated by electron transfer in the Eu^{3+} sample. The electron transfer may have occurred to Eu^{3+} at the inside surface or to traces of Eu^{3+} that remained at the outside. However, because a large amount of Eu^{3+} should be present at the inside surface of these vesicles the latter explanation seems more plausible. For at least a fraction of car$^-$, transfer of electrons to Eu^{3+} at the inner surface was not possible.

Photochemically Produced Electron Transfer

Photoionization. The simplest photochemically produced electron transfer is the photoionization of an organic molecule, such as pyrene or tetramethylbenzidene, in vesicles. Significant laser induced photoionization that has been observed (31, 32) is quite similar to data in nonionic micelles such as Triton X-100. The yield of photoionization increases rapidly at the phase transition temperature (T_c), but no particular significance can be placed on this observation because the yield of excited states also shows a similar increase at T_c. The increasing clarity of the sample at T_c transmits more light to produce more photolytic products.

Electron Transfer Between Two Reactants. Vesicles have received close attention from photochemists because of their similarity to biological membranes (8, 33–40), and numerous photophysical studies have been carried out by biochemists to determine various parameters of these systems. The close proximity of reaction partners in vesicles and liposomes, or multilayer vesicles, has been demonstrated by Förster-type singlet–singlet energy transfer from alloxazines to isoalloxazines (37). Both vesicle structure and fluidity affect the energy-transfer process. Below the phase change T_c, the only mechanism of energy transfer is via a Förster mechanism. Above T_c the mean separation of reactants increases as the vesicle size increases, and the efficiency of energy transfer is impaired.

The vesicular property of situating two reactants in close proximity yet separated by a suitable barrier (i.e., on either side of the vesicle wall), with one reactant in the inner water phase and the other in the bulk medium, is particularly attractive for electron-transfer studies. Several systems have been studied, in particular the trisbipyridylruthenium–methyl viologen (RuII–MV^{2+}) system, with a repair agent such as ethylenediaminetetraacetate, EDTA (33, 34, 36). In one instance, the MV^{2+} is located at the vesicle wall by using a viologen with a long chain, such as cetylmethyl viologen (CV^{2+}), and RuII and EDTA may be located exclusively in the inner water region; this preparation places MV^{2+} or CV^{2+} in the outer wall toward the aqueous bulk. Photolysis of the system leads to reduced methyl viologen (MV^+). The suggested reactions are

$$RuII \xrightarrow{\ h\nu\ } (RuII)^* \qquad \text{inner region}$$

$$(RuII)^* + MV^{2+} \rightarrow RuIII + MV^+ \qquad \text{outer wall}$$

$$\text{vesicle across wall}$$

$$RuIII + EDTA \rightarrow RuII + \text{radical} \qquad \text{inner region}$$

Electrons may tunnel from $(RuII)^*$ to MV^{2+} across the lipid vesicle wall.

In some cases an electron mediator is required to obtain efficient electron transport across the cell wall (34). For example, photoinduced electron transfer from porphinatozinc complex to disodium 9,10-anthraquinonedisulfonate is more efficient in vesicle systems if 1,3-dibutylalloxazine is used as an electron carrier.

Photoinduced electron transfer from N-alkylphenothiazene derivatives (PH) to electron acceptors such as RuII in vesicles have been reported (35, 38). The RuII complex was located at the cell wall by using a long alkyl-chain derivative and the phenothiazene was intercalated into the lipid layer. Pulsed laser experiments show that RuI and PH$^+$ are formed on photolysis of this system. With methylphenothiazene, a portion of the photogenerated cation rapidly picks up an electron from a neighboring RuI; some cations escape into the main water pool, and the remainder escape into the aqueous bulk. If the cation is retained at the vesicle by using a long-chain N-phenothiazene, then the cation recombines rapidly with RuI. These systems are very suitable and provocative models for naturally occurring chloroplasts in which membrane-mediated electron-transfer processes are observed following light absorption by chlorophyll.

Monolayer Systems

Earlier systems have indicated the extent to which hydrophobic interactions, such as those occurring between hydrocarbon chains, can effect the structure taken by an amphiphile in solution. For example, long-chain fatty acids can form micelles when dissolved in aqueous solution. However, if these materials are placed on a polar liquid surface, such as water, then monolayers may be found. Figure 3 represents this process in a diagrammatical form where a single, or monolayer, of fatty acid is shown on a water surface. Details of the methods of preparation of these assemblies and the care needed to exclude impurities can be found in the literature (41–49). This section will merely outline procedures that are necessary to discuss monolayers in a context with other assemblies.

Preparation and Properties. A standard freshman laboratory experiment involves the spreading of an oleic acid film on water by gradual addition of a concentrated oleic acid–benzene solution to the water surface. Evaporation of the benzene leaves a film of fatty acid on the water surface. At first sight this relatively easy preparation of a hydrophobic film on a polar surface is encouraging to prospective workers in the field; however, the experiment is misleading. The two prime problems are the preparation of clean smooth surfaces and the exclusion of im-

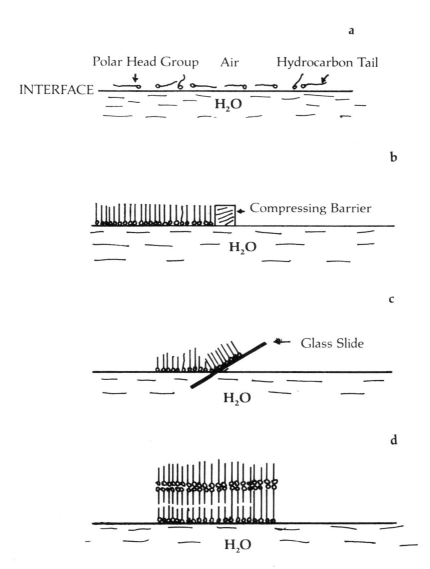

Figure 3. Adsorption of surface-active molecules at interfaces. a, water–air (random); b, water–air (compressed and ordered); c, transfer to glass slide; and d, multilayer.

purities. The amount of film spread on a surface is very small, and dust can soon contribute a major role to film properties. Nevertheless, the simple spreading of a film on a polar surface by immersion of the surface in a solution of the material to be spread is widely used (49–52). Adsorption of materials from organic solutions onto polar solid surfaces, such as glass, may produce homogeneous compact monolayers, and the process is claimed to be reproducible.

A versatile technique developed by Kuhn et al. (47, 48), uses the Langmuir balance method. As with the earlier spreading methods, the monolayer material is introduced onto an aqueous surface from solution in a volatile solvent (e.g., $CHCl_3$ or benzene). Evaporation of the solvent leaves the molecules on the water surface. The aqueous surface is enclosed by means of a trough, one side of which is mobile and can be swept across the water surface. This action compresses the film and is the basis of many famous experiments by Langmuir and Blodgett. By sensing the force resisting the movement of the barrier compressing the film, one can judge when the film is a truly organized monolayer on the water surface, as depicted in Figure 3. Techniques have been developed to manipulate the film. Figure 3 depicts how the film can be relocated on any other surface, such as a microscope slide. This relocation is achieved by dipping the slide into the monolayer followed by careful withdrawal. Repetition of this process leads to multilayer assemblies on the support as shown. As many layers as are required can be built onto the slide support. Introduction of guest molecules into the water surface film at different stages in this process can be used to space guest molecules at known distances from each in the assembly. Several variations can be used to organize reactants precisely in a series of monolayers in assembly (47, 48).

Physical and Chemical Reactions in Monolayer Assemblies. The monolayer technique has been used to investigate a variety of phenomena.

ENERGY TRANSFER. The monolayer assemblies, although strictly located in space, nevertheless show some degree of organizational arrangement. This organization is presumed to occur via openings or pores in the films, which allow the film components to diffuse laterally on the film. Some rapid but limited movement around a small region of space is also evident. For example, a single octadecane film on a glass substrate containing dye I shows absorption and emission spectra typical of monomeric dye. However, a multilayer system, although still showing absorption spectra typical of monomeric dye, nevertheless now shows fluorescence identified with excimers of the excited dye. Therefore, neighboring dye molecules show rapid readjustment following excitation to produce excimers. The movement producing excimers is slight and does not involve gross movement. However, significant lateral movement is observed in oleic acid monolayers on water (54). Excimers of pyrenedodecanoic acid are formed in this film, and a two-dimensional lateral diffusion coefficient of 1.7×10^{-6} cm^2/s can explain the data. Thus, some limited movement is possible to facilitate kinetic processes such as energy transfer. However, for the most part Förster-type energy transfer has been studied, where energy is transferred efficiently between fixed molecules at some distance apart (as much as 50 Å or greater).

The nature of the fatty acid and the counterion used also affects the degree of movement in a monolayer. For cadmium arachidate monolayers no significant dye transport is reported from one layer to another, and energy transfer between molecules is completely via a Förster mechanism. However, with cadmium stearate monolayers enhanced energy transfer over and above that noted in the previous films may be due to dye penetration from one layer to another. This effect is even larger for barium stearate. Also, small molecules such as diazonium and azo salts diffuse through even arachidic acid monolayers fairly readily. To eliminate destruction of the assembled system by reactant movement, cadmium arachidate is used for the monolayer, and long alkane chains (C_{18} and up) are attached to the dye molecules to restrict their movement.

FÖRSTER ENERGY TRANSFER. Monolayers of arachidic acid have been used to maintain known separations of two dye molecules, one of which is excited and donates energy to an acceptor dye. A typical donor-type dye is dye I, which absorbs in the UV and fluoresces in the blue part of the spectrum; and a suitable acceptor is dye II, which absorbs in the

dye I

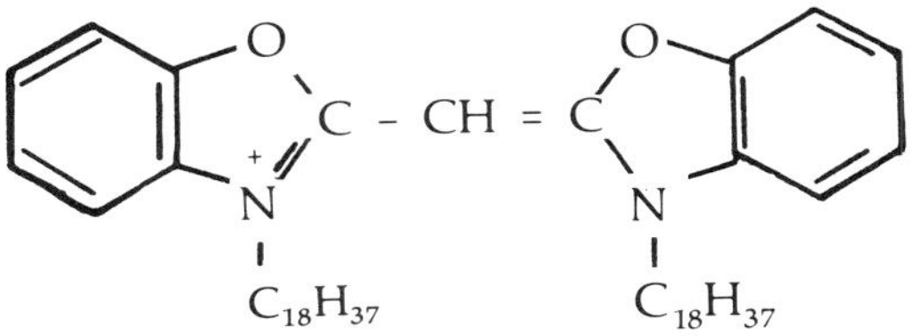

$$C - CH = C$$

$$C_{18}H_{37} \qquad C_{18}H_{37}$$

dye II

$$C - CH = CH - CH = C$$

$$C_{18}H_{37} \qquad C_{18}H_{37}$$

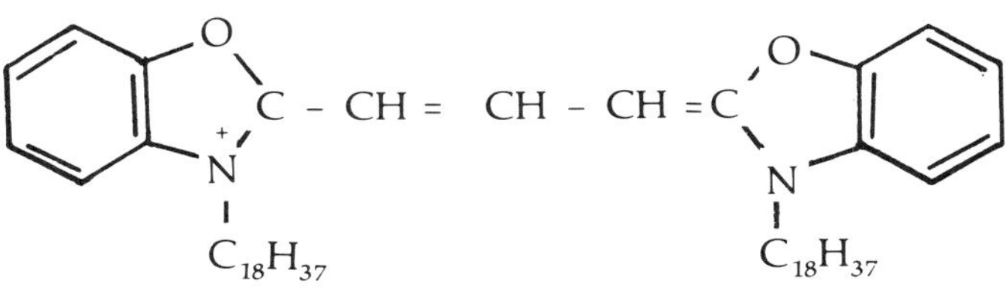

blue and fluoresces in the yellow. Excitation by UV light leads only to excitation of dye I because dye II has little absorption in this part of the spectrum. Introduction of dye II into the system at known separations from excited dye I causes a quenching of the dye I fluorescence as the separation d decreases. The following relationship holds quantitatively in this system and others like it

$$\left(\frac{I_d}{I_D}\right)_s = \left[1 + \left(\frac{d_o}{d}\right)^4\right]^{-1}$$

where I_d is the fluorescence of dye I for a separation of d, I_D is the fluorescence for infinite separation, and d_o is the separation at which 50% of excited dye I loses its energy by energy transfer to dye II and produces yellow fluorescence. Typical data are shown in Figure 4 for such systems. This technique may be used to gain information on the excited state properties of dyes that do not luminesce. The energy of the nonluminescing dyes may be transferred to a luminescing acceptor type dye by arranging them in close, exact proximity. Energy transfer to the acceptor monitors the behavior of the nonluminiscent dye. In all reactions, the energy-transfer process leads to a decrease in the lifetime of the excited state donating the energy.

Multipole Nature of Excited States

The luminiscence of organic molecules corresponds to an electric dipole emission. However, the exact nature of the multipole process is uncertain because the excited molecule may act as an electric quadrupole or as an electric magnetic dipole emitter. These processes all show a different distance dependence on Förster energy transfer and may be conveniently studied by the monolayer technique.

For a magnetic dipole, the previous dipole energy-transfer equation takes on the form

$$\frac{I_d}{I_D} = \left[1 + \left(\frac{d_o}{d}\right)^2\right]^{-1}$$

For an electric quadrupole emitter, the form is

$$\frac{I_d}{I_D} = \left[1 + \left(\frac{d_o}{d}\right)^6\right]^{-1}$$

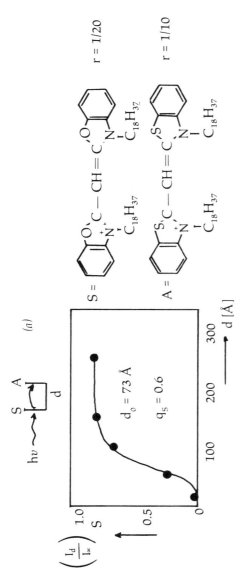

Figure 4. Energy transfer in layer systems.

System of monolayers of S and A at distance d (curves a–d) and S, A, and A' at distance d and d' (curve e). Experimental specifications are given with each curve with the values for the quantum yield q_S of the sensitizer to calculate d_o (curves a–d) and q_A to calculate d_A' (curve e). Curve a, fluorescence intensity of S vs. d; and curve b, fluorescence intensity of S and A vs. d. The fluorescence of A, where S and A are in direct contact, is larger than expected. Curve c is the decay time of luminescence of S vs. d (curve follows $(\tau_d/\tau_\infty)_s = (I_d/I_\infty)_s$). Curve d is the velocity of photochemical bleaching of A vs. d (curve follows $(v_d/v_o)_A =$ $(I_d/I_o)_A$). Curve e is the fluorescence intensity of A vs. distance d' between A and A'; and distance d between S and A. Key: □, 3 Å; △, 27 Å; and ○, 53 Å (47, 48). Continued on next page.

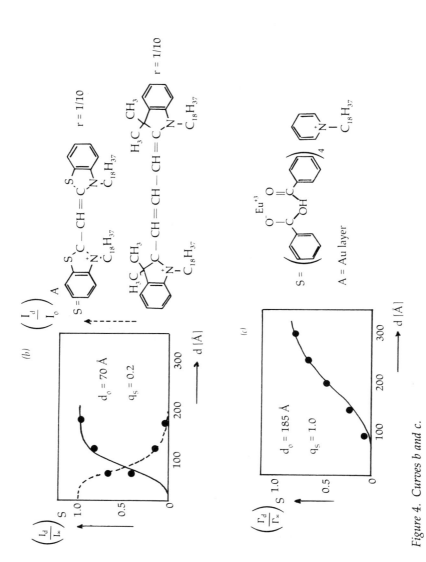

Figure 4. Curves b and c.

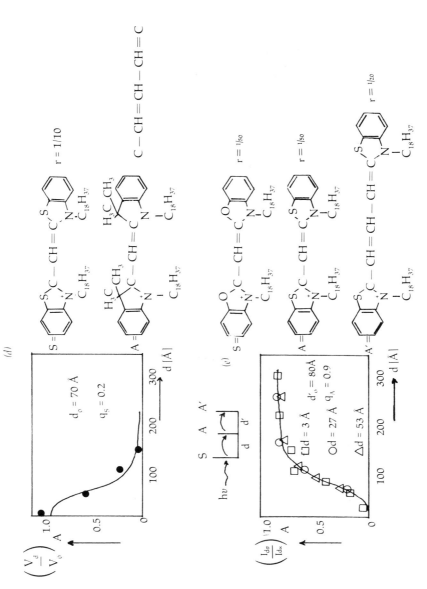

Figure 4. Curves d and e.

The dependence of I_d/I_D on d can usually distinguish the exact situation in question. In principle, the standing wave technique can also distinguish between the various processes.

Standing Wave Technique. The fatty acid monolayers may be deposited on a mirror to create a series of steps (Figure 3). When light is shone on this system, standing waves are produced that may exhibit nodes or antinodes at the layers. If a dye molecule is situated at a node, then little absorption of light occurs compared to one at an antinode; this step is dark because it does not luminesce well. This effect is a direct consequence of the fact that the absorption power is proportional to the square of the local electric field strength. However, in the case of an electric quadrupole or magnetic dipole oscillator, the absorption power is proportional to the square of the electric field gradient. Hence, maximum absorption should occur at the nodes.

Complex I

The position of a minor reflector near the excited state can also decrease the emission lifetime of the state. Typical data are shown in Figure 5 for the luminiscence of complex I at various distances from a silver mirror and from a fatty acid–air interface. The field determining the probability of observing a light quantum may be represented as a classical oscillator antenna. The mirror, or interface, reflects the wave associated with the field and causes it to interact with the source. The phase of the echo field relative to the motion of the oscillator may impart an accelerating or retarding effect in the decay time. These effects are quite clearly visible in Figure 5 in monolayer systems; however, to date they have not been observed in micelles, microemulsions, or vesicles.

Electron Transport in Monolayers. The monolayer surfaces may be encased in metallic conductors or electrodes. Excitation of a dye in the monolayer system produces a current in an external circuit connected

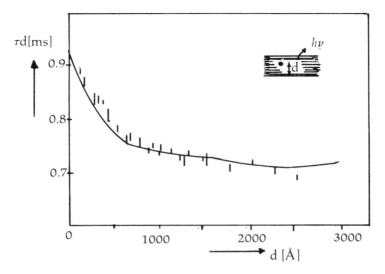

Figure 5. Europium complex I at distance d *from fatty acid–air interface. Decay time of luminescence* (τ_d) *as a function of* d. *Theoretical curve and experimental points.*

to the electrodes if a potential is applied to the system. The excited electron in the dye and monolayer insulator has not sufficient energy to climb the energy barrier imposed on it. However, the wave nature of electrons allows a small probablility for the electron to exist in the other side of the barrier. This probability is about 10^{-14} for a barrier of 1 eV and a thickness of 30 Å. The tunneling process, as the probability is called, leads to an electric current in the external circuit, or electron transfer to a suitable acceptor located near to the excited molecule. This process is particularly interesting to biosystems such as photosynthesis, where adsorption of light leads to charge separation in an assembly system.

Literature Cited

1. Israelachvili, T. N.; Mitchell, J.; Ninham, B. W. *J. Chem. Soc., Faraday Trans. 2* **1976**, *72*, 1525.
2. Israelachvili, T. N.; Mitchell, J.; Ninham, B. W. *Biochim. Biophys. Acta* **1977**, *470*, 185.
3. Tanford, C. *Science* **1978**, *200*, 1012.
4. Huang, C. *Biochemistry*, **1969**, *8*, 344.
5. Bangham, A. D. In "Progress in Biophysics and Molecular Biology"; Butler, J. A. V., Ed.; Noble: London, 1968; p. 29.
6. Kremer, J. M.; Esker, M. W. J.; Pathmanorahan, C.; Wiersma, P. H. *Biochemistry* **1977**, *16*, 3932.
7. Batzri, S.; Korn, E. D. *Biochim. Biophys. Acta* **1973**, *298*, 1015.
8. Fendler, J. H. *Acc. Chem. Res.* **1980**, *13*, 7.

9. Träuble, H. *Naturwissenschaften* **1971**, *18*, 177.
10. Morris, D. A. N.; Thomas, J. K. In "Micellization, Solubilization, and Microemulsions"; Mittal, K. L., Ed.; Plenum Press: New York, 1976, p. 913.
11. Almgren, M. *Chem. Phys. Lett.* **1980**, *71*, 539.
12. Almgren, M. *J. Am. Chem. Soc.* **1980**, *102*, 7882.
13. Almgren, M.; Grieser, F. Thomas, J. K. *J. Am. Chem. Soc.* **1979**, *101*, 279.
14. Morris, D. A. N.; McNeil, R.; Castellino, F. J.; Thomas, J. K. *Biochim. Biophys. Acta* **1980**, *599*, 380.
15. Cogan, U.; Shinitzky, M.; Weber, G.; Nisida, T. *Biochemistry* **1973**, *12*, 521.
16. Galla, H. J.; Schwartz, E. *Ber. Bunsenges. Phys. Chem.* **1974**, *78*, 949.
17. Cheng, S.; Thomas, J. K. *Radiat. Res.* **1974**, *60*, 268.
18. Vandekooi, T. M.; Callis, J. B. *Biochemistry* **1974**, *13*, 4000.
19. Pownall, H. J.; Smith, L. C. *J. Am. Chem. Soc.* **1973**, *95*, 3136.
20. Morris, D. A. N. Ph.D. Thesis, University of Notre Dame, Ind. 1977.
21. Hayden, D. A.; Taylor, J. *J. Biol. Chem.* **1963**, *4*, 281.
22. Mandersloot, J. G.; Reman, F. C.; VanDiener, C. C. M.; DeGier, J. *Biochim. Biophys. Acta* **1975**, *382*, 22.
23. Lucy, T. A. *Nature* **1970**, *227*, 815.
24. Rideal, E.; Taylor, F. H. *Proc. R. Soc.* **1958**, *B148*, 450.
25. Reman, T. C. Thesis, Utrecht, Netherlands, 1971.
26. Taylor, C. L.; March, R. E. *Can. J. Chem.* **1973**, *51*, 216.
27. Schnecke, W.; Grätzel, M.; Henglein, A. *Ber. Bunsenges. Phys. Chem.* **1977**, *81*, 821.
28. Henglein, A.; Schnabel, W. Ein. Duhning in die Strahlenchemie Weinheld/ Bergste Valag. Chemie 1969.
29. Barber, D. J. W.; Thomas, J. K. *Radiat. Res.* **1978**, *74*, 51.
30. Almgren, M.; Thomas, J. K. *Photochem. Photobiol.* **1980**, *81*, 329.
31. Morris, D. A. N.; Barber, D.; Thomas, J. K. *Chem. Phys. Lett.* **1976**, *37*, 481.
32. Escabi-Perez, J.; Romero, A.; Lukac, S.; Fendler, J. H. *J. Am. Chem. Soc.* **1979**, *101*, 2231.
33. Matsuo, T.; Nagamura, T.; Itoh, K.; Nishijima, T. *Mem. Fac. Eng. Kyushu Univ.* **1980**, *40*, 25.
34. Matsuo, T.; Itoh, K.; Takuma, K.; Hashimoto, K.; Nagamua, T. *Chem. Lett.* **1980**, 1009.
35. Infelta, P. P.; Grätzel, M.; Fendler, J. H. *J. Am. Chem. Soc.* **1980**, *102*, 1479.
36. Ford, W. E.; Otvos, J. W.; Calvin, M. *Nature* **1978**, *224*, 507; *Proc. Nat. Acad. Sci. USA* **1979**, *76*, 3590.
37. Aso, Y.; Kano, K.; Matsuo, T. *Biochim. Biophys. Acta* **1980**, *599*, 403.
38. Takayanagi, T.; Nagamana, T.; Matsuo, T. *Ber. Bunsenges. Phys. Chim.* **1980**, *84*, 1125.
39. Romero, A.; Sunamoto, J.; Fendler, J. H. *Life Sci.* **1976**, *18*, 1453.
40. Kano, K., Fendler, J. H. *Chem. Phys. Lipids* **1979**, *23*, 189.
41. Lord Rayleigh. *Philos. Mag.* **1899**, *48*, 337.
42. Langmuir, I. *J. Am. Chem. Soc.* **1917**, *39*, 1848.
43. Blodgett, K. B. *J. Am. Chem. Soc.* **1935**, *57*, 1007.
44. Adam, N. K. "The Physics and Chemistry of Surfaces"; Oxford Univ. Press: Oxford, **1941**.
45. Gaines, G. L. "Insoluble Monolayers at Liquid Gas Interfaces"; Interscience: New York, **1966**.
46. Adamson, A. W. "The Physical Chemistry of Surfaces"; Wiley Interscience: New York, **1976**.
47. Kühn, H.; Möbius, D. *Angew Chem.* **1971**, *10*, 620.
48. Kühn, H.; Möbius, D.; Bucher, H. "Physical Methods for Chemistry"; Weissberger, A.; Rossiter, B. W., Eds.; Waky: New York, 1972; Vol. I, p. 577, Part III B.
49. Sagiv, J. *J. Am. Chem. Soc.* **1980**, *102*, 92.
50. Sagiv, J. *Isr. J. Chem.* **1979**, *18*, 339.

51. Scamehora, J. F.; Schechter, R. S.; Wade, W. H. *J. Coll. Interface Sci.* **1982**, *85*, 463.
52. *Ibid.* **1982**, *85*, 479.
53. *Ibid.* **1982**, *85*, 494.
54. Loughran, T.; Hatlee, M. D.; Patterson, L. K.; Kozak, J. J. *J. Chem. Phys.* **1980**, *72*, 5791.

11

Specialized Systems

In earlier chapters we have discussed the specialized effects of various organic structures (micellar microemulsions, vesicles, etc.) on the photochemistry of molecules lodged in, or on them. Many other colloidal systems are available, and we will discuss some of them in terms similar to those used in micellar systems.

Colloidal Silica

The effect of aqueous colloidal SiO_2 particles on the electron exchange between (RuII)* and a dipropionate derivative of MV^{2+} (PMV) has received detailed attention (1). The colloidal SiO_2 system is more effective than sodium lauryl sulfate (NaLS) and other systems in providing a high quantum yield for reduced PMV. The main reason for the large yield of reduced PMV may be the high negative charge in the SiO_2 particles due to adsorbed OH^- and to ionized silanol groups (SiO^-). Trisbipyridylruthenium (RuII) may be adsorbed on the particles in preference to the Na^+ counterion; and the e^- transfer from (RuII)* to PMV may occur at the SiO_2 surface and produce reduced PMV, which, because it is negatively charged, is repelled from the particle and away from the RuIII.

$$(RuII)* + PMV \rightarrow Ru(III) + PMV^-$$

A repair agent, such as triethanolamine (TEA), then donates an e^- to RuIII and re-forms RuII; the net photochemical reaction is the transfer of e^- from TEA to PMV. Perhaps the resultant negative charge on the colloidal SiO_2 is larger than that on the NaLS micelle. This difference could explain the larger yield of PMV^- in the SiO_2 system. However, the NaLS micelle also solubilizes PMV, an organic molecule, to a greater extent than does the SiO_2 particle. Thus, the escape of PMV^- from NaLS micelles is hindered compared to SiO_2 particles. The nature of the e^- transfer process could also be different on the SiO_2 surface compared to micellar NaLS. However, the SiO_2 system promotes e^- transfer from (RuII)* to PMV to a greater extent than NaLS micelles.

0065-7719/84/0181-0267$06.50/1
© 1984 American Chemical Society

Other probe studies provide more information on this process (2). Colloidal SiO_2 particles, similar to those used in the earlier study (1), did not solubilize neutral organic molecules; cationic species were solubilized on these particles.

Two probe molecules, Ru(II) and 4-(1-pyrenyl)butyltrimethylammonium bromide (PN^+), have been used to investigate the nature of colloidal silica particles in water. The fluorescence spectra of the two probes show that the silica surface is very polar and provides an environment for the probes that is similar to water.

Quenching studies (Tables I and II) on the excited states of RuII and PN^+ by anionic quenching molecules show that the particles are negatively charged; however, the charge is not as effective as that on NaLS micelles. For example, both negative ions $Fe(CN)_6^-$ and 3,5-dinitrobenzoate quench (RuII)* more slowly on silica compared to water; however, the rates are not as slow as in NaLS, where the rates were too slow to measure accurately. The decreased reaction rates are due to repulsion of the anionic quenchers by the anionic silica particles or NaLS micelles. The Debye modification of the Smoluckowski equation may be used to explain these data. The equation indicates that

Table I.
Quenching of Excited RuII Observed at $\lambda = 610$ nm

Quencher	System, k $(LM^{-1} s^{-1})$		
	Water	NaLS Micelle	Silica
None	1.67×10^6	1.3×10^6	1.4×10^{6a}
			$(1.5 \times 10^6)^b$
			$[1.5 \times 10^6]^c$
O_2	2.8×10^9	2.1×10^9	1.3×10^{9a}
			$[1.2 \times 10^9]^c$
Nitrobenzene	3×10^9	1×10^9	7.6×10^{8a}
			$(8.8 \times 10^8)^b$
			$[7.5 \times 10^8]^c$
$Fe(CN)_6^{3-}$	3.8×10^{10}	$<10^7$	1.0×10^{8a}
			$(3 \times 10^9)^b$
			$[1.0 \times 10^8]^c$
3,5-Dinitrobenzoate	7.8×10^9	$<10^7$	1.5×10^{8a}
Cu^{2+}	5.3×10^7	1.8×10^8	$(1.6 \times 10^7)^b$
Heptyl viologen	7.4×10^8	8.75×10^8	3.43×10^{8b}
			$(3.6 \times 10^8)^b$

[a] Small; pH 10.4; $r = 40$ Å.   [b] Acidic; $r = 200$ Å.   [c] Large; pH 9.0; $r = 200$ Å.

Table II.

Quenching of Excited Cationic Pyrene Probe (PN$^+$) Observed at λ = 375 nm

	System, k $(LM^{-1} s^{-1})$		
Quencher	Water	NaLS Micelle	Silica
None	6.8×10^6	4.9×10^6	7.06×10^6
O$_2$	1.0×10^{10}	7.0×10^9	1.8×10^9
CH$_3$NO$_2$	3.8×10^9	2.6×10^9	1.1×10^9
Dimethylaniline	4.0×10^9	(Poisson)	1.1×10^9
Tl$^+$	3.2×10^9	1.7×10^{10}	1.6×10^{10}
Cu^{2+}	1.947×10^9	1.8×10^9	5.4×10^8 (2% acidic Si)
3-Nitropropionate	6.7×10^9	$<10^7$	2.0×10^8

the diffusion-controlled rate constant, k_1 for reaction of two ions is given by

$$k = \frac{4\pi r}{1000} \frac{D N}{rEkT} \frac{Z_1Z_2e^2}{rEkT} \bigg/ \left[\exp\left\{ \frac{Z_1Z_2e^2}{rEkT} \right\} - 1 \right]$$

where N is Avogadro's number, r is the interaction radius, E is the dielectric constant of the medium, Z_1e and Z_2e are the charges of the two reactants, and D is the sum of diffusion constants of probe and quencher. The RuII data can be explained quantitatively if charges of -8 to -10 units/particle are used for silica; a much larger charge (-20) is necessary to explain the NaLS data. The exact position of the probe in the surface is important in this calculation; the data could also indicate that for reaction to occur the anionic quenchers do not have to penetrate as close to SiO$_2$ as NaLS.

Dimethylaniline (DMA) rapidly quenches (PN$^+$)* on silica and produces the anion of PN$^+$, (PN$^+$)$^-$, and the DMA cation, DMA$^+$. The ions are short-lived because DMA$^+$ and (PN$^+$)$^-$ do not escape from the particle rapidly enough to prevent back electron transfer. This reaction has also been observed in anionic NaLS micelles (3).

Laser excitation of RuII leads to (RuII)* and to a bleaching of the RuII ground-state absorption in the region of 4600 Å. Heptyl and methyl viologen (HV^{2+} and MV^{2+}) rapidly quench the (RuII)*, but the well-established release of long-lived intermediates such as reduced MV$^+$ is not observed: (RuII)* + MV^{2+} → (RuIII) + MV$^+$. This observation is similar to the PN$^+$–DMA system where the anionic silica surface binds the cationic products and promotes back e^- transfer before the product ions can be separated.

Interestingly, Ag^+ reacts with (RuII)* on silica and produces long-lived (several seconds) bleaching of RuII and forms colloidal silver.

$$(RuII)^* + Ag^+ \rightarrow (RuIII) + Ag^0 \quad \text{Colloidal silver}$$

The back reaction of Ag^0 + (RuIII) is rapid in water but is strongly retarded in silica because Ag^0 is ejected from the vicinity of (RuIII), which is strongly bound to the silica particle. Such a long-lived separation of products is not usually observed in homogeneous aqueous solution.

Quenching studies with cationic quenchers show that the cations are bound strongly to the silica particles but do not move as readily around the surface as on anionic micelles. A small steric effect is observed with neutral quenchers. Several charge-transfer reactions, including photoionization, are strongly affected by the silica particles. The studies show many similarities to anionic micelles. They differ from micelles in two important aspects, however; they do not solubilize neutral organic molecules, and cationic organic molecules, such as PN^+, hexadecyltrimethylammonium bromide, and hexadecylpyridinium chloride, cluster on the silica surface rather than disperse uniformly around it, as with ionic micelles.

Porous Colloidal Silica

Colloidal silica particles with porous properties have been prepared by polymerization of silicic acid around small (20 Å) seed particles of SiO_2 (4). The particles formed have radii several hundred ångströms in diameter, as determined by dynamic light scatter and electron microscopy. These particles do not absorb neutral organic molecules. However, Ru(II) is absorbed onto the particles; and, immediately following adsorption, its photophysical and kinetic properties are similar to those in the earlier SiO_2 particles. With time (several hours), the fluorescence lifetime of (RuII)* increases from 0.5 to 2.0 μs, and the emission spectrum exhibits a blue shift. The data indicate considerable restriction of the (RuII)* in its surroundings and suggest that it may now be occluded in the polymerized particle. This observation is confirmed via quenching studies where quenchers, such as O_2 and nitrobenzene, are found to be ineffective in quenching the (RuII)* fluorescence. These quenchers are very effective in homogeneous solution and when (RuII)* is located on the surface of the particle. The quenching by MV^{2+} is slower in the polymerized system and may occur via an e^--tunneling process. The steady-state yield of MV^+ in this system is much larger than in other systems. Therefore, efficient escape of MV^+ from (RuIII) is indicated, again in agreement with the concept that RuIII and MV^+ are formed while still some distance apart.

Colloidal Clay Systems

Clays, or aluminosilicates, are abundantly distributed and widely used. Crushed clays can be readily suspended in aqueous media (5). At first sight, one may be tempted to relate any photochemical peculiarities of molecules on colloidal clays to similar events in colloidal silica particles. Many added complications of clays exist, however, in particular the occurrence of significant amounts of reduced or oxidized material (e.g., Fe^{2+}, Fe^{3+}) in the clay structure. The basic preparation of a clay colloid is simple; the material is ground to a fine powder (about 1 μm in diameter) and then suspended in aqueous media by rapid mixing.

    At this point we should consider the known features of clay structures.

Clay Structure. Many studies on clay structures and properties, including thermal reactions promoted on clays, are available (6–9). The theoretical formula for montmorillonite clay may be $(OH)_4Si_8Al_4O_{20}\cdot nH_2O$, where the nH_2O is the interlayer water. Clays have layered structures composed of units made up of two silica tetrahedral sheets with a central alumina octahedral sheet. Figure 1 shows a typical clay structure with the tips of the tetrahedrons pointing toward the center of the unit; both tetrahedral and octahedral sheets are arranged so that the tetrahedron tips of each silica sheet and one of the octahedral-sheet hydroxyl layers form a common layer. The individual sheets take up molecular species where these guest molecules are intercalated between the clay sheets. Substitution of Si^{4+} by Al^{3+} at the tetrahedral sites yields residual negative charge on the layer, and the structure adsorbs Na^+ or Ca^{2+} ions to achieve electrical neutrality. The Na^+ or Ca^{2+} cations are interchangeable with other cations in a cationic exchangelike process. The incorporation of large cationic surfactant molecules (e.g., octadecyl ammonium) at the Na^+ sites leads to separation of the clay layers, which has been observed by X-ray techniques (6, 10–14). This unique property of clays to adsorb organic molecules and enclose them in unique inorganic environments may be used in photochemical experiments. Also, clays promote certain specific reactions of chemical interest and may be considered catalysts (7, 10–19).

Adsorption of Molecules on Clays. Cations are readily adsorbed on clays by replacing the Na^+ counterions. For a typical montmorillonite clay, about 100 meq of cations can be adsorbed per 100 g of clay. The

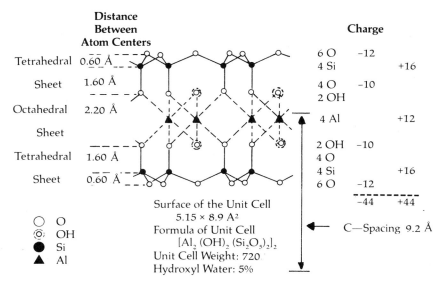

Figure 1a. Structure of clay mineral. Reproduced with permission from "An Introduction to Clay Colloid Chemistry." Copyright 1963 Interscience Publishers.

ions are adsorbed both on the surface and in the layers. The adsorption of noncharged molecules at low concentrations on aqueous colloidal clays is small and occurs at Lewis acid and base sites. However, colloidal organoclays readily solubilize hydrophobic noncharged molecules. Adsorption of small amounts of cetyltrimethylammonium bromide (CTAB), well below the cation exchange capacity, leads to adsorbed surfactant molecules that tend to lie flat within the layers. Increasing adsorption gradually leads to a monolayer type of arrangement, which progresses to a bilayer with further adsorption. Figure 1 shows the processes diagrammatically. X-Ray-diffraction data on the separation of the layers confirm the simple picture shown. Such layer-type processes readily occur on montmorillonite clays, but are absent on other clays, such as kaolin, where only surface adsorption can occur. A comparison of the chemistry in the two systems can be instructive with regard to statements of reactions on clay surfaces as against reaction in clay inner layers. A fully CTAB-exchanged clay is hydrophobic and is readily dispersed in hydrophobic solvents such as benzene. These organoclays may be dispersed in water by using excess CTAB; further absorption of surfactant leads to a charged double layer at the clay surface, hence rendering it hydrophilic. Fluorescent cationic molecules almost invariably fail to fluoresce strongly on clays. An exception is RuII; however, the fluorescence is severely curtailed.

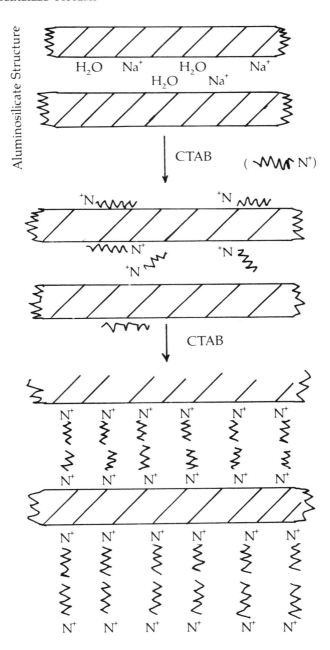

Figure 1b. Illustration of interaction of alkyl quaternary ammonium compound with the clay structure.

Reactions on Clays. Numerous organic reactions exist that are induced by clay systems (7, 14, 20–22). These organic systems range from the oxidation of amines, such as benzidine, to mono- and divalent cations to the polymerization of monomers such as styrene and methyl methacrylate. Data obtained from the effects of clay type (e.g., montmorillonite or kaolin) and clay treatment (i.e., Na^+-exchanged and polyphosphate-treated) enable the following statements to be made with respect to clay reactivity:

1. Ions, such as ferric and ferrous, that may comprise part of the clay structure often undergo electron exchange with adsorbed organic compounds. For example, complete reduction of ferric ions in montmorillonite to the ferrous form using hydrazine eliminates the ability of the clay to oxidize benzidine. These data suggest that the Fe^{3+} of the natural clay undergoes e^- transfer with benzidene.

2. Aluminum at the crystal edges can also catalyze the reaction. The edges possess positive charge, and treatment of the edges with anionic polyphosphate reduces this type of reactivity. A monomer such as styrene binds to the Al edge site via a Lewis-base mechanism, a condition that leads to polymerization (Structure 1).

Structure 1

Polar solvents, such as alcohols, reduce this type of reactivity, because they are stronger Lewis bases than styrene and, hence, replace the styrene at the active clay site. The ability of different clays to decompose organic molecules depends critically on the oxidation state of multivalent ions, such as ferric/ferrous, the level of Lewis acid and Lewis base sites, and the availability of these sites in the structure.

Photochemical Reactions. These previously discussed factors, which influence thermal reactions on clays, can also influence photochemical reactions. An additional source of reactivity could be the clustering of adsorbed molecules on the clay surfaces. This clustering, at low light intensities, leads to excited-state quenching by the ground state and, at high intensities, to bimolecular excited-state annihilation. These effects have been observed in colloid clay systems (23, 24) but not in clay films (25). Many organic molecules fail to fluoresce when absorbed on either montmorillonite or kaolin clays. Typical examples are the cationic dyes, such as rhodamine B and 6G, and quaternary ammonium derivatives of simple chromophores, such as PN^+. In the latter case, coadsorption of CTAB onto the clay leads to development of fluorescence. This result is due to colonies of adsorption of CTAB and PN^+ on the clay where the CTAB insulates and protects PN^+ from deactivation by the clay sites. CTAB has no effect on the rhodamine dyes, probably because the dye chromophore is located directly on the clay surface and cannot be protected by CTAB. Addition of cetylpyridinum chloride (CP^+), to the CTAB–PN^+–clay system leads to a static quenching of PN^+ fluorescence without a change in lifetime. This quenching is due to coadsorption of PN^+ and CP^+ in close proximity. Other molecules, such as dimethylaniline and nitrobenzene, illustrate quenching of PN^+ by a dynamic mechanism because the PN^+ fluorescence lifetime decreases in the presence of these molecules. Comparison of such experiments on kaolin and montmorillonite indicate surface quenching in kaolin and surface and internal-layer quenching takes place on the montmorillonite.

Trisbipyridylruthenium luminiscence is slightly quenched on clays at low light intensities. Two distinct decay times are observed; therefore, at least two adsorption sites exist for the molecule on the clay. At high light intensities, the quenching increases rapidly, and the luminescence lifetime shortens. These observations are due to excited-state annihilation because of the close proximity of the adsorbed molecules. Much research remains to be done regarding photochemistry on clay materials. It is an important area in nature for the alteration and destruction by sunlight of organic molecules such as pesticides, which adsorb into ground clay.

Colloidal Metals

Most metals may be obtained in a colloidal form in aqueous solution. Silver, gold, and platinum have received special attention by photochemists and radiation chemists.

Colloidal Silver. Techniques exist for preparing colloidal silver that utilize the thermal reduction of silver cations (26). Preparation of pure silver sols is possible by the radiolysis of silver salts in water in the presence of a surfactant, such as NaLS or polyvinyl alcohol, both of which act as stabilizers. The net reaction is

$$H_2O \rightarrow e_{aq}^- + H^+ + OH$$
$$e_{aq}^- + Ag^+ \rightarrow Ag \rightarrow sol$$

The hydrated electron may be used to produce a reducing species (e.g., Cu^+),

$$e_{aq}^- + Cu^{2+} \rightarrow Cu^+$$

which also reduces Ag^+. The advantage of this method is the clean lines of the system, because, unlike thermal preparations, an added reducing agent is not needed and quite small particles (radii $\sim$ 100 Å) are formed. Similar methods may be used to produce gold sols (27). The silver and gold sols formed are quite stable and may be used for subsequent chemistry (27, 28–30).

Radicals of high negative potential, such as α-alcohol radicals, formed in the radiolysis of the alcohol or water,

$$OH + RCH_2OH \rightarrow RCHOH + H_2O$$

readily transfer electrons to colloidal silver particles (28–30). The particles become charged by the donated electrons and eventually produce H_2 by reducing water. The electrons may also react with suitable acceptors that approach the charged particle.

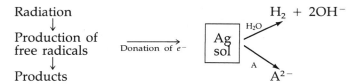

The Ag and Au sols provide an added dimension to free-radical chemistry by promoting reactions that lead to reduction of water and other compounds by free radicals, in deference to simple radical–radical recombination or disproportionation. The reduction of water to H_2 is a particularly interesting factor of these systems. However, the platinum system has been used much more extensively for this reaction (31–33).

Colloidal Platinum Catalysts. Colloidal platinum catalysts have been used to promote a variety of reactions. The classical method of producing the catalyst is to heat chloroplatinic acid in water with organic compounds such as citric acid. A brown sol with a relatively small radius ($\sim$ 30 Å) forms (33); and the quantum yield of H_2 from these colloids used in photosystems is as high as 0.3 (34). Production of these sols is possible by photolysis of the chloroplatinic acid (35). This reaction is particularly useful in systems where the Pt catalyst has to be coated onto another colloid, such as a colloidal semiconductor (35). The most commonly studied reaction is the transfer of electrons from reduced methyl viologen (MV^+) to the Pt colloid,

$$MV^+ \xrightarrow[\text{transfer}]{e^-} \boxed{Pt}^{\,n-} + MV^{2+}$$

followed by reduction of water at the Pt-colloid surface,

$$\boxed{Pt}^{\,n-} \xrightarrow{H_2O} H_2 + 2OH^-$$

The first reaction is fairly rapid and occurs with a first-order rate constant of 5.7×10^4 s^{-1} with 12.5×10^{-4} M Pt. Other radicals such as the benzophenone radical, $Ph_2\dot{C}\text{-}OH$, also react rapidly with Pt sols. These reactions are similar to those observed with Ag and Au sols. Pt sols also catalyze the reaction of ceric ions with water to produce O_2

$$4Ce^{4+} + \boxed{Pt} \xrightarrow{2H_2O} 4Ce^{3+} + O_2 + 4H^+$$

The chemistry of these rather complex systems may be studied to advantage by pulse laser and pulse radiolysis techniques.

Metal Oxides and Semiconductors. Several metal oxides are used as catalysts in industrial processes, but few have been studied photochemically in the colloidal state. Methods of producing colloidal oxides are well established (36). From the photochemical point of view, suspended powdered or colloidal titanium dioxide, a semiconductor, has received the most attention. The powdered or colloidal TiO_2 absorbs light at $\lambda < 4000$ Å and, when excited, can initiate a whole variety of chemical events using solutes present in the aqueous phase. This effect is particularly true if the TiO_2 particle is coated with platinum or adsorbs an electron acceptor (e.g., Cu^{2+} or MV^{2+}). In both cases, excitation of TiO_2 leads to charge separation, and the electron goes to Pt or an acceptor.

$$TiO_2 \xrightarrow{\;h\nu\;} [TiO_2^+ + e^-]$$

The electron acceptors capture e^- and release the positive holes in the excited semiconductors for subsequent chemistry. For Pt sols, H_2 is produced (35, 36). With MV^{2+}, which is also adsorbed at the TiO_2 surface, MV^+ is formed, which can also react with Pt to produce H_2 (37). To obtain large yields of H_2, electrons must be injected back into the semiconductor to repair the positive hole. Sacrificial agents, such as EDTA, are used; however, almost any carbon source is successful (36). For acids (e.g., CH_3COOH), the products following e^- transfer are CO_2 and CH_4 or C_2H_6.

$$TiO_2^+ + CH_3COO^- \rightarrow (TiO_2) + CH_3COO \rightarrow CH_3 + CO_2$$

Such systems can produce significant quantities of amino acids when an N_2 source is also included (38).

In the synthesis of particles containing TiO_2, RuO_2, and Pt (39), the Pt again acts as an electron acceptor from excited TiO_2 and generates H_2, and the RuO_2 acts as a catalyst to promote O_2 formation from the positive holes. Quantum yields of $H_2 > 0.2$ are observed in these systems (40).

Pulsed laser experiments show that I_2^- (41) and CO_3^- (42) are formed when colloidal TiO_2 is excited in the presence of I^- and CO_3^- ions, respectively. These species are formed rapidly within the laser pulse and are due to e^- transfer from the adsorbed anion to the TiO_2 positive hole.

Cadmium Sulfide. A well-studied semiconductor with many ideal photoproperties, cadmium sulfide (CdS) possesses a strong visible absorption spectrum up to about 5000 Å. This material readily transfers e^- to many acceptors. CdS has similar properties in aqueous suspension and in the colloidal form. Visible irradiation of aqueous CdS suspensions produced O_2^- (43) and reduced MV^{2+} (44); the yield MV^{2+} increased in the presence of EDTA. H_2 was also efficiently produced on irradiation of colloidal CdS + MV^{2+} + Pt in the presence of a sacrificial agent such as EDTA (45, 46). Photoinduced e^- transfer takes place giving MV^+, which is converted to H_2 and MV^{2+} on the Pt. The CdS positive hole is repaired by EDTA. The CdS system may be codispersed with RuO_2 and Pt in water. This dispersion produces a mixture that gives H_2 and O_2 on irradiation (47). The RuO_2 promotes the formation of O_2 from interaction of CdS positive holes and water. Pulsed irradiation of CdS colloids in the presence of MV^{2+} leads to MV^+, which is formed rapidly and within a 3-ns laser pulse (48). Electron transfer from CdS to MV^{2+} only occurs with surface-bound MV^{2+}. The MV^+ thus formed exhibits an initial rapid decay (half-life $\sim$ 0.5 μs) that gradually decreases, so that some of the MV^+ exists for hours. The gradual decrease is due to a geminate recombination of MV^+ with the CdS positive hole on the CdS surface and, at high laser intensities, to dimerization of MV^+ on the particle surface.

Colloidal semiconductor systems are a particularly fruitful area of photochemical study because of their potential interest as energy storage systems and also the intriguing reactions of the positive holes and electrons that occur under unique conditions on the particle surface.

Cyclodextrins

Cyclodextrins are doughnut-shaped cyclic oligosaccharides that are soluble in water. Three types are in common use: α-cyclodextrin (cyclohexaamylose), β-cyclodextrin (cycloheptaamylose), and δ-cyclodextrin (49, 50). These compounds solubilize hydrophobic organic molecules in the dextrin core away from the bulk water phase. The guest molecule is screened from certain types of reactivity with reactants because of its location in the dextrin; this screening is used in the construction of an enzymelike catalysis in these systems (51, 52).

Pyrene is readily solubilized in β-cyclodextrins up to 10^{-3} M in 10^{-2} M dextrin, and the pyrene is occluded within the doughnut hole of the dextrin (54). The pyrene fluorescence spectrum indicates a polar environment at low pyrene/dextrin ratios (<104/1); this environment shows increasing hydrophobicity as the pyrene content increases (53). The increase is due to multiple occupancy of a dextrin by more than one pyrene molecule (54). Multiple occupancy is also observed in other systems involving dextrins, acetonitrile (55, 56), and surfactants (53), which can share the dextrin cavity with an aromatic molecule; other examples are reported (57, 58). Solubilization of pyrene in the dextrin severely limits its fluorescence quenching by O_2, I^-, and CH_3NO_2; a decrease in quenching-rate constant of $>1 \times 10^{-1}$ is quite common. Similarly, the phosphorescence of 1-bromonaphthalene and 1-chloronaphthalene is readily observed in dextrin system; and phosphorescence quenching by NO_2^-, which is diffusion controlled in homogeneous solution, is much reduced (55). Similar effects are observed for the inclusion of several 1,3-bichromophoric systems, such as 1,3-dinaphthylpropane (56) (Structure 2), in which O_2 quenching of the fluorescence of these compounds is severely reduced. Conversely, the excimer emission of the bichromic systems is greatly increased relative to homogeneous solution, because of the inclusion of both chromophores in the dextrin cavity in close proximity.

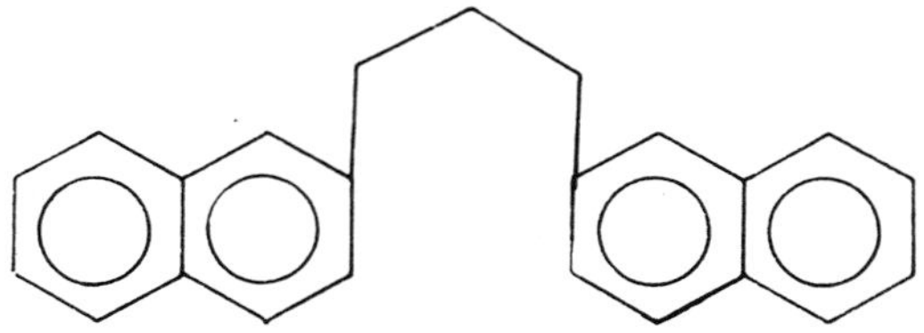

Structure 2

Dextrins may provide interesting systems for comparison to micelles. Their property of solubilizing molecules may be used to partially expose a portion of the guest molecule to a reactant in the aqueous phase. This effect could provide important information on the molecular details of many excited state quenching reactions.

Macrocyclic Compounds

Macrocyclic compounds, such as crown ethers, bear some resemblance to cyclodextrins, at least as far as photochemical interests are concerned (59). The crown ether dibenzo-18-crown-6 (Structure 3) readily complexes metals, such as Tl^+, and hence may introduce geometric restrictions of interest into a reaction where a complexed Tl^+ quenches an excited state. The salient features of metal ions binding to various macrocyclic compounds have been described (60).

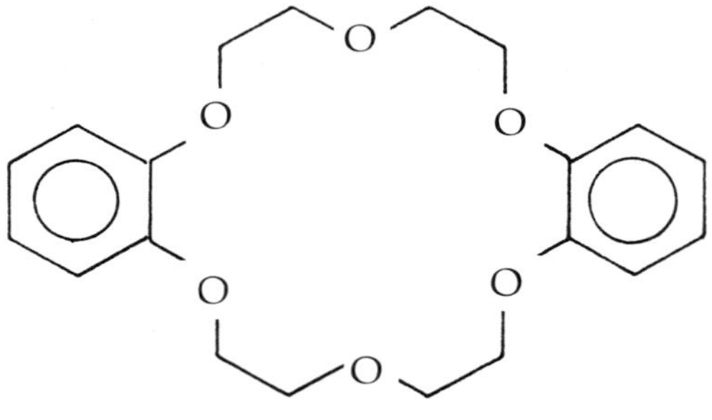

Structure 3

For dibenzo-18-crown-6, Tl^+ binding efficiently reduces the fluorescence of the benzo group (61). The fluorescence lifetime in the absence of Tl^+ is 2–6 ns, and bulk $[Tl^+]$ as low as 10^{-3} M efficiently quenches the fluorescence without a measurable change in fluorescence lifetime. The quenching is static and approaches unit efficiency for a molecule with a Tl^+ bound to it. Analysis of the data shows that the binding constant, for the process, crown ether + Tl $\rightleftharpoons$ (crown ether,

Tl$^+$), is $4.3 \pm 0.2 \times 10^3$ M^{-1} at 22 °C. Proton-NMR data show that the Tl$^+$ ions in the crown cavity interact strongly with the benzene moieties of the ether. Complexed Tl$^+$ is a factor of 10 less efficient than free Tl$^+$ in quenching excited states such as pyrene, presumably because the crown ether restricts the molecular association of Tl$^+$ and excited pyrene, a condition necessary for reaction to take place. The reduction of complexed Tl$^+$ by solvated electrons to produce Tl^0 is reduced about fivefold over that for Tl$^+$ in aqueous solution.

$$Tl^+ + e_{aq}^- \rightarrow Tl$$

Macrocyclic compounds with long hydrocarbon chains have been synthesized (*62*) such as crown 2-tetradecyl-1,4,10,13-tetraoxa-7,16-diazacylooctadecane (Structure 4). The preceding compound strongly absorbs Ag$^+$ ($k_{eq} \sim 10^8$) and also readily forms micelles (CMC = 3×10^{-4} M). Excitation of a dye molecule in the crown–Ag$^+$ micelle leads to rapid reduction of Ag$^+$ to Ag^0, which remains complexed with the ring ether structure (*63*). Back reaction of the oxidized dye and Ag^0 does not take place. The main use of macrocyclic complexing agents may be the ability of such agents to obstruct chemical reactions of bound ions in an understandable fashion and, thus illuminate molecule mechanisms. Preservation of a product such as Ag^0 is also of use in solar-energy storage or in photoprocessing and copying.

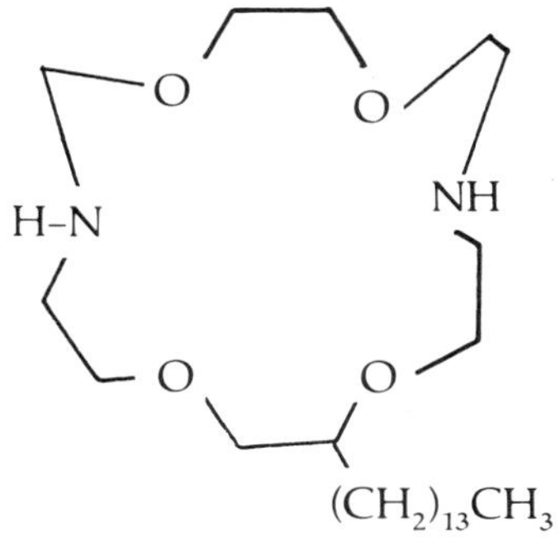

$(CH_2)_{13}CH_3$

Structure 4

Gel Structures and Polyelectrolytes

Polyelectrolytes can often markedly affect the rheological properties of solvents in which they reside, especially the macroviscosity of the medium and the diffusion of ions associated with the polyelectrolyte (*64–66*). In particular, several polyelectrolytes form closely interlinking gel-like structures in aqueous solutions. κ-Carrageenan, carboxymethylcellulose, and agarose belong to this class of polyelectrolytes; as little as 1% of the polyelectrolyte forms extremely viscous solutions at room temperature (67). Structure 5 is typical of a κ- and λ-carrageenan gel, which is a polysaccharide containing α-D-galactopyranosyl sulfate residues (68).

R = H for a κ-carrageenan
R = SO_3^- for an λ-carrageenan

Structure 5

The gel is formed by partial cross-linking of the anion groups by added cations; a loose polysaccharide network is formed throughout the aqueous phase. The resulting product is macroscopically rigid, although fluid, and has a waterlike microviscosity. This fact is quite apparent from measured rate constants in these gels for reactions of e_{aq}^- with added solutes (69, 70), which are almost identical with those observed in bulk water. The rate constants for these reactions are much lower in viscous solvents, such as glycerol, where the microviscosity is high.

The gel systems alone do not solubilize hydrophobic molecules such as pyrene. However, solubilization is achieved if surfactants such as NaLS are present (71). The molecular environments of the guest molecules, such as pyrene or N-phenylnaphthylamine (NPN), are similar to those of micelles or slightly more polar. Fluorescence polarization studies show that the structures formed are more fluid than micelles and that the onset of guest molecule solubilization takes place well below the CMC of the surfactant alone. Quenching of the excited states of guest molecules by CH_3NO_2 or Tl^+ is similar to that observed in micellar systems alone. Therefore, the surfactants form clusters on the gel structure that can solubilize guest molecules. The rigidity and order of these structures is not high as shown by polarization studies.

Polyacrylic and Polymethacrylic Acids

Both polyacrylic acid (PA) and polymethacrylic acid (PMA) show marked pH-dependent viscosity behavior in aqueous solution (72–74). Close to neutral pH, the carboxylic acid groups are almost completely ionized, and the polymer is stretched out as a result of the repulsion of these anionic groups. At lower pH, where the free polymerized acid is formed, the polymer collapses into a tight hydrophobic coil. Accordingly, the viscosities of aqueous PA and PMA are very high at neutral pH and low at low pH. At low pH the globular polymer can solubilize hydrophobic molecules, such as pyrene, and in fact behaves as a "static" anionic micelle (75). The polymers can also capture cationic organic dyes such as crystal violet (76, 77).

Pyrene solubilized in PMA at low pH exhibits a fluorescence fine structure indicative of a nonpolar environment, much less polar than micelles. Increasing the pH causes an abrupt decrease in the intensity of the fluorescence of pH 4–5, and the fluorescence fine structure shows a polar environment similar to water for pyrene in PMA at pH > 6. The data are shown in Figure 2 for PMA and PA. The pH titration curves in Figure 2 are identical to other measurements that show an opening of the polymer on neutralization at a pH of 4. When the polymer opens, the pyrene is released to the aqueous phase where its fluorescence yield decreases. Stop-flow experiments (75) on this system using pyrene as a probe of the polymer uncoiling process show that the rate of uncoiling varies as $(time)^{1/3}$, and for PMA occurs from microseconds to seconds. The quenching of pyrene fluorescence in PMA at low pH (<3) is considerably slower than that observed in homogeneous solution. For example, the quenching-rate constants for excited pyrene by O_2, CH_3NO_2, and Tl^+ in PMA compared to homogeneous solution are 2×10^{-2}, 5×10^{-3}, and 2×10^{-3}, respectively. These data show that the coiled polymer exerts a considerable protective effect on excited pyrene, probably by restricting the movement of quencher molecules to it.

Polymethacrylic acid also binds perylene and 1-anilino-8-naphthalenesulfonate anion (77); neutralization of the polymer leads to a decrease in the solubilizing power of the polymer (e.g., pyrene as a guest molecule). Fluorescence polarization measurements show that perylene is held very tightly in the polymer coil; the only observed rotation is that of the polymer itself.

Cationic dyes, such as crystal violet, also bind strongly to PMA (76) and apparently broaden the pH-induced transition from coiled states of the polymer, although binding is stronger to the compact coiled state.

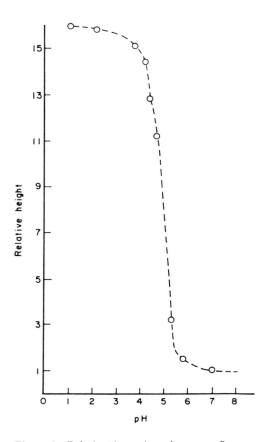

Figure 2. Relative intensity of pyrene fluorescence in 0.1 M PMA as a function of pH (75).

Polyvinyl Sulfate

Polyvinyl sulfate promotes the photoinduced electron transfer from
excited RuII to Cu^{2+} because both these cationic species are bound to
the anionic polymer; thus, the proximity effect of enhanced reaction is
promoted (78–81). The anionic polymer aids in the separation of the
ionic products if the electron acceptor is neutral; thus, e^- transfer produces an anion that is repelled away from the polymer and the site of
RuIII. Suitable electron acceptors for e^- transfer from excited RuII are
iron nitrilotriacetate and cobalt(III) acetylacetonate. For efficient charge
separation to take place, one of the reactants must be repelled from the
sphere of reaction of the other product. This fact is apparently true for
quenching of excited RuII by ferricyanide on cationic polybrene, poly-
(N,N,N',N'-tetramethyl-N-trimethylenehexamethylenediammonium di-

bromide) (*82*). Fe(CN)$_6^{3-}$ is bound to the polymer and e^- transfer from (RuII)^4 occurs to form Fe(CN)$_6^{4-}$. RuIII is promoted and Fe(CN)$_6^{4-}$ is repelled from the cationic polymer.

Photochemistry in Cellulose

The type of matrix provided by a gel or polyelectrolyte might be taken one step further; thus, cellulose is introduced as a medium for photochemical reactions. This field has important implications for the fabric industry, where dye molecules incorporated into cellulose materials can lead to fabric degeneration in sunlight (*83–85*).

Many organic molecules may be adsorbed from solution into cellulose or cellophane (*86–88*). Aqueous solutions may be used if sulfonic acid derivatives of the probe molecules are used (*88*). If the adsorbed sample is subsequently dried, oxygen is occluded from the matrix; and quenching of probe fluorescence and even phosphorescence is minimized. The extent of phosphorescence quenching depends on the nature of the cellulose material and reflects on the mobility and presence of O$_2$ in these samples. Adsorbed water increases the penetration of O$_2$ into the sample and decreases the phosphorescence lifetime.

These data indicate the rigid nature of the cellulose matrix. However, photochemical reactions may be produced in this medium. An example is the photolysis of the RuII–MV^{2+} system in cellulose, which gives rise to reduced MV^{2+} and gives the cellulose a deep blue color (*89–91*). Electron transfer occurs from excited (RuII)* to MV^{2+} and produces MV$^+$; the subsequent reaction of O$_2$ and MV$^+$, which occurs readily in solution, is prohibited by the cellulose. The e^- transfer reaction may occur via an e^- tunneling process (*91*) because the reactants are held in close proximity. RuIII, one product of the photochemistry, is not found. However, RuIII disappears rapidly by reaction with methylcellulose in solution, a feature that must also play a role in solid-phase cellulose system. Cellulose provides a convenient room temperature rigid matrix for many photochemical processes and, as for the (RuII)*–MV^{2+} system, provides a medium for storage of light energy as ionic products.

Literature Cited

1. Willner, J.; Yang, Jer-Ming; Loane, Colja; Otvos, J. W.; Calvin, M. *J. Phys. Chem.* **1981**, *85*, 3277.
2. Wheeler, J.; Thomas, J. K. In "Inorganic Reactions in Organized Media"; ACS Symposium Series No. 177, American Chemical Society: Washington, 1982; p. 97.
3. Katsusin-Razem, B.; Wong, M.; Thomas, J. K. *J. Am. Chem. Soc.* **1978**, *100*, 1679.
4. Wheeler, J.; Thomas, J. K. *J. Phys. Chem.*, in press.

5. Bulletins of the Georgia Kaolin Company.
6. Grim, R. E. "An Introduction to Clay Colloid Chemistry"; Interscience: New York, 1963.
7. Theng, B. K. G. "The Chemistry of Clay Organic Reactions." Halsted Press, 1974.
8. Solomon, D. H. *Clays Clay Minerals* **1968**, *16*, 31; Adam, Hilger, London 1974.
9. Thomas, J. M.; Adams, J. M.; Walters, J. M. *J. Chem. Soc., Dalton Trans.* **1978**, *1459*.
10. Brindley, G. W. "X-ray Identification and Crystal Structure of Clay Minerals"; Mineralogical Society: London, 1951.
11. Platikinov, D.; Weiss, A.; Tagaly, G. *Colloid Polym. Sci.* **1977**, *255*, 907.
12. Greene-Kelly, R. *Trans. Faraday Soc.* **1955**, *51*, 412.
13. Adams, J. M. *J. Chem. Soc., Dalton Trans.* **1974**, 2286.
14. Tennakoon, T. B.; Thomas, J. K.; Triker, M. J.; Williams, J. O. *J. Chem. Soc., Dalton Trans.* **1974**, 2207.
15. Weiss, A. *Chem. Ber.* **1958**, *91*, 487.
16. Hasegawa, M. *J. Phys. Chem.* **1962**, *66*, 834.
17. Mortland, M. M.: Pinnavaia, T. J. *Nature London Phys. Sci.* **1971**, *229*, 75.
18. Tennakoon, T. B.; Thomas, J. M.; Tricker, M. J.; Graham, S. H. *J. Chem. Soc., Chem. Commun* **1974**, 124.
19. Doner, H. E.; Mortland, M. M. *Science* **1969**, *166*, 1406.
20. Solomon, D. H. *Clays Clay Minerals* **1968**, *16*, 31.
21. Hakusui, A.; Motsunaga, Y.; Nonehara, K. *Bull. Chem. Soc. Jpn* **1970**, *43*, 709.
22. Lahav, N. *Isr. J. Chem.* **1972**, *10*, 925.
23. DellaGuardia, R.; Thomas, J. K. "Proceedings of the Chemical Systems Laboratory Conference on Decontamination"; Aberdeen: Maryland, 1981.
24. DellaGuardia, R.; Thomas, J. K. Unpublished data.
25. Krenske, D.; Abdo, S.; Van Damme, H.; Cruz, M.; Fripiat, J. J. *J. Phys. Chem.* **1980**, *84*, 2447.
26. "Melens Handbook der Anorganischem Chemie"; *61*, 113, p. 183 Verlag Chemie: Weinheim, 1975.
27. Meisel, D. *J. Am. Chem. Soc.* **1979**, *101*, 6133.
28. Henglein, A. *J. Phys. Chem.* **1979**, *83*, 2209.
29. Tausch-Tremel, R.; Henglein, A.; Lillie, J. *Ber. Bunsenges. Phys. Chem.* **1978**, *82*, 1335.
30. Henglein, A. *Angew. Chem.* **1979**, *91*, 449.
31. Lehn, J. M.; Sauvage, J. P. *Nouv. J. Chim.* **1977**, *1*, 449.
32. Kiwi, J.; Grätzel, M. *J. Am. Chem. Soc.* **1979**, *101*, 7214.
33. Turkevich, J.; Aika, K.; Ban, L. L.; Okura, I.; Namba, S. *J. Res. Inst. Catal Hokkaido Univ.* **1976**, *24*, 54.
34. Harriman, A.; Porter, G.; Richoux, M. C. *J. Chem. Soc., Faraday Trans. 2* **1980**, 3.
35a. Kiwi, J.; Kalyanasundaram, K.; Grätzel, M. Structure and Bonding, 1981, *49*, 37.
35b. Kawai, T.; Sakata, T. *Chem. Phys. Lett.* **1980**, *72*, 87.
36a. Matijevic, E. *Acc. Chem. Res.* **1981**, *14*, 22.
36b. Kawai, T.; Sakata, T. *Nature* **1979**, *282*, 283; *Ibid.* **1980**, *286*, 474.
37. Duonghong, D.; Borgarello, E.; Grätzel, M. *J. Am. Chem. Soc.* **1981**, *103*, 4685.
38. Dunn, W. W.; Aikawa, Y.; Bard, A. J. *J. Am. Chem. Soc.* **1981**, *81*, 6893.
39. Borgarello, E.; Kiwi, J.; Pellizzetti, E.; Viscea, M.; Grätzel, M. *J. Am. Chem. Soc.* **1981**, *103*, 6324.
40. Grätzel, M. *Acc. Chem. Res.* **1981**, *14*, 376.
41. Henglein, A. *Ber. Bunsenges Phys. Chem.* **1982**, *86*, 241.
42. Chandrasekaran, K.; Thomas, J. K. Unpublished data.

43. Harbour, J. R.; Hair, M. L. *J. Phys. Chem.* **1977**, *81*, 1791.
44. Harbour, J. R.; Wolkow, R.; Hair, M. L. *J. Phys. Chem.* **1981**, *85*, 4026.
45. Darwent, J. R.; Porter, G. *J. Chem. Soc., Chem. Commun.* **1981**, 145.
46. Grätzel, M. *Ber. Bunsenges Phys. Chem.* **1980**, *84*, 981.
47. Kalyanasundaram, K.; Borgarello, E.; Grätzel, M. *Helv. Chim. Acta.* **1981**, *64*, 362.
48. Kuczynski, J.; Thomas, J. K. *Chem. Phys. Lett.* **1982**, *88*, 445.
49. Bender, M. L.; Domiyama, M. "Cyclodextrin Chemistry"; Springer-Verlag: Berlin, 1978.
50. Cramer, F. W.; Saenger, W.; Spatz, H. C. *J. Am. Chem. Soc.* **1967**, *89*, 14.
51. Breslow, R. *Acc. Chem. Res.* **1980**, *13*, 170.
52. Tabushi, I. *Acc. Chem. Res.* **1982**, *15*, 66.
53. Edwards, H. E.; Thomas, J. K. *Carbohydr. Res.* **1978**, *65*, 173.
54. Atik, S. S.; Thomas, J. K. Unpublished data.
55. Turro, N. J.; Bolt, J. D.; Kuroda, Y.; Tabushi, I. *Photochem. Photobiol.* **1982**, *35*, 69.
56. Turro, N. J.; Okubo, T.; Weed, G. K. *Photochem. Photobiol.* **1982**, *35*, 325.
57. Ueno, A. K.; Takahashi, I.; Hiro, Y.; Osa, T. *J. Chem. Soc., Chem. Commun.* **1980**, 921; and **1981**, 194.
58. Rideout, D. C.; Breslow, R. *J. Am. Chem. Soc.* **1980**, *102*, 7816.
59. Cram, D. J.; Cram, J. M. *Acc. Chem. Res.* **1978**, *11*, 8.
60. Pedersen, C. J.; Frensdorf, K. H. *Angew. Chem.* **1972**, *11*, 16.
61. Platzner, T.; Grätzel, M.; Thomas, J. K. *Z. Naturforsch* **1978**, *33B*, 614.
62. Cinquini, M.; Montarani, F.; Tundo, P. *J. Chem. Soc., Chem. Commun.* **1975**, 393.
63. Humphrey-Baker, R.; Grätzel, M.; Tundo, P.; Pelizzetti, E. *Angew. Chem.* **1979**, *91*, 669.
64. Magdelenat, H.; Turq, P.; Chemea, M. *Biopolymers* **1974**, *13*, 1535.
65. *Ibid.* **1976**, *15*, 175.
66. Kowblansky, A.; Sasso, R.; Spagnuola, V.; Ander, P. *Macromolecules* **1977**, *10*, 78.
67. Vink, H. *Polymer Commun.* Peking **1982**, *23*, 6.
68. Anderson, N. S.; Dolan, T. C. S.; Rees, D. A. *J. Chem. Soc., Perkin Trans. 1* **1973**, 2173.
69. Philips, G. O.; Wedlock, D. J.; Micic, O. I.; Milosauljevic. B. H.; Thomas, J. K. *Int. J. Radiat. Chem. Phys.* **1979**, *15*, 187.
70. Micic, O. I.; Milosavljevic, B.; Phillips, G. O.; Wedlock, D. J. *J. Chem. Soc., Faraday Trans. 1* **1979**, *75*, 1142.
71. Wedlock, D. J.; Phillips, G. O.; Thomas, J. K. *Polym. J.* **1979**, *11*, 681.
72. Katchalsky, A.; Eisenberg, H. *J. Polym. Sci.* **1952**, *6*, 145.
73. Okamoto, H.; Wada, Y. *J. Polym. Sci.* **1974**, *12*, 2413.
74. Arnold, R.; Overbeek, J. T. G. *Rec. Trav. Chem.* **1950**, *69*, 192.
75. Chen, T.; Thomas, J. K. *J. Polym. Sci.* **1979**, *17*, 1103.
76. Stork, W. H. J.; Hasseth, P. L. de; Lippits, G. J. M.; Mandel, M. *J. Phys. Chem.* **1973**, *77*, 1778.
77. Treloar, F. E. *Chem. Scr.* **1976**, *10*, 219.
78. Meisel, D.; Matheson, M. S. *J. Am. Chem. Soc.* **1977**, *99*, 6577.
79. Meyerstein, D.; Rabani, J.; Matheson, M. S.; Meisel, D. *J. Phys. Chem.* **1978**, *82*, 1879.
80. Thornton, A. T.; Laurence, G. S. *J. Chem. Soc., Chem. Commun.* **1978**, 408.
81. Taha, I.; Morawetz, H. *J. Am. Chem. Soc.* **1971**, *93*, 829.
82. Sassoon, R. E.; Rabani, J. *J. Phys. Chem.* **1980**, *84*, 1319.
83. Egerton, G. S.; Morgan, J. *J. Soc. Dyers Colour.* **1970**, *86*, 79; **1970**, *86*, 242; **1971**, *87*, 223; and **1971**, *87*, 268.
84. Bentley, P.; McKellar, J. F.; Phillips, G. O. *J. Soc. Dyers Colour. 5*, 33, **1974**.
85. Baugh, P.; Phillips, G. O.; Worthington, N. W. *J. Soc. Dyers and Colourists*, **1969**, *85*, 241.

86. Pringsheim, P.; Vogels, H. *J. Chim. Phys.* **1936**, *33*, 345.
87. Szent-Györgyi, A. *Science* **1957**, *126*, 751.
88. Schulman, E. M.; Walling, C. *J. Phys. Chem.* **1973**, *77*, 902.
89. Kaneko, M.; Motoyoshi, J.; Yamada, A. *Science* **1980**, *285*, 468.
90. Kaneko, M.; Yamada, A. *Macromol. Chem.* **1981**, *182*, 1111.
91. Milosaulavic, B.; Atik, S.; Thomas, J. K. Unpublished data.

Biosystems of Interest

MICELLES, MICROEMULSIONS, AND VESICLES are directly related to biological systems; therefore, a discussion of photoinduced reactions in biosystems from the viewpoint of information established for model assemblies will be useful. We will consider several photochemical and photophysical studies first in membranes, then in proteins, and finally in cells. Much information has been written about light in biosystems and processes such as photosynthesis and vision, but these very important photobiological systems cannot be given true justice in such a short treatise.

Probe Studies in Proteins

The high sensitivity of fluorescence techniques applies well to biological studies, where quite often only small quantities of material are available. Biological materials often have a native fluorescence of their own (*1, 2*), but the deliberate addition of a selected fluorescent molecule with unique properties to the biostructure is more convenient. Many reviews are available that describe a host of such systems (*3–7*). The reviews discuss in detail several fluorescent molecules that bind at selected sites in biosystems and, furthermore, show selected photophysical properties that help to describe the binding site. The basic ideas of these works should be discussed in the same manner and used for the discussion of organic assemblies.

Anilinonaphthalenesulfonic Acid. Anilinonaphthalenesulfonic acid (ANS) is a particularly useful probe for many systems because its fluorescence is very low in polar media and increases with decreasing polarity of the medium. The maximum of the ANS fluorescence also shows a red shift with increasing polarity (*8*). The anionic ANS binds strongly to many proteins, a process that is accompanied by a sharp increase in the fluorescence yield (*9–11*). The polarization of the fluorescence also increases when the probe binds to proteins. This property may be used to determine the effective volume of a protein and any conformational changes that may take place on varying the conditions of the system (*12*). Concentration effects on the fluorescence polarization can be interpreted to give average separations of probe molecules on proteins (i.e., separations of binding sites) (*11, 13, 14*).

0065-7719/84/0181-0289$06.00/1

A major disadvantage of ANS is the short fluorescence lifetime of a few nanoseconds, which varies with the polarity of the probe environment. These disadvantages are at least partially overcome by using pyrenebutyric acid or pyrenesulfonic acid; the fluorescence lifetime of the former is almost 150 ns; however, the precise value depends on the environment (15). This probe binds strongly to serum albumin conjugates and to human immunoglobin M.

Effect of External Quenchers. The foregoing experiments illustrate a direct comparison between fluorescence probe studies in proteins and the earlier studies in organic assemblies. The comparison goes even further when we consider quenching studies, which are of prime interest in organic assemblies and contain some interest for biochemists. One of the earliest studies is the quenching of tryptophan and tyrosine fluorescence in several proteins (16–18). In particular, oxygen quenching has claimed attention because of its biointerest. Some problems arise because the fluorescence lifetimes of tryptophan (Try) and tyrosine (Tyr) are < 6 ns and depend on probe environment. The short excited-state lifetime means that oxygen pressures up to 100 atm have to be used to achieve measurable quenching. These high pressures have little, if any, effect on the physical states of the proteins as determined by bioassay and by the relative invariance of the wavelength position of the fluorescence. Changes in environment usually markedly shift the fluorescence of Try and Tyr; an increase in polarity produces a red shift (19, 20). According to the data, oxygen efficiently quenches Tyr and Try fluorescence in proteins; typical data are shown in Tables I–III. Rate constants, measured from Stern–Volmer plots of O_2 quenching of the fluorescence, and the fluorescence lifetimes are, for some proteins, about an order of magnitude smaller than those measured in homogeneous solution. However, the data indicate that O_2 moves freely in these proteins. The rate constants for I^- quenching of Tyr and Try fluorescence are smaller than those for O_2 and, unlike the O_2 data, depend on the ionic charge of the protein.

As indicated earlier, various probe molecules readily bind to proteins. This fact is illustrated by the probes pyrene (P), pyrenesulfonic acid (PSA), pyrenebutyric acid (PBA) and pyreneduodecanoic acid (PDA), which show long-lived fluorescences in bovine serum albumin that are quenched by a variety of molecules (21). The fluorescence properties of the probes (e.g., τ) are also dependent on the physical state of the system. The fluorescence lifetime (τ) of pyrene changes from 390 ns at 20 °C to 150 ns at 80 °C in benzenesulfonic acid (BA) and from 220 ns at 2 °C to 120 ns at 70 °C in ribonuclease (RNase). Discontinuities were observed at 30 °C, in BA, and at 10 °C and 60 °C, in RNase. An unfolding of RNase occurs at 60 °C that allows pyrene to move more freely in the

Table I.

Oxygen Quenching Constants and Fluorescence Lifetimes for Proteins

Proteins	$K(M^{-1})$	$\tau_o{}^a$ (ns)	$K/\tau_0 \times 10^{-10}$ (M^{-1} s^{-1})	Fluoresence Emission max
α-Chymotrypsin[b]	4.1	2.1	0.20	332
Bovin serum albumin[c]	15.2	6.2	0.24	342
Human serum albumin[c]	14.7	6.0	0.25	342
Edestin[d]	8.0	3.2	0.25	328
Carbonic anhydrase[e]	11.4	4.4	0.26	341
Aldolase[b]	6.1	2.3	0.27	328
Azurin[c,h]	10	3.4	0.30	
Carboxypeptidase A[e]	6.4	1.7	0.38	330
Trypsinogen[b]	8.7	2.0	0.43	334
IgG[c]	10.1	2.2	0.46	332
Lysozyme[c]	7.4	1.5	0.48	340
Pepsin[f]	28.6	5.5	0.52	342
Trypsin[b]	10.2	1.9	0.54	335
Ribonuclease[g]	8.6	1.2	0.71	304

NOTE: 280 ± 2 nm excitation was used for all proteins. In determining the Stern–Volmer quenching constant (K), the fluorescence intensity was monitored at the maximum of emission (±3 nm).
[a] For lifetime measurements a Corning 7–54 filter was used to block the parasitic visible light that is particularly evident with the 0.25-m Jarrell-Ash monochromator used for excitation. The fluorescence emission was observed through a Corning 0–54 filter.
[b] 0.001 M HCl, pH 3.
[c] 0.1 M sodium phosphate, pH 7.0.
[d] 0.05 M sodium phosphate + 1 M NaCl, pH 7.0.
[e] 0.025 M Tris-HCl + 0.5 M NaCl, pH 7.5.
[f] 0.01 M HCl.
[g] 0.005 M sodium phosphate, pH 7.14.
[h] The fluorescence intensity of Try was followed at 355 nm, which does not correspond to the fluorescence emission maximum.
SOURCE: Ref. 16

protein and also leads to an increased quenching by reactive groups in the protein. Similar effects are observed for the effect of pH on the fluorescence lifetimes of pyrene and PSA in BA, where the lifetimes show a sharp decrease at pH < 4.0; similar effects are reported for dyes absorbed on BA (22, 23). Decreasing the pH may loosen the protein and thereby increase the mobility of guest molecules.

Further confirmation of increased mobility is observed in the quenching-rate constants for pyrene and PSA fluorescence by O_2 and CH_3NO_2. In Figure 1 the quenching rate constants at pH > 4.0 are nearly 1/100 of those found in homogeneous solutions. These facts illustrate the lack of penetration of O_2 and CH_3NO_2 to the site where the probes lie in the protein. Below a pH value of 4 the rate constants

Table II.
Iodide Quenching Constants for Proteins

Protein	$K(M^{-1})$	$K/\tau_0 \times 10^{-10}$ $(M^{-1}\ s^{-1})$
Trypsinogen	0.67	0.034
Carboxypeptidase A	0.54	0.032
Carbonic anhydrase	0.85	0.019
Pepsin	11.2–7.7[a]	0.21–0.14[a]

NOTE: 290 ± 2 nm excitation was used. The buffers described in Table I were used except for [KCl] + [KI] = 0.5 M and ca. 10^{-4} M $Na_2S_2O_3$.
[a] Downward curvature was seen in the Stern–Volmer plot.

Table III.
Quenching Rate Constants k_q ($M^{-1}\ s^{-1}$) for the Pyrene Excited Singlet State in BSA Solutions

Probe	Lifetime (ns) in Absence of Quencher	Quencher			
		O_2	I^-	CH_3NO_2	Tl^+
Pyrene	344	3.0×10^8	1.2×10^9	1.6×10^8	1.6×10^8
PSA	103	1.5×10^9	1.6×10^8	1.7×10^8	4.5×10^7
PBA	190	7.8×10^8	2.19×10^7	2.13×10^8	$<10^7$
PBDA	180	1.1×10^8	4.3×10^7	1.75×10^8	$<10^7$

SOURCE: Ref. 21

increase by nearly 10-fold, which indicates a marked increase in mobility of molecules in the protein. The quenching rate constants decrease again below pH 2–2.5, an effect that is quite similar to data obtained via fluorescence polarization (22) and that is attributed to break up of the protein into subunits.

Tl^+ quenching of pyrene probes in BA indicates that Tl^+ is bound to the polymer; however, quenching-rate constants are still only 1/100 of those observed in homogeneous solution. Iodide ion is a very effective quencher; however, the rate constant approaches that in homogeneous solution. Similar effects are observed for quenching of Try fluorescence in BA (17); the site of solubilization of pyrene may be close to the Try units of the protein. Other pyrene probes (PSA, PBA, and PDA) are much less reactive with I^- than pyrene and are bound at some other location in the protein. The data suggest the presence of cationic groups in the vicinity of pyrene and Try where I^- may bind and hence quench efficiently. These cationic groups repel Tl^+ ions; this repulsion accounts for its low fluorescence quenching efficiency in BA.

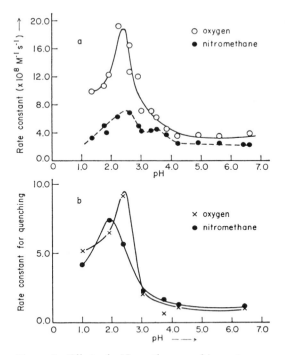

Figure 1. Effect of pH on the quenching-rate constants of pyrene and PSA fluorescence in BSA solutions. Quenchers, oxygen (○, ×) and nitromethane (●). (a) BSA, 1.5 × 10⁻⁵ M; pyrene 10⁻⁵ M. (b) BSA, 1.5 × 10⁻⁵ M; PSA 10⁻⁵ M (21).

The fluorescence-quenching technique can be used to comment on the binding of various nonfluorescent molecules to BA and other proteins. Both sodium lauryl sulfate (NaLS) and cetyltrimethylammonium bromide (CTAB) bind to BA and increase the quenching rate constant of pyrene fluorescence by O_2 and CH_3NO_2. CTAB also increases the I^--quenching rate, but NaLS causes a decrease presumably because of anionic repulsion. The amphiphiles NaLS and CTAB cause conformational changes in proteins (24, 25) that increase the permeability of the protein to O_2 and CH_3NO_2. Structural changes in the proteins caused by the binding of large drug molecules such as penicillin G also show up in fluorescence-quenching data. A maximum in the quenching rate of pyrene fluorescence by O_2 is observed at a BA to penicillin ratio of 1:4; smaller effects are observed on CH_3NO_2 and I^- quenching. Similar effects have been observed for drugs on the rates of reaction of radiolytically produced e_{aq}^- with BA (25, 26). The reactivity of BA with e_{aq}^- is thought to be via sulfur-containing and lysine amino acid residues of the proteins (27, 28). Conformational changes in the protein, which are caused by drug binding, make these sites more or less available to e_{aq}^-. Penicillin binds to BA by ionic interaction between NH_3^+ groups on the amino acids and carboxylic acid groups of the penicillin. This blocking of the NH_3^+ leads to a decreased reactivity of BA with e_{aq}^-. Cephalosporin C and 6-aminopenicillanic acid binds at other sites of BA. Increased reactivity is noted with this type of binding, because –S–S– groups are exposed to e_{aq}^- attack. Interpretation of the e_{aq}^- rate data are in accord with other physical measurements that reflect on the architecture of the protein.

The foregoing studies indicate certain specific effects noted in protein systems by fluorescence probing or e_{aq}^--reactivity techniques. This type of work is illustrated by the binding of arene probe molecules at specific sites on DNA, where the probes are bound as sandwich complexes between two base pairs of the double helix (29–31). Fluorescence methods are particularly useful in observing the arene–DNA complexes (32, 33). The fluorescences of intercalated benzo[a]pyrene (aP) and benzo[e]pyrene (eP) are severely affected on binding to calf thymus DNA (33). About 50% of the probe molecules are located at sites where the radiative decay is small, probably at adenosine–thymine (AT)-intercalation sites. The fluorescent lifetimes of the bound molecules are also short (τ = 3.2 and 10 ns for aP, and 10 and 33 ns for eP). Silver ions quench up to 70% of the arene fluorescence; this fact indicates that the binding sites are guanine–cytosine (GC) sites, a known site for efficient silver binding.

The short-probe fluorescence lifetime in these systems tends to limit quenching experiments. However, triplet states of the arenes are formed, even at the AT-intercalation sites, and the triplet lifetimes are quite long, 35 ms for aP and 155 ms for eP. Oxygen quenches the triplet states with rate constants of 1–2×10^8 $M^{-1}s^{-1}$ (i.e., 10 to 20 times slower than in homogeneous solution); this situation was noted earlier in globular proteins.

Many probe molecules have been designed to study specific effects in biosystems; a few are listed in Table IV.

Quite often, energy related processes in biology are connected with interfacial potential and proton gradients. Therefore, much work has gone into designing probes that monitor this very event (6, 34). For pH gradients the basic concept in the design of such probes is the change in acidity of the probe molecule on excitation, a situation that may give rise to two excited states. An example of this reaction is β-naphthol, which gives a fluorescence at $\lambda_{max} \sim 340$ nm with the un-ionized molecule and at $\lambda_{max} > 360$ nm form the ionized form.

Table IV.

Probes and Their Approximate Location and Function in Assemblies

Probe	Location[a]	Function[b]	Reference[c]
Pyrene	H	P, LRM, Q	70
Pyrenecarboxaldehyde	I, H	P	70
Pyrenealkylcarboxylates	I	LRM, Q	70
Pyranine	I, P	P, H	34
Dipyrene	H	SRM	70
N-Phenyl-1-naphthylamine	I	P	5
Anilinonaphthalenesulfonate	I	P	5
Ethydium bromide	Base pairs, I	Binding site	5
Cyanine dyes	I	Potential changes	6
Merocyanine dyes	I	Potential changes	6
Oxonol dyes	I	Potential changes	6
Eu^{3+}, Tb^{2+}	I	Degree of binding to sites	5

[a] H, hydrophobic; P, polar; and I, interface
[b] P, polarity; LRM, long range motion; SRM, short range motion
[c] Q, access of quencher molecules (e.g., O_2) to binding site.

$\lambda_{max} = 340$ nm $\lambda_{max} > 380$ nm

An alternative approach is to use the difference in absorption spectra of the ionized and un-ionized molecule. An example is (34)

$\lambda_{max} = 400$ nm $\lambda_{max} = 450$ nm

The fluorescence emission of pyranine in water is centered at $\lambda_{max} =$ 510 nm and in most cases is due to the ionized form, which is a much stronger acid than the un-ionized form. The yield of fluorescence at 510 nm on excitation at 400 nm and 450 nm gives the relative distribution of nonionized and ionized forms of the ground state of the molecule. These data in turn can be related to the pH of the medium, which controls the degree of ionization with a pK_a of 7.22. This probe has been used to investigate pH gradients in the vicinity of vesicles.

Cyanine and other related dyes have been successfully used to monitor membrane potentials of various biological suspensions, red blood cells, mitrochondria, synaptosomes, bacterial cells, and purple-membrane containing vesicles (6). The dyes, which are ionic in nature, stain the biomaterial of interest. The partition of the dye between the membrane and adjacent polar regions controls the level of observed fluorescence; the partitioning is strongly affected by the membrane potential. This type of fluorescence probing is particularly useful in nerve membranes, where potential changes accompany the bioevent. The fluorescence dyes respond rapidly to the nerve membrane changes and accurately mimic bioevents.

Before progressing to lipid systems, we should draw attention to fluorescence studies of ion binding to proteins, a feature of interest in enzymology. Several metal ions are fluorescent in water (e.g., Eu^{3+} and Tb^{3+}). These ions exhibit little spectral absorption in the 300-nm region, a region of the spectrum where Try and Tyr absorb. Excitation of Try or Tyr gives rise to short-lived excited states, and only extremely high quencher concentration can effectively react with these excited states. However, Förster energy transfer may occur quite efficiently in these systems and over large reactant separations. Hence, if the fluorescent ions Eu^{3+} or Tb^{3+} are bound to the protein and the Try and Tyr residues are excited, then Eu^{3+} or Tb^{3+} fluorescence is observed from those ions bound to the protein; nonbound ions are not affected (35, 36). The binding of fluorescent ions can be measured conveniently by this technique. The binding of nonfluorescent ions (e.g., Ca^{2+} and Mg^{2+}) can also be measured by experimentally arranging for competitive binding of Tb^{3+} or Eu^{3+} with the nonfluorescent ion. The sensitivity of the technique is well suited to biosystems, where only small quantities of material are available.

Fluorescent Probing of Membranes. Many of the fluorescent probes used in protein studies have also been used to study membranes. The studies often mimic those carried out with vesicles, which were described earlier. Four features are of most concern: temperature- and ion-induced phase changes; movement in membranes, both lateral and transverse; location of molecules in membranes and their availability to outside agents; and specialized functions, such as ion transport in nerve membranes. Photophysical measurements, such as those carried out in vesicles, are more complex in biomembranes because the probe location may be difficult to ascertain; for example, the probe may be in the lipid or protein portions of the membrane or in both. This problem has been addressed in some publications; but for the most part, it is ignored or merely hinted at. The extent that a lipid or protein participates in membrane changes can be estimated by fluorescence studies. Using the fluorescence probes 1-ANS and N-phenyl-1-naphthylamine (NPN) at high concentrations, which extend to the saturation limit of the probes in *E. coli* membranes, we concluded that 42.5% of the fluorescent probe, or of the membrane, takes part in phase changes with temperature (37). Therefore, each protein molecule in the membrane is surrounded by 100 lipid molecules.

Fluorescent probes have been used extensively to investigate the nature of movement in membranes. On the basis of this work, movement across membranes may be via thermal motion of the lipid chains whereby kinks are produced; these kinks facilitate movement of guest molecules (*38*). The diffusion constant of such a lipid chain link could be quite high (i.e., $\sim 10^{-4}$ cm^2/s). At first sight, the concept of volume change is a useful way of looking at membrane transport phenomena. An alternative approach is to consider possible phase changes induced in the membranes by external means. For example, increased fluidity associated with phase changes in charged lipids is accompanied by a decrease in the electrostatic free energy of the membranes (*39*). The Gouy–Chapman theory also predicts a decrease in the phase transition temperature (T_c) with an increase in charge density. Bilayers of phosphatidic acid, which have a change in charge from 1 to 2 over the pH range 7–9, provide reasonable evidence for the validity of these suggestions. A pH change from 7 to 9 lowers T_c by almost 20 °C as predicted by theory; Mg^{2+} and Ca^{2+} ions increase T_c because they bind to the anionic group of the bilayer and reduce the membrane charge. Monovalent cations (e.g., K^+ and Li^+) reduce T_c because of increased charge on the bilayer surface. Such model data would seem to provide clues to the function of nerve membrane excitation.

Transport across a membrane is often measured by radioactive tracer techniques, which are quite suited to slow transport phenomena. More rapid transport requires fluorescence methods, and tagging the molecule of interest by covalently binding a fluorescent probe to it is often quite convenient. A good example of this technique has been provided by observing galactose transport in *E. coli* membranes (*40*). The fluorescent probe used was 2-(*N*-dansyl)aminoethyl-β-D-thiogalactoside (DG), which, although not actively transported, nevertheless completely inhibits lactose transport by *E. coli* membrane vesicles. The fluorescence of DG in an aqueous suspension of vesicles and probe increases on addition of D-lactate. This increase is due to binding of

DG to the membrane on activation of the membrane by D-lactate. The binding is such that the dansyl group of DG experiences a more hydrophobic environment. Figure 2 illustrates the time evolution of increase of dansyl fluorescence on addition of D-lactate and the decrease on addition of lactose, which reverses the activation of the transport system by D-lactate. Vesicles that do not contain the β-galactoside transport system do not cause this effect, thus the binding of DG to the transport system is established. Further, on binding, DG shows fluorescence on excitation at 290 nm, into the tryptophan band; this result is absent if DG is not bound to the vesicle. The 290-nm excitation excites the aromatic amino acids, which transfer energy to DG. Similar results are observed with whole cells.

Pyrene is readily adsorbed into *E. coli* whole cells and into extracted membranes. This adsorption facilitates the design of probe studies that are quite similar to those carried out in micelles and microemulsions and, with caution, may be interpreted in a similar fashion. Two membranes are readily separated from *E. coli* bacteria; an inner membrane (IM) and an outer membrane (OM) or cell wall. This latter membrane contains a lipopolysaccharide (LPS) coat in addition to lipids and pro-

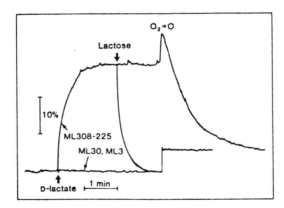

Figure 2. Time course of changes in dansylgalactoside fluorescence dependent on D-lactate. Lithium D-lactate (20 mM) was added to a cuvette containing dansylgalactoside (0.033 mM) and membrane vesicles (ML 308–255) and the fluorescence at 500 nm was recorded (excitation, 340 nm). Similar experiments were conducted with membranes prepared from uninduced E. coli (ML 30) or from E. coli (ML 3). In a second experiment with ML 308–255 membranes, 0.1 M lactose was added to a final concentration of 1 mM (arrow). In the absence of D-lactate, lactose had little or no effect on dansylgalactoside fluorescence (40).

teins. The behavior of pyrene is quite different in each membrane; the fluorescence lifetime and quantum yield is lower in the IM (41, 42). The inner membrane shows a continuous decrease in fluorescence lifetime (τ) and yield (ϕ) with increasing temperature (0–40 °C). This result indicates a continuous decrease in microviscosity of the IM with increasing temperature, which leads to a greater mobility of the probe pyrene. This is reflected in a greater extent of quenching by active groups in the membrane. The τ and ϕ of pyrene in the OM show two sharp decreases at ~14 °C and 28 °C; these temperatures are also noted as phase changes from cell growth studies.

Pyrene and pyrene derivatives, such as pyrenesulfonic acid, incorporated into E. coli membranes and living cells show quenching rates that are markedly different from those found in homogeneous solution (42). Comparison of quenching rates with membranes containing LPS coatings to a mutant strain without LPS suggests the following conclusions:

1. Cell membranes containing LPS showed significantly lower quenching rates for pyrene and pyrenesulfonic acid than cell membranes without LPS.

2. These two probes are located near and under the LPS and close to membrane proteins.

3. No effect of LPS membrane content was observed for the probes pyrenebutyric acid and pyrenebutyroyldecanoic acid. These probes may be located primarily in the lipid part of the membrane.

Disruption of energy transduction in whole cells has been followed by fluorescence probe techniques that use pyrene and N-phenyl-1-naphthylamine. Both of these probes show enhanced fluorescence on leaving the polar aqueous phase and entering into hydrophobic cell membranes (43).

The probes are partitioned between the aqueous phase and a suspension of E. coli cells. Addition of colicin Ia to the suspensions increases the fluorescence yield twofold; the fluorescence lifetime increases; the spectrum exhibits a blue shift; and the polarization of fluorescence increases. These results correspond to increased binding of the probe to the cells. Other additives, such as malonate, amytal, cyanide, and the uncouplers CCCD and azide, show similar effects. They are all deenergizers for the cell membrane and are inhibitors of electron transport. Depleting the cells of oxygen and indigeneous substrates gives similar data; the latter effects are reversed on addition of substrate or O_2.

The transformation of healthy mammalian cells by viruses has been monitored by various fluorescent probes (44–46). Quite often, the mi-

croviscosity of the membrane of a transformed cell, observed by fluorescence polarization, is lower than that of a normal cell. This observation supports the concept that a transformed cell lacks certain membrane
elements that are present in a normal cell.

The fluorescent probes pyrene, pyrenebutyric acid, and N-phenyl-
1-naphthylamine have been used to investigate the changes that accompany in vitro transformation of a baby hamster kidney cell line with
Rous sarcoma virus (RSV) (44). The fluorescent probes residing in the
membrane were used to compare the changes in microviscosity and
polarity of the membranes of normal cells with two transformed cell
lines. The spectrofluorimetric data indicate that, following transformation, the probe N-phenyl-1-naphthylamine resides in a more polar
environment. However, with the pyrene probe, the yield of excimer
indicates decreased mobility of this probe in the membrane of transformed cells. The data also indicate differences between the two transformed cell lines. Laser photolysis was used to study the lifetime of
the pyrene probes and the quenching of the pyrene fluorescence in the
membrane by several different quenching molecules. The data indicate
differences between the three cell lines and suggest that transformation
decreases movement within the cell membrane.

The fluorescent probes pyrene, pyrenebutyric acid, and N-phenyl-
1-naphthylamine were used to study membranes of normal cells, RSV-
transformed cells, cells treated with a proteolytic enzyme, and cells
persistently infected with lymphocytic choriomeningitis virus (45). The
lifetime of excited pyrene and pyrenebutyric acid showed only minor
changes when these probes were in normal, transformed, trypsinized,
or persistently infected cells. However, the pyrene, but not pyrenebutyric acid, fluorescent lifetime is shorter in cell membranes than in
homogeneous solvents. The quenching of excited pyrene in cells by
quencher molecules was slower than corresponding reactions in homogeneous solutions; therefore, the probe was screened from the
quenchers by the membrane. However, quenching reactions with the
pyrenebutyric acid probe were similar in cells and homogeneous solvents. This result indicates that pyrene and pyrenebutyric acid reside
in different lipid regions of the membranes. Transformed and trypsinized cells showed increased membrane fluidity compared to normal and
persistently infected cells. Membrane fluidity was determined from the
excimer/monomer fluorescence ratios of pyrene and by the polarization
of N-phenyl-1-naphthylamine fluorescence. Several techniques distinguished between normal and transformed or trypsinized cells; however,
the only parameter unique to viral transformation was a blue shift of
the fluorescence maxima of N-phenyl-1-naphthylamine. This shift reflected a less polar environment for N-phenyl-1-naphthylamine in virus-
transformed cells.

Fluorescent probing may be used to investigate any biosystem that will take up the probe. We may conclude by indicating briefly a few systems, other than simple proteins, membranes, or cells; lipoproteins are good examples for discussion (47, 48).

Very low density lipoproteins (VLDL) from the serum of rabbits at various stages of hypercholesterolemia (95–1665 mg cholesterol/100 mL of serum) show quite marked differences to normal VLDL. The most notable chemical change in hypercholesterolemic (hc) VLDL was the greatly increased content of cholesteryl esters and the greatly decreased content of triglycerides, compared to normal (n), VLDL. Structurally, the lipid region of nVLDL possessed a much lower microviscosity than did hcVLDL, when analyzed by fluorescence polarization and pyrene eximer methods. The microviscosity of the redispersed nVLDL lipid extract was considerably greater than that observed in nVLDL; but less than that of hcVLDL. Incorporation of pyrene into the lipid region of nVLDL and VLDL allowed assessment of various properties of the surface and hydrocarbon regions of these lipoproteins. Only slight differences were found in the pyrene monomer 3:1 fluorescence emission peak ratios and in the rate constant for quenching of pyrene by O_2. However, the quenching rate constant of pyrene by I^- and iodoheptane were different for each lipoprotein.

The biochemical literature abounds with many examples of fluorescence probe measurements for a variety of biosystems. The examples chosen are most akin in experimental execution and interpretation to the earlier reports on purely chemical assemblies.

In previous sections we reported the use of pulse radiolysis to investigate the binding of drugs to proteins. Similar studies have been reported for erythrocyte membranes (49). The red blood cell membranes were suspended in water, and the reactivity of the membranes with various radiolytically produced radicals investigated. In particular, Br_2^- was investigated; it reacts rather specifically with Tyr and Try residues and not with the membrane. Little reactivity was observed because the reactive residues are buried in the membrane, and Br_2^- does not readily penetrate to these sites from the aqueous phase. Solubilization of the membrane with NaLS or alkali exposes proteins at the active sites, and a marked increased reactivity with Br_2^- is observed. This technique is mainly used to investigate the presence or absence of reactive groups on a membrane surface.

Radiolysis of Enzymes

The activity of an enzyme depends on a specific active site of the molecule. This property is similar to locating a probe in the large molecule or assembly and then watching its reactivity with additives. The additives may be free radicals, which are produced photochemically or radiolytically. The systems in the latter case are of particular interest because an activation of an enzyme by high-energy radiation is the essence of food preservation by radiolysis (50). Several excellent reviews are available (51, 52). Enzyme inactivation studies are usually carried out in aqueous solution, where the free radicals produced in the aqueous phase react with the enzyme. For the most part, the extent of inactivation is exponentially related to the number of radicals produced (the absorbed dose), and the yield of inactivation is low. The exponential relationship arises because the free radicals may react with an enzyme already inactivated as well as with active ones; and, because many sites of reactivity exist in the enzyme, the yield of inactivation is low. In addition, initial free-radical attack can cause a conformational change in the enzyme, which opens up more effective attack in the active site (53). Studies with radicals of specific reactivity with certain groups show that damage to the enzyme is attributed to attack at sulfur groups (–SH or –S–S–) via reactions of the following types: $-Sh + OH \rightarrow -S^{\cdot} + H_2O$ and $-S-S- + CO_2^- \rightarrow -S^- + S^- + CO_2$. In some situations, the free radical may be repaired and enzyme inactivation is lost; for example,

$$-S^{\cdot} + RSH \rightarrow -SH + RS^{\cdot}$$

$$(\text{cysteine})$$

The ionic species may also be repelled away by adsorbing polar surfactants on the protein (54); hence, the rate of reaction is decreased. It is also possible to produce radicals or molecules of biological importance and monitor their reactivity, for example, free radicals of nicotinamide nucleotides, which are coenzymes for dehydrogenases, and nicotinamide adenine dinucleotide (56, 57).

The superoxide free radical (O_2^- or HO_2)

$$O_2^- + H^+ \rightleftharpoons HO_2$$

$$pk_a = 4.6$$

is an important species in biosystems, as well as the enzyme superoxide dismutase (SD), which catalyzes inactivation of the free radical via (58) $2O_2^- + 2H^+ \rightarrow H_2O_2$. The O_2^- and HO_2 free radicals are readily produced in pulse-radiolysis experiments and can be monitored by their absorption spectra at $\lambda \sim 250$ nm. The bimolecular rate constant for reaction of O_2^- with SD is $2.37 \pm 0.18 \times 10^9$ $M^{-1}s^{-1}$ at 25 °C (59). In addition, the enzyme can handle larger quantities of O_2^-, indicating the stability of the enzyme during catalysis. This feature is important for high-energy radiation studies, which, unlike photochemical probe studies, are destructive in nature.

Radiolysis of Cells

The high-energy inactivation of tumor cells is one of the major medical techniques for treating cancer; cell death caused by inactivation shrinks or kills the tumor. The action of the radiation occurs all over the cell; however, many different studies show that most damage occurs at the nucleus of the cell and that the phenomena is via a free-radical mechanism (5, 60–62). The free radicals may be generated by direct action of the radiation in the vital part of the cell or nucleus (the direct-action theory), or by the secondary reaction of free radicals produced in the cell bulk with the nucleus (the indirect-action theory). Such radiation-induced chemical events lead to cell damage to such an extent that the cell can no longer reproduce. Typical data are shown in Figure 3 for the survival of Chinese hamster cells in 0.02 M hydroxyethylpiperazine-ethanesulfonic acid (HEPES) buffer on irradiation with X-rays. The survival parameter (S), is defined as

$$S = \frac{\text{Number of colonies formed}}{\text{Number of single cells plated} \times \text{(PE)}}$$

where PE is the number of colonies formed per number of single cells plated without irradiation. The above definition becomes clear on considering the experimental procedures. Cells, irradiated or nonirradiated, are counted on a solid growth medium after incubation for a period of time. Cell reproduction leads to the formation of colonies, which are observed visually on the growth medium.

Two points are worthy of note: 1, in the presence of oxygen or air, the irradiation is relatively ineffective at low doses below about 200 rads, but rapidly accelerates beyond this point; and 2, oxygenated cells are more sensitive than anoxic cells.

The first point arises from the ability of the cell to repair some of the damage inflicted on it by irradiation. Point 2 indicates the free-radical nature of the process because O_2 reacts rapidly with radicals prior to their recombination or repair and thus extends the radiation damage.

$$R_1 + R_2 \rightarrow R_1R_2 \qquad \text{(no damage)}$$

$$R_1 \text{ or } R_2 + O_2 \rightarrow R_1O_2 \text{ or } R_2O_2 \qquad \text{(damage)}$$

The free radicals involved in the process live for many microseconds, as demonstrated by several different types of pulsed studies (63, 64).

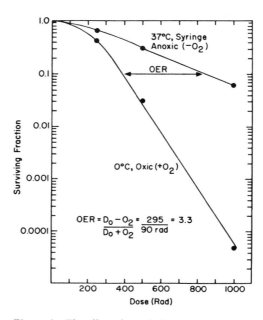

Figure 3. *The effect of metabolically produced hypoxia on the radiation response in dense suspensions of Chinese hamster ovary cells. In the top curve the 2×10^8 CHO cells/mL in 0.02 M HEPES buffered media were drawn up into a glass syringe, immediately radiated at 37 °C, and diluted and assayed for clonogenic survival. In the lower curve the same density of cells was spread as a layer of cells on the surface of a T30 flask, gassed with humidified CO_2, irradiated at 0 °C, and diluted and assayed for colony-forming ability (60).*

Simulation of the O_2 enhancement of radiation damage by other chemical agents is important, particularly because tumor cells tend to be anoxic and, hence, more radiation resistant than healthy oxic cells. Considerable effort has gone into the design and mode of operation of several chemical radiation sensitizers (65–67). At the present time nitroimidazoles are considered to be the most efficient sensitizers of radiation action in cells. A correlation exists between the electron affinity of the sensitizer and its radiation sensitizing effectiveness (68).

Once again the situation is similar to other studies involved with the penetration of a reactant to a reactive center in the assembly. The active site is a free radical in the vicinity of the nucleus, and the reactants (O_2 or nitroimidazoles) have to penetrate to and react with this center. For a cell, the main concerns are the actual concentration of sensitizer at the reactive site; the restrictions placed on it at this site (e.g., rigidity or polarity); and the applicability of knowledge gained from studies carried out in the homogeneous phase in the reaction in vitro. The reasonable correspondence reported between cell survival and the physical properties of the sensitizers would indicate that these concerns do not present a problem. However, the physical correlation of the properties of the sensitizers with the bioevents could reflect on the degree of partitioning of the sensitizers between the aqueous phase and the cell medium.

Compounds of the thiol-type can lead to protection of the cells from radiation damage. The mechanism is thought to be simple H-atom donation to the radical from the thiol, as indicated earlier. The factors affecting the effectiveness of the thiol are exactly those discussed earlier for sensitizers.

Fluorescence probe studies of the type used earlier in cells and membranes could provide useful answers to the questions raised regarding the mode of action of sensitizers in cells. One report exists of using pyrene derivatives in cells and monitoring the action of O_2 in the cell via the pyrene probe (69).

Literature Cited

1. Konev, S. V. "Fluorescence and Phosphorescence of Proteins and Nucleic Acids"; Udenforend, S., Ed.; Plenum Press: New York, 1967.
2. Dandliker, W. B.; Portman, A. J. In "Excited States of Proteins and Nucleic Acids"; Steiner, R. F., Ed.; Plenum Press: New York, 1971; p. 199.
3. Steiner, R. F.; Edelhoch, H. Chem. Rev. 1962, 62, 457.
4. Weber, G.; Teale, F. W. In "The Proteins"; Neurath, H., Ed.; Acad. Press: New York, 1965; Vol. 3, 445.
5. Brand, L.; Gohlke, J. R. Ann. Rev. Biochem. 1972, 41, 843.
6. Waggoner, A. J. Membr. Biol. 1976. 27, 317.
7. See Peter Froehlich, p. 49; Bradley, R. A., p. 91; and Churchich, Jorge E., p. 217 in "Modern Fluorescence Spectroscopy"; Wehry, E. L. Ed.; Plenum Press: New York, 1976; Vol. 2.
8. Weber, G.; Laurence, D. J. R. Biochemistry 1954, 56, XXXI.
9. Stryer, L. J. Mol. Biol. 1965, 13, 482.
10. Daniel, E.; Weber, G. Biochemistry 1966, 5, 1893.
11. Weber, G.; Daniel, E. Biochemistry 1966, 5, 1900.
12. Castellino, F. J.; Brockway, W. J.; Thomas, J. K.; Tiao, H.-T.; Rawitch, A. B. Biochemistry 1973, 12, 2787.
13. Weber, G. Trans. Faraday Soc. 1954, 50, 552.
14. Weber, G. Biochemistry 1952, 51, 145.
15. Knopp, J. A.; Weber, G. J. Biol. Chem. 1969, 244, 6309.
16. Lakowicz, J. R.; Weber, G. Biochemistry 1973, 12, 4161.
17. Lehrer, S. S. Biochemistry 1971, 10, 3254.
18. Lakowicz, J. R.; Weber, G. Biochemistry 1973, 12, 4171.
19. Weber, G. In "Light and Life"; McElroy, W. D.; Glass, B.; Eds.; John Hopkins Press: Baltimore, 1961; p. 82.
20. Lippert, E. Z. Elektrochem. 1957, 61, 962.
21. Cooper, M.; Thomas, J. K. Rad. Res. 1977, 70, 312.
22. Weber, G. J. Biochem. 1952, 51, 155.
23. Chen, R. In "Fluorescence"; Guilbault, G., Ed.; Dekker: New York, 1967; p. 443.

306

24. Aoki, K.; Foster, J. F. *J. Am. Chem. Soc.* **1956**, *78*, 3538.
25. Tanford, C. "The Hydrophobic Effect;" Waley: New York. 1973.
26. Phillips, G. O.; Power, D. M.; Richards, J. T. *Isr. J. Chem.* **1973**, *11*, 517.
26a. Phillips, G. O.; Power, D. M.; Davies, J. V. In "Fast Processes in Radiation Chemistry and Biology"; Adams; Fielden; Michael, Eds.; Wiley: London, 1974; p. 180.
27. Adams, G. E.; Wilson, R. L.; Aldrich, J. E.; Cundall, R. B. *Int. J. Radiat. Biol.* **1969**, 333.
28. Phillips, G. O.; Power, D. M.; Robinson, C.; Davies, J. V. *Biochim. Biophys. Acta* **1970**, *215*, 491.
29. Boyland, E.; Green, B.; Liu, S. L.; *Biochim. Biophys. Acta* **1964**, *87*, 653.
30. Nagata, C.; Kodama, M.; Tagashia, Y.; Imamura, A. *Biopolymers* **1966**, *4*, 409.
31. Green, B.; McCarter, J. A. *J. Mol. Biol.* **1967**, *29*, 447.
32. Kodama, M.; Nagata, C. *Biochemistry* **1975**, *14*, 4645.
33. Geacintov, N. E.; Prusik, T.; Khosrofian, J. M. *J. Am. Chem. Soc.* **1976**, *98*, 6444.
34. Kano, K.; Fendler, J. H. *Biochem. Biophys. Acta* **1978**, *509*, 289.
35. Eisinger, J.; Lamola, A. *Biochem. Biophys. Acta* **1971**, *240*, 299; *Ibid.* **1971**, *240*, 313.
36. Luk, C. K. *Biochemistry* **1971**, *10*, 2838.
37. Träuble, H.; Overath, P. *Biochem. Biophys. Acta* **1973**, *307*, 491.
38. Träuble, H.; *J. Mol. Biol.* **1971**, *4*, 193.
39. Täuble, H. *Proc. Natl. Acad. Sci.* **1974**, *71*, 214.
40. Reeves, J. P.; Shechter, E.; Weil, R.; Kaback, H. R. *Proc. Natl. Acad. Sci.* **1973**, *70*, 2722.
41. Cheng, S.; Thomas, J. K.; Kulpa, C. F. *Biochemistry* **1974**, *13*, 1135.
42. Wong, M.; Kulpa, C. F.; Thomas, J. K. *Biochem. Biophys. Acta* **1976**, *426*, 711.
43. Nieva-Gomez, D.; Konisky, J.; Gennis, R. B. *Biochemistry* **1976**, *15*, 2747.
44. Edwards, H. E.; Thomas, J. K.; Burleson, G. R.; Kulpa, C. F. *Biochem. Biophys. Acta* **1976**, *448*, 451.
45. Burleson, G. R.; Kulpa, C. F.; Edwards, H. E.; Thomas, J. K. *Experimental Cell Biology* **1978**, *116*, 291.
46. Inbar, M.; Shinitzky, M.; Sachs, L. *FEBS Lett.* **1974**, *38*, 268.
47. Castellino, F. J.; Thomas, J. K.; Ploplis, V. A. *Biochem. Biophys. Res. Commun.* **1977**, *75*, 857.
48. Ploplis, V. A.; Thomas, J. K.; Castellino, F. J. *Chem. Phys. Lipids* **1979**, *23*, 49.
49. Bisby, R. H.; Cundall, R. B.; Wardman, P. *Biochem. Biophys. Acta* **1975**, *389*, 137.
50. Taub, I. A. *J. Chem. Educ.* **1981**, *58*, 162.
51. Adams, G. E.; Wardman, P. In "Free Radicals in Biology"; Acad. Press: New York, 1977; Chapter 2.
52. Redpath, L. *J. Chem. Educ.* **1981**, *58*, 131.
53. Bisby, R. H.; Cundall, R. B.; Redpath, J. L.; Adams, G. E.; *J. Chem. Soc., Faraday Trans. 1* **1976**, *72*, 51.
54. Cooper, M.; Grätzel, M.; Thomas, J. K. In "Radiation Research"; Nygrard, Ed.; Acad. Press: New York, 1976; p. 511.
55. Eadsforth, C. V.; Power, D. M.; Thomas, E. W.; Davies, J. V. *Int. J. Radiat. Biol.* **1976**, *30*, 449.
56. Land, E. J.; Swallow, A. J. *Biochem. Biophys. Acta* **1968**, *162*, 327.
57. Land, E. J.; Swallow, A. J. *J. Biochem.* **1976**, *157*, 781.
58. Fridovich, I. *Acc. Chem. Res.* **1972**, *5*, 321.
59. Fielden, M.; Roberts, P. B.; Bray, R. C.; Lowe, D. J.; Mautner, G. N.; Rotilio, G.; Calabrese, L. *J. Biochem. Tokyo.* **1974**, *139*, 49.
60. Biaglow, J. E. *J. Chem. Educ.* **1981**, *58*, 144.
61. Greenstock, C. L. *J. Chem. Educ.* **1981**, *58*, 156.

62. Ward, J. F. *J. Chem. Educ.* **1981**, *58*, 135.
63. Adams, G. E.; Michael, B. D.; Asquith, J. C.; M. A. Shenoy, M. A.; Watts, M. E.; Whillans, D. W. In "Radiation Research"; Nygrard, Ed.; Acad. Press: New York, 1975; p. 478.
64. Epp, E. R.; Weiss, H.; Kessaris, N. D.; Santomasso, H.; Heslin, T.; Long, C. C. *Rad. Res.* **1973**, *54*, 171.
65. Adams, G. E.; Cooke, M. S. *Int. J. Radiat. Biol. Relat. Stud. Phys. Chem. Med.* **1969**, *15*, 457.
66. Adams, G. E.; Dewey, D. L. *Biochem. Biophys. Res. Commun.* **1963**, *12*, 473.
67. Raleigh, J. A.; Chapman, J. D.; Bossa, J.; Kremers, W.; Reuvers, A. P. *Int. J. Radiat. Biol. Sel. Stad. Phys. Chem. Md.* **1973**, *23*, 377.
68. Adams, G. E.; Flockhart, J. R.; Smithen, C. E.; Stratford, J. J.; Wardman, P.; Watts, M. E. *Rad. Res.* **1976**, *67*, 9.
69. O'Loughlin, N. A.; Willins, D. W.; Hunt, J. W.; *Rad. Res.* **1980**, *84*, 477.
70. Thomas, J. K. *Chem. Rev.* **1980**, *80*, 283.

INDEX

INDEX

311

Copy editors: Deborah Corson and Susan Robinson
Indexer: Susan Robinson
Production editor: Anne Riesberg
Jacket artist: Anne G. Bigler
Managing editor: Janet S. Dodd

Elements typeset by EPS Group Inc., Baltimore, Md. and Hot Type Ltd., Washington, D.C.
Printed and bound by Maple Press Co., York, Pa.